Satellite Signal Propagation, Impairments and Mitigation

Satellite Signal Propagation, Impairments and Mitigation

Rajat Acharya

Space Applications Centre,
Indian Space Research Organisation (ISRO),
Ahmedabad, India

ACADEMIC PRESS

An imprint of Elsevier

Academic Press is an imprint of Elsevier
125 London Wall, London EC2Y 5AS, United Kingdom
525 B Street, Suite 1800, San Diego, CA 92101-4495, United States
50 Hampshire Street, 5th Floor, Cambridge, MA 02139, United States
The Boulevard, Langford Lane, Kidlington, Oxford OX5 1GB, United Kingdom

Notices

Knowledge and best practice in this field are constantly changing. As new research and experience
broaden our understanding, changes in research methods, professional practices, or medical treatment may
become necessary.

Practitioners and researchers must always rely on their own experience and knowledge in evaluating
and using any information, methods, compounds, or experiments described herein. In using such
information or methods they should be mindful of their own safety and the safety of others, including
parties for whom they have a professional responsibility.

To the fullest extent of the law, neither the Publisher nor the authors, contributors, or editors, assume
any liability for any injury and/or damage to persons or property as a matter of products liability,
negligence or otherwise, or from any use or operation of any methods, products, instructions, or
ideas contained in the material herein.

Library of Congress Cataloging-in-Publication Data
A catalog record for this book is available from the Library of Congress

British Library Cataloguing-in-Publication Data
A catalogue record for this book is available from the British Library

ISBN 978-0-12-809732-8

For information on all Syngress publications
visit our website at https://www.elsevier.com/books-and-journals

Working together
to grow libraries in
developing countries

www.elsevier.com • www.bookaid.org

Publisher: Matthew Deans
Acquisition Editor: Carrie Bolger
Editorial Project Manager: Carrie Bolger
Production Project Manager: Mohana Natarajan
Designer: Vicky Pearson

Typeset by SPi Global, India

Dedicated to Maa, Baba and Moni

Contents

Ancillary material for this book is available at http://booksite.elsevier.com/ISBN/
9780128097328

Preface

My first book titled 'Understanding Satellite Navigation' was widely circulated amongst the students of GNSS across the world. It was a delightful experience when I started getting mails and calls from the readers, some of them just appreciating the book while others suggesting improvements. This probably was the inspiration which had driven me to engage myself once again in writing this book on satellite signal propagation.

In today's world, in which corporate decisions are being taken through videoconferencing, while medical treatments are being carried out remotely over satellite links, virtual classrooms connecting students over the world, natural disasters being predicted using satellites and above all, common men using internet and navigation tools fervently, the role of the satellites in extending the communication, navigation, and similar applications is inexorably extensive. In the backdrop of this fact, this book, 'Satellite Signal Propagation Impairments and Mitigation' aims to create a foundation of signal propagation through atmospheric media, and it tries to do so with the help of elementary physics and rational sense. It discusses the physical processes at work in the atmosphere and attempts to bring out the pragmatic issues pertinent to any satellite service—the propagation impairments and how to alleviate them.

This book focuses on three major areas. The initial chapters involve the physics of propagation of waves, including the basic electromagnetic theory. It develops the basis for the later chapters. The following chapters apply the propagation effects to the satellite links. They include aspects of mathematical modeling the impairments and also describe ways of mitigating them. Finally, it develops an idea of designing a satellite link.

Throughout the book, the basic working principles are explained right from the foundation level, keeping in mind all those readers who are being introduced to the subject for the first time. At many places, various complexities are avoided intentionally to provide the reader an ease of understanding. However, the real niche in this book is the visualization of the topics under discussion using the MATLAB programs. Such programs are provided for every important new concept introduced in this book. Here, the reader will have the choice to alter the values of the parameters and see the results to appreciate the effects and look deep inside the problem under different scenarios. This is a novel idea that has been continued in this book from my previous one.

It is important for a teacher in any subject to elucidate the basic theory, in a simple manner from the first principle. That is what I attempted to provide in this book. I believe, this book will not only serve the readers to get a full grasp of the subject but will also be a source of enjoyment of learning.

Finally, I assert that the views and the opinions presented here are those of my own and do not necessarily reflect the views of my employer, nor of the Government of India.

Rajat Acharya

Acknowledgement

The publication of this book has become possible due to the contribution and efforts of many people. First of all, I would like to thank all my teachers, who made me qualified to author this book. My teachers Gaurhari Bhuyan, Prosanto Bose, Shankar Bhattacharya, Parimal Chatterjee, Narayan Das, B.N. Das need special mention with many others. My knowledge in the subject is all due to my professors. Prof. Ashish Dasgupta, Prof. Animesh Maitra, Prof. Apurbo Datta, Prof. Bijoy Banerjee and Prof. S.K. Chatterjee, Dr. Ramaranjan Mukherjee are some amongst them. It is due time to acknowledge their selfless and dedicated contributions.

Thanks are due to Kalyan Bandyopadhyay, M.R. Sivaraman, Joel Lemorton, Laurent Castanet and Xavier Boulanger for sharing their professional knowledge with me. I wish to extend my sincere gratitude to Dr. Mohanchur Sarkar, Dr. Surendra Sunda, Dr. Chandrashekhar, Mr Nilesh Desai and Mr. Atul Shukla for their patronage and the generous gestures they have extended to me. I also thank the whole team of Elsevier for their untiring support during the production of this book. The thoughtful suggestions by the reviewers to improve this book are also very much appreciated.

I am indebted to my parents who shaped up my life while recognition is also due to my family and friends for their interminable cooperation. I must specially mention here my niece Purbita and my brother Dr. Bijoy Roy for their support and encouragement. I would also not forget my dogs Chiku, Charki and Gunda and all those members of the animal kind, who gave me unconditional love and motivated me during my difficult times. Last but not the least, I sincerely thank my wife Chandrani and my son Anubrata, for their support inspirations and for the sacrifice they have made during the writing of this book to make this effort a success.

Introduction

1

Communication is essentially the exchange of information between two entities. It is probably the oldest of the technologies that is continuing from the early ages of mankind. However, the art of communication predated the advent of mankind. Even before human race evolved, the primitive kinds of the life forms also conveyed their minds by their own means.

Almost every introductory book on telecommunication portrays a brief history of the subject, describing the means by which early men used to communicate with others over distance. They tell us about how the fire signs or smoke puff signals were used for sending information to a distant observer or how the beats of the drums used to carry meaningful messages. We all read them with great amazement. However, in early days, communication over a large distance was not at all an easy task. There were impediments in every facet of such activities. How did those primitive men use the fire signal when it was raining? Were the smoke plumes correctly interpreted during heavy winds or fog? Were the message-bearing drum beats heard really loud enough to be intelligible to the listener even amid other noises? Did they have any alternative methods for communication in such trying situations? Someone asking such questions may sound weird and perhaps none would be interested today to find the answers. Yet, one must realize that the difficulties that men used to face in such conditions were enormous. Today, it is even more difficult to perceive the difficulties to their precise extent. Even Fig. 1.1 understates the grim of the situation.

Incidentally, the means and the techniques of communications have changed over the ages, but the problems have not. Rains, fogs, noise, etc. still put hindrances in today's communication systems. Therefore, it is equally interesting to understand, even today, what marring effects the signals face while propagating through different media, why such problems arise and how they are eliminated, moderated or appropriately taken care of. My professional 'experience', fortunately, falls in line with this topic. Although it is very aptly said that, 'Experience is the name wise men give to their mistakes' (William Wordsworth), yet that is what has made me eligible to write this book. So, in this book we shall delve into the foundations of physics and the aspects of engineering behind the simple propagation of the satellite signals and their apparently challenging impairments.

At the outset, we shall discuss as a prerequisite, the preliminary ideas in this chapter, which will form the basis for understanding many of the subsequent discussions presented in this book.

Satellite Signal Propagation, Impairments and Mitigation. http://dx.doi.org/10.1016/B978-0-12-809732-8.00001-6

FIG. 1.1

Early forms of telecommunication and their impairments.

1.1 **INTRODUCTION TO ELECTRONIC COMMUNICATION**

Modern electronic telecommunication, in general, is the technique of sending and receiving any information between two distant points using electromagnetic waves. The information is converted into discrete codes, the units of which are called Bits. Bits are transmitted over the medium ridden over a sinusoidal 'Carrier' wave, in which every different kind of bit is represented by a predefined state of the carrier amplitude, phase or frequency. When this is done, the carrier is said to be 'Modulated' by the bits. The information bearing carrier is basically an electromagnetic wave of comparatively much higher frequency than the rate of change of the bits. The modulated carrier is generated and transmitted by the source and propagates through the medium towards the receiver, where it is received and the individual bits are identified in sequence to retrieve the information back. The basic theory of telecommunications can be understood from any good book on the topic. Considering the scope of this book, here we concentrate only on the different aspects of the propagation of the satellite signals through the atmospheric medium.

There are a number of mechanisms by which radio waves may travel from a transmitter to a receiver. The most important of these, in which propagation takes place through the open atmospheric medium are, (i) the ground wave, (ii) the sky wave and (iii) the space wave. These are discussed in brief below (Terman, 1955; Kennedy and Davis, 1999; Chatterjee, 1963; Dolukhanov, 1971).

1.1.1 **GROUND WAVES**

Ground waves are those waves which passes between the transmitting and the receiving antenna situated close to the earth surface. These waves are essentially vertically polarized, i.e. the electric field of the wave is directed vertically over the ground. These are typically low and medium frequency signals which spread out from the transmitter, and as they propagate over the earth, induce charges on the earth surface. These induced charges travel with the wave and constitute a current. As a result, instead of just travelling in a straight line, the signals tend to follow the curvature of the Earth along the path the currents are induced like a guided wave. Therefore, the waves can move beyond the horizon. However, due to the finite conductivity of the earth, the conducting current gets absorbed and thus the energy carried by the wave is attenuated. Consequently, the ground wave gets weakened with distance from the transmitter. Due to the same reason, the ground wave moves more efficiently over the conducting sea surfaces. Ground wave radio propagation is used to provide local radio broadcasting.

1.1.2 **SKY WAVE**

Sky waves are those in which the waves reach the receiver from the transmitter as a result of a bending of the wave path due to the refraction by the ionized region of the atmosphere. This ionized region, termed as the ionosphere, has abundance of free electrons and ions, coexisting in dynamic equilibrium with the neutral particles. The ionosphere extends from 50 km to more than 1000 km above the earth surface. Practically, all earlier very long distance communications were made using sky waves.

The distance to which the reflected ray will bounce back over the earth surface and the maximum frequency up to which the ionospheric reflection will occur for a given transmission angle both depend upon the amount of ionization. More precisely, they depend upon the free electron density and the height of the peak density of the ionosphere, both of which in turn, vary diurnally, with seasons and years. We shall learn about the ionosphere and its constituents later in Chapter 4 and shall develop the theory behind the reflection of the waves in the ionosphere leading to the sky wave propagation. However, there we shall look into it, more in context of the transition between the waves from those getting transmitted through the ionosphere to those reflecting back to the earth. The high frequency band operates almost exclusively with sky waves. For a given location, time and elevation angle of transmission, the useful transmission band remains below a maximum usable frequency (MUF). Frequencies above the MUF are transmitted into space to constitute the space wave, which we shall discuss below. Sky wave propagation takes place over few thousand kilometres of range.

1.1.3 **SPACE WAVES**

Space waves, also known as 'direct waves', are radio waves which travel directly from the transmitter to the receiver, such that the propagation occurs in a line of sight path between them even without any supporting medium. When both the transmitting and the receiving antenna are located on the earth, the maximum distance between two antennas along the line of sight depends on the height of each antenna and the curvature of the earth. The propagation of the satellite signals is also an interesting example of direct waves. Because a typical transmission path of such a wave passes through the atmosphere and earthly ambience like hills, buildings and other obstacles, it is possible for radio waves to get absorbed or get reflected, refracted, diffracted or scattered by them. This results in the waves arriving at the receive antenna from several different directions causing major distortion to these signals in the form of fading, multipath, scintillations, depolarization, etc. space wave propagation is carried out in bands having very high frequencies, e.g. V.H.F, U.H.F or in even higher bands. The following Fig. 1.2 illustrates the different types of propagation.

1.2 SATELLITE SIGNALS AND PERFORMANCE
1.2.1 SATELLITE APPLICATIONS

The literary meaning of satellite is any natural or artificial body moving around her planet. The idea of artificial satellite was first conceived by the British scientist Arthur C Clarke (Clarke, 1945) who proposed that an artificial satellite moving around the earth with the same angular velocity with it can act as a super terrestrial relay station for long distance communication system. The first artificial satellite, Sputnik, was put into orbit by USSR in the year 1957. This created a revolutionary change in the field of electronic telecommunication system which was till then dependent primarily upon the sky waves. Since then many countries in the world have send their own satellites in space, not for just communication purposes, but also for varied other applications, including remote sensing of the earth, navigation and astronomical studies.

Today's modern satellite-based telecommunication comprised of broadband transmission of high speed multimedia data to the user, directly to their fixed, mobile or portable devices. However, modern applications of satellite system have moved far beyond telecommunication. Satellites are being used for environmental monitoring. It includes the study of structure of atmosphere, formation and movement of winds and clouds, obtaining ocean and land parameters like ocean surface temperature, heights, cartographic parameters, monitoring glacial movement, volcanic activities, soil moisture and what not. Besides, such satellites also find applications in Geology, Agriculture and Forestry. Satellites are also used for Navigation for finding position of a user on the earth surface or near it along with accurate time. Satellites used for astronomical researches probe into near and distant space. They help to find the answers to the eternal questions of the birth, evolution and the character of the universe. These satellites carry transponders for communication or sounders, imagers and other sensors for monitoring earth's environment or they carry payloads for their application-specific purposes. The information generated by these

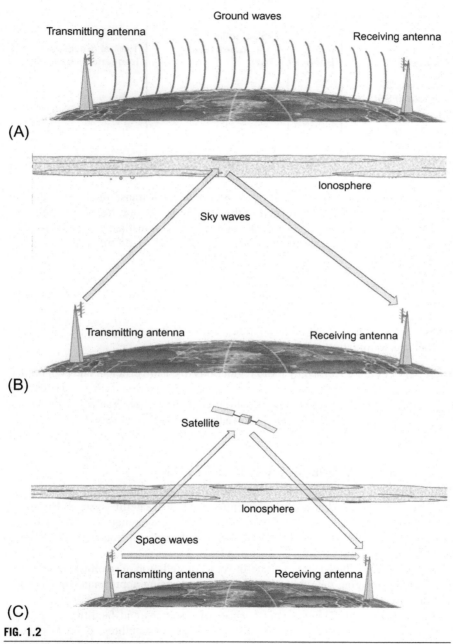

FIG. 1.2

(A) Ground wave, (B) sky wave and (C) space wave propagation.

Table 1.1 Frequency allocations

Services	Band	Typical uplink frequencies used (GHz)	Typical downlink frequencies used (GHz)
Fixed satellite services (FSS)	C	6	4
	Ku	14	12
	Ka	30	20
Mobile satellite services (MSS)	L	1.6	1.5
	Ka	30	20

sensors in the satellite or received by its transponders is transmitted to the ground station along with other data related to the satellite behaviour, health and control.

The satellite services may be broadly divided into few distinct categories for definite applications. These are namely (Maral and Bousquet, 2006),

- Fixed satellite services (FSS)
- Mobile satellite services (MSS)
- Broadcasting satellite services (BSS)
- Earth exploration satellite services (EES)
- Space research and operation satellite services (SRS)
- Radio determination satellite services (RSS)

These services are governed by radio regulations to ensure efficient and economical use of the limited radio frequency spectrum. Frequency bands are allocated to each of the services mentioned previously to allow compatible use. Table 1.1 shows the frequency allocations for FSS and MSS radio-communication services.

1.2.2 ARCHITECTURE OF SATELLITE SYSTEM

To understand how different types of information are exchanged from the satellites to the ground, we need to know the architecture of a typical satellite system. The architecture of the satellite system consists of three distinct segments, viz. the space segment, the control segment and the user segment (Maral and Bousquet, 2006; Pratt et al., 1986).

Space segment: It consists of a satellite or a constellation of satellites orbiting the earth and exchanging data with the other segments using appropriate signals such that the targeted applications are fulfilled.

Control segment: This segment consists of resources and facilities on ground that monitor, control and maintain the space segment satellites. It sends control and configuration commands to the space segment units to carry out any specific task, reallocate the resource onboard, maintain or discipline the satellite on its orbit or merely for the management of the data traffic. It also receives the health, housekeeping and other telemetric data from it.

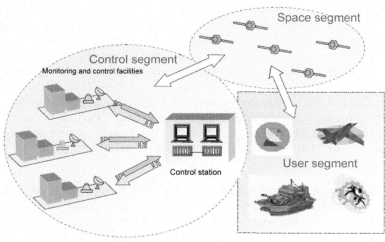

FIG. 1.3

Segments of a satellite system.

User segment: It consists of the users who interact with the satellite and exchange data with it through the signals to accomplish the intended applications.

The schematic of the segments and their interactions are shown in Fig. 1.3.

Today, there are innumerable scientific, societal, strategic and other applications which can be carried out using a satellite system. But, whatever be the application of the satellite, the basic architecture of the satellite system is comprised of the above-mentioned distinct segments. All the satellites in the space segment are connected to the control segments, either in simplex or in duplex mode through radio wave signals.

1.2.3 PERFORMANCE OF SATELLITE SYSTEM

As user to any of the satellite-based services, many of us have experienced the frustrating situation of suddenly getting disconnected just at the most critical moment of the communication or during the most important segment of the service transaction. Although many of the satellite-based applications can tolerate such link break or unavailability, there are many other satellite services which invariably need uninterrupted links, e.g. telemedicine (Relgin and Relgin, 2001). It is always undesirable, and sometimes even disastrous to have a break in such a service. Therefore, it is necessary to understand why a snap in the satellite link actually happens. To the matter in a systematic way, let us first define what the performance metrics of the satellite services are. Then we shall explore the reasons which cause degradation of the service and thus affect the performance.

1.2.3.1 Performance metrics

This is one of the most important sections of the whole book. This is important because it relates the physics affecting a propagating signal (that we shall be learning in the following chapters) with the associated engineering parameters defining the performance of a telecommunication system. Although the subject may be elaborated to the length of a full chapter, considering the scope of this book we confine this to only in the few following paragraphs. But, these few paragraphs are of immense importance upon which the understanding and the appreciation of the rest of the book stand.

We know that the satellite communication system basically transfers the information between the participating entities, viz. satellite and the user or control segment. Communication system performances are mainly determined by few important factors—Fidelity, Availability and Continuity.

a. Fidelity

The literary meaning of fidelity is the degree of conformity, i.e. the exactness with which something is reproduced. In satellite communication, this is not a popular term in use. But, since we are considering here, in generic sense, transfer of information in any arbitrary application, I prefer to use this term 'Fidelity' to represent the measure of correctness of the received information by the receiver compared to what has been sent by the transmitter. In plain words, it states how truthfully the information has been communicated. Since the information-bearing units are called bits, Fidelity is quantitatively represented by a more obvious and popular parameter, bit error rate (BER), also sometimes called as bit error ratio. BER is the ratio of the number of bits in error in a stream of N received bits. Lesser the BER, better the fidelity and hence better the quality of communication. This is directly related to the bit error probability (BEP), which is the statistical probability of a received information bit to be detected in error (Maral and Bousquet, 2006). While BER is the more factual quantitative measure of the effect of bits received erroneously, BEP is a long-term statistical index of performance and is dependent upon how well the transmitted signal reaches the receiver. The conformity between the transmitted and received signals depends upon factors like the transmitted signal quality, propagation effects, goodness of receiver quality, etc. We shall discuss the propagation effects on the propagating satellite signals in the following chapters, while also having a vivid discussion on the noise effects.

b. Availability

Availability is defined by the probability that the communication system will make the signal resources accessible to the user within a specified coverage area and over a specified time with warranted quality. So the time that simultaneously meets the condition of signal accessibility and the signal quality in terms of BER better than a predefined threshold, is only accounted in calculating the availability of a system.

c. Continuity

Continuity is the probability that a service, once available, will continue to be available and performing over a predefined duration. The number of times that the service fails to sustain over the total period of use after the user starts using the service adds to the degradation in continuity.

1.2.3.2 Signal parameters

From the above discussions, we come to know about the metrics which quantitatively define the quality of service of a satellite system. Here, we shall discuss about the signal parameters which directly affect these terms.

The BER, Availability and Continuity terms are determined by the ability with which the receiver can acquire the signal and can correctly identify and track the current states of the signal parameters, viz. amplitude, phase, frequency, etc., which bears the information in form of bits. Only on correct identification of the bits that the receiver can faithfully recreate the information embedded in it. It is obvious that the propagation factors will distort the signal at the receiver from its condition at the transmitter. The signal is designed and transmitted in such a way that the performance targets are still achieved with the designated receivers for the expected conditions of propagation. So, under nominal condition, there is no scope for any difficulty to happen in carrying out the communication processes with expected performance. But, the pragmatic scenario may be different and the signal may actually be received in much more distorted form than what is otherwise expected with consequent divergence from its predicted performance. In fact, there are candidate factors which distort the signal character and thus deteriorate the signal quality as it passes from the transmitter to the receiver. The signal power may be excessively reduced, some meaningless random voltages, called Noise may get added with the actual signal levels or similar degrading effects may happen. The effects may be so severe that the information in the signal may get totally unintelligible at the receiver. However, the more rigorously the propagation factors are considered in the design, the more rugged the signal is to the propagation impairments. Here, we are now going to learn about the signal parameters which have explicit bearing upon the performance.

The information-bearing units in the signal are called bits. Binary bits can assume values '0' or '1'. Bits, in a carrier signal, are represented by predetermined state of the signal parameter. For example, in quadrature phase shift keying (QPSK) type of modulation, the four possible codes, corresponding to two bits of information, which are (0,0), (0,1), (1,0) and (1,1), can be represented by four different phase states of the signal. The states are typically separated by equal phase differences. So, with reference to an arbitrary zero phase of the unmodulated carrier signal, two of the codes may be represented by the phase $+\pi/4$ and $+3\pi/4$, while the other two codes may be represented by phase $-\pi/4$ and $-3\pi/4$ of the signal.

At the receiver, these bits are identified from the signal states. Correct identification of the bits is determined by the ability of the receiver to distinguish between these states of the signal representing different bits. This distinction is made by comparing the signal state with a predetermined reference state. For a given separation of the states, the receiver is able to correctly distinguish them and decide the correct bit when more energy is carried by the actual signal within the duration of a bit. Moreover, increased contribution of the noise in the signal increases the likelihood of incorrect identification of the information bit. Therefore, the overall probability of a bit to be identified correctly depends upon how much the different states are separated from each other, and also upon how much energy is carried by the signal within the interval of a bit (E_b). It is also inversely proportional to the variance of the noise. The total added noise is a linear function of the noise power spectral density, N_0. Hence, the ratio E_b/N_0, is one of the most important parameters in satellite communication system which determines the BEP or BER. We can write this term as

$$\frac{E_b}{N_0} = \frac{S \times T_b}{N/B}$$
$$= \frac{S}{N} \times \frac{B}{R_b} \tag{1.1}$$

where, S and N are the power of the signal and the noise, respectively, and T_b ($=1/R_b$) is the bit duration. R_b is the bit rate and B is the bandwidth of the signal. The ratio S/N is called the signal to noise power ratio or simply signal to noise ratio (SNR). Here, we have considered the fact that the noise power spectral density N_0 is constant over the bandwidth, B. Rearranging, Eq. 1.1 can also be written as

$$\frac{S}{N} = \frac{E_b}{N_0} \times \frac{R_b}{B} \tag{1.2}$$

Therefore, S/N and E_b/N_0 are related by the ratio R_b/B, called as the bandwidth efficiency. It represents the bit rate that can be achieved given a definite bandwidth of the signal and depends upon the type of modulation. So, for the same requirement of E_b/N_0, the S/N requirement will be different depending upon the modulation. Further, as the signal to noise ratio is derived from the RF signal, with C and N as the respective powers in the carrier and noise, we can write,

$$\frac{E_b}{N_0} = \frac{C/N}{R_b/B}$$
$$= \frac{C}{N_0} \times \frac{1}{R_b} \tag{1.3}$$

If, there are more than one information channel in the same carrier, the total carrier power is accordingly apportioned. This division of power is accredited to individual channel at the time of transmission. E_b/N_0 of each channel is then estimated with its respective carrier power and can be written as

$$\left(\frac{E_b}{N_0}\right)_j = \left(\frac{C}{N_0}\right)_j \times \left(\frac{1}{R_b}\right)_j \tag{1.4}$$

As BER improves with increasing E_b/N_0, the improvement is observed with both increased (C/N_0) and with increased T_b, the latter being equivalent to reduction in R_b. While BER is a measured number, the relationship of E_b/N_0 exists with the BEP. The exact mathematical expression relating the BEP with the E_b/N_0 can be obtained from any good book on digital communication (Proakis and Salehi, 2008; Chakrabarty and Datta, 2007; Mutugi, 2012). Table 1.2 provides the final expression for the BEP in terms of the E_b/N_0 (Maral and Bousquet, 2006; Mutugi, 2012).

It is worth reiterating that, in order to retrieve the data, the receiver needs to acquire the signal first. Acquisition is the detection of the approximate carrier states and adjustments of the local reference signal of the receiver for demodulation. Once the signal is acquired, it is to be tracked throughout the period of its operation. Tracking is nothing but following the changes in the incoming signal states and accordingly adjusting the local reference signal so that the demodulation process is correctly continued.

Table 1.2 Expressions for BEP for different modulations

Modulation	Required theoretical E_b/N_0 (dB, approx)		
	BER $= 10^{-3}$	BER $= 10^{-5}$	BER $= 10^{-7}$
BPSK	6.75	9.57	11.3
QPSK	10.32	12.90	14.52
16-PSK	17.60	20.05	21.63

Both acquisition and tracking performance, which directly affect the availability and continuity of the services, improve with better values for C/N_0.

It is evident, thus, that higher the C/N_0 of the received carrier, better the probability of the receiver to acquire and track a signal and identify the information-bearing bits in it correctly. Hence, the quality of the satellite services, which depends upon the correctness of the received signal, is determined by the received carrier power to the noise power density.

The term C/N_0 of the received signal is thus found to be the most important parameter influencing the performance of a communication system. Now, let us understand how this factor is modified during propagation. We shall deal with modest details here and will recall some of the fundamental concepts discussed here again in Chapter 9. The power of a signal received by a receiver depends upon the transmitted power P_T. This power when transmitted by a directive antenna, having directive gain G, the total power is concentrated along the direction of the gain of the antenna. This enhanced power in the direction of the antenna gain is called the effective isotropic radiated power (*EIRP*) of the signal. Therefore,

$$EIRP = P_T \times G \tag{1.5}$$

The enhanced power is then transmitted towards the targeted direction. On propagating through the space, the power of the signal gets weaker as it spreads over larger and larger area. At a distance R, the power passing per unit cross sectional area is the power flux density, given by

$$\Phi(R) = \frac{EIRP}{4\pi R^2} \tag{1.6}$$

Every receiving antenna has an effective receiving area determined by its gain such that the power incident over this area can be received by it. If G_R is the gain of the receiving antenna and A_e is its effective area, then the relationship is given by

$$G_R = \frac{4\pi}{\lambda^2} A_e$$
$$A_e = \frac{G\lambda^2}{4\pi} \tag{1.7}$$

Here λ is the wavelength of the signal. Therefore, the power received by the receiving antenna with gain G_R is

$$P_R = \Phi(R) A_e$$
$$= \frac{EIRP}{4\pi R^2} \times G_R \frac{\lambda^2}{4\pi}$$
$$= \frac{EIRP}{(4\pi R/\lambda)^2} \times G_R \tag{1.8}$$

Calling $(4\pi R/\lambda)^2$ as free space path loss (L_{FS}), the received power after being enhanced by the receiving antenna gain becomes,

$$P_R = \frac{EIRP}{L_{FS}} \times G_R \qquad (1.9a)$$

The absolute power level variation from the *EIRP* to the received power is very large, and hence, is more conveniently represented in logarithmic scales. So, in logarithmic (decibel) scale,

$$P_R = EIRP - L_{FS} + G_R \text{ in dB} \qquad (1.9b)$$

In addition to the free space loss occurring due to the obvious spreading of the power with distance, there is also some associated loss due to the passage of the signal through the atmosphere. It is called the Atmospheric propagation loss, L_A. Considering the latter, the received power becomes

$$P_R = EIRP - L_{FS} - L_A + G_R \text{ in dB} \qquad (1.9c)$$

To this received power, some unwanted and uncorrelated power from random noise will get added. These are generated due to the emissions of incoherent electromagnetic wave by the atmospheric elements at finite temperature and which are picked up by the antenna. Further added to this will be the noise generated by the random motion of the electronic charges in the antenna and the receiver hardware. This noise component gets adhered with the actual signal and degrades the signal quality and thereby reducing the ability of the receiver to correctly identify the signal state.

The occurrence of noise in time domain is fully random and its probability density follows Gaussian distribution with zero-mean. The power spectrum of noise is white, i.e. noise has almost the same spectral power over the frequency band width of interest. A black body at a finite temperature will also emit noise over all frequency bands. The spectrum of the received power from such emitted noise is approximately uniform in the radio frequency range and is only dependent upon the temperature of the body. Therefore, the noise added to the signal may be assumed to be transmitted by a black body at a finite temperature, T. This is called the equivalent noise temperature. Since the power contribution over the entire spectral range is equal, the noise is characterized and better represented by noise spectral density N_0. This is the total noise power over a unit bandwidth. As this is proportional to the noise temperature, we can write $N_0 = kT$. The proportionality constant 'k' is called the Boltzmann constant and its value is 1.38×10^{-23} J/K or -228.6 dB. Thus, the terms N_0 and T may be used interchangeably. Therefore, the ratio of the carrier power to the receiver noise density may be expressed as

$$\begin{aligned} C/N_0 &= EIRP - L_{FS} - L_A + G_R - kT \text{ in dB} \\ &= EIRP - L_{FS} - L_A + G_R/T - k \text{ in dB} \end{aligned} \qquad (1.10)$$

All other parameters being factors of transmitter and propagation medium, the performance of the receiver terminal is measured by the term G_R/T, the ratio of the receive antenna gain and the system noise temperature. It is called the figure of merit of the receiver.

Box 1.1 MATLAB Exercise

The MATLAB Link.m was run to generate the following variations of the received C/N_0 for different frequencies of the satellite signal. The given input and the program output for this particular run are shown below. The difference in C/N_0 is essentially due to the difference in path loss, all other factors remaining the same. A considerable difference is observed between Ku and Ka band signals.

Signal parameters:
Input the transmission frequency (GHz): (12:2:24)

Transmitting ground station parameters:
Input the transmission power (W): 6
Input the ground station antenna diameter (m): 2.4
Input the ground station antenna efficiency: 0.55
Input the ground station latitude (deg; E: +ve, W: −ve): 22
Input the ground station longitude (deg; N: +ve, W: −ve): 88

Receiving satellite parameters:
Input the satellite antenna diameter (m): 1.2
Input the satellite antenna efficiency: 0.6
Input the satellite longitude (deg; E: +ve, W: −ve): 83
Input the noise temperature at receiver (K): 450

The estimated parameters are:
The ground station antenna gain for the range of input frequencies is respectively (dB$_i$)

46.992   48.331   49.4908   50.5139   51.429   52.2569   53.0126

The ground station *EIRP* for the range of input frequencies is respectively (dB$_w$)

54.7736   56.1125   57.2723   58.2954   59.2105   60.0384   60.7942

The satellite-ground station distance is 36,568.7602 km
The power flux density for the range of input frequencies is respectively (dB$_w$/m^2)

−107.4807   −106.1418   −104.982   −103.9589   −103.0438   −102.2159   −101.4601

The satellite received power for the range of input frequencies is respectively (dB$_w$)

−109.1647   −107.8258   −106.6659   −105.6429   −104.7277   −103.8999   −103.1441

The noise at the satellite is −202.0679 dB$_w$
The carrier to noise ratio for the range of input frequencies is respectively (dBHz)

92.9032   94.2421   95.4019   96.425   97.3401   98.168   98.9238 dB

Input the minimum E_b/N_0 required for your modulation (dBHz): 14
The maximum bit rate possible for the range of input frequencies is respectively (Mbps)

77.68138   105.733   138.1002   174.7831   215.7816   261.0957   310.7255

Continued

Box 1.1 MATLAB EXERCISE—Cont'd

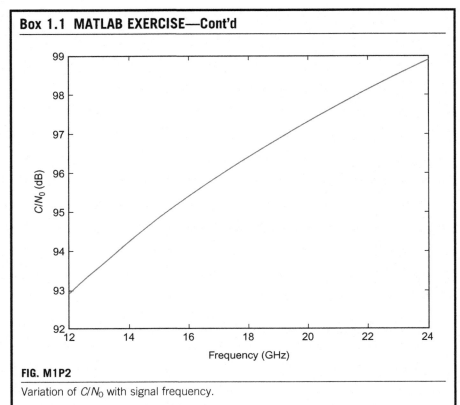

FIG. M1P2

Variation of C/N_0 with signal frequency.

Run the program for different positions of the satellite, different G/T and for different $EIRP$ and compare the graphs. The atmospheric loss is considered to be absent here. The program takes only frequency values in an array.

For satellite-based broadcasting, navigation, remote sensing and certain other applications, communication of data takes place in one direction between the satellite and the ground terminal. Therefore, either the uplink or the downlink is in consideration. There, the link performance is obtained from the one way estimate of C/N_0 as given in the equation mentioned previously. Nevertheless, in a communication system, where the link is established between two ground terminals through a transparent satellite, the total link between these two ground terminals will consist of an uplink and a downlink connection. The satellite will receive the signal from the transmitting ground station and convert it to a different frequency before transmitting it again towards the receiving ground station after amplification. For such a communication link, the total C/N_0 of the signal over the complete path at the receiving ground terminal, represented by $(C/N_0)_T$, is given by

$$\left(\frac{C}{N_0}\right)_T = \left[\left(\frac{C}{N_0}\right)_U^{-1} + \left(\frac{C}{N_0}\right)_D^{-1}\right]^{-1} \tag{1.11}$$

where $(C/N_0)_U$ is the carrier to noise density ratio for the uplink and $(C/N_0)_D$ is the carrier to noise density ratio for the downlink part of the total connection, considered individually.

From the expression in Eq. (1.1), it is clear that $(C/N_0)_T$ is even lower than the lowest among the values of $(C/N_0)_U$ and $(C/N_0)_D$. Therefore, if only one of these values degrades to fall below the threshold needed to ensure a certain performance level, the effective $(C/N_0)_T$ cannot be raised above this threshold just by increasing the C/N_0 of the other side. Therefore, to keep the $(C/N_0)_T$ above a particular required value, both the uplink and the downlink components of it are individually required to be maintained above it. However, if one of these components is kept at a sufficiently larger values than the other, then the $(C/N_0)_T$ becomes approximately equal to the other component. The following Focus 1.2 describes how to calculate the C/N_0 of a link considering all the pertinent factors. This will help in understanding where the propagation effects are involved and by how much. Similar problems will be dealt with in Chapter 9.

Focus 1.2 Satellite Link C/N_0 Calculation

We have already seen how the signal transmitted from the ground is received by the satellite and is retransmitted back to the ground after amplification. Given the ground station and the satellite trans-receive parameters and their distances, let us find the C/N_0 values at the destination receiver.

The system design parameters are:

Frequency	$f = 30$ GHz
Satellite to earth station distance	$d = 40,000$ km

Among the ground station parameters, assumed to be identical for transmitting and receiving:

Transmission power	$P = 10$ W
Antenna diameter	$D_g = 4$ m
Antenna efficiency	$\eta = 0.6$ m
Noise temperature at the receiver	$T_g = 250$ K

The satellite design parameters are

Satellite antenna diameter	$D_s = 2$ m
Satellite antenna efficiency	$\eta_s = 0.7$
Noise temperature at the satellite receiver	$T_s = 450$ K
Satellite transmission power	$P_s = 6$ W

The wavelength for the frequency used is

$$\lambda = 3 \times 10^8 / 30 \times 10^9$$
$$= 10^{-2} \text{m}$$

The transmitting antenna gain is given by

$$G_T = 4\pi/\lambda^2 * (\pi D^2/4) * \eta$$
$$= 9.4748 \times 10^5$$
$$= 59.76 \text{dB}_j$$

Therefore, the *EIRP* is given by,

$$EIRP = P \times G_T$$
$$= 9.4748 \times 10^6 \, W$$
$$= 69.76 \, dB_W$$

The power flux density at the distance of the satellite is given by

$$\varphi = EIRP / (4\pi d^2)$$
$$= 4.7124 \times 10^{-10} \, W/m^2$$

The effective area of the satellite receiving antenna is

$$A_{es} = (\pi D_s{}^2 / 4) \times \eta_s$$
$$= 2.1991 \, m^2$$

The carrier power received by the satellite is

$$P_s = \varphi \times A_{es}$$
$$= 1.0363 \times 10^{-9} \, W$$
$$= -89.84 \, dB_W$$

Again, using the constant value of k, the noise density at the satellite receiver is

$$Ns = k \times Ts$$
$$= 6.2117 \times 10^{-21}$$
$$= -202.07 \, dB$$

Therefore, the $(C/N_0)_U$ becomes

$$(C/N_0)_U = Ps/Ns$$
$$= 1.668 \times 10^{11}$$
$$= 112.22 \, dBHz$$

In this variable, the propagation factor is reflected in the received power flux density.

Now, the signal is amplified at the satellite transponder after the carrier frequency is appropriately adjusted. But, once the noise gets added to the signal, the signal to noise ratio cannot be improved just by amplifying the combined signal. This is because the noise also gets simultaneously amplified as a result, keeping the ratio unaltered. Therefore, if the carrier is amplified by factor A at the satellite transmitter to become $C \times A$, the noise power density there due to the already added noise during the uplink is also $N_0 \times A$. Taking satellite transmitting antenna same as the receiving antenna, with gain G_s,

$$G_s = 4\pi / \lambda^2 \times A_{es}$$
$$= 2.763 \times 10^5$$
$$= 54.41 \, dB_i$$

$$EIRPs = 6 \times 2.763 \times 10^5$$
$$= 16.57 \times 10^5 \, W$$
$$= 62.19 \, dB$$

The nominal C/N_0 at the ground receiver with $T_g = 250 \, K$ is

$$(C/N_0)_g = 1.8 \times 10^{11}$$
$$= 112.55 \, dBHz$$

Now, the received carrier power at the destination ground station receiver

$$C_d = C \times A \times G_s / d^2 \times A_{eg}$$

where d is the radial distance from the satellite and A_{eg} is the effective antenna area of the ground station. Similarly, the noise power received here due to the noise already added with the signal during uplink is

$$N_0u = N_0 \times A \times Gs/d^2 \times A_{eg}$$

If the noise temperature at the destination ground station is $T_g = 250$ K, the C/N_0 for the downlink is

$$(C/N_0)d = \left[(C \times A \times Gs)/d^2 \times A_{eg} \right]/kT_g$$

Now, during the RIP from the satellite to the ground, the signal also acquired some noise power. The power density of this noise being N_0d, we get the total noise power density at the ground receiver as

$$N_T = \left[(N_0u \times A \times Gs)/d^2 \times A_{eg} \right] + N_0d$$

Therefore, the C/N_0 after the round trip of the signal is

$$(C/N_0)_T = \left[(C \times A \times Gs)/d^2 \times A_{eg} \right]/\left[\{(N_0u \times A \times Gs)/d^2 \times A_{eg}\} + N_0d \right]$$
$$\text{Or, } (C/N_0)_{T^{-1}} = \left[\{(N_0u \times A \times Gs)/d^2 \times A_{eg}\} + N_0d \right]/\left[(C \times A \times Gs)/d^2 \times A_{eg} \right]$$
$$= (N_0u/C) + \left(d^2 \times A_{eg} \times N_0d \right)/(Cu \times A \times Gs)$$
$$= (C/N_0)u^{-1} + (N_0d/C_d)$$
$$= (C/N_0)u^{-1} + (C/N_0)d^{-1}$$

Using the obtained values, we get

$$(C/N_0)_T = 8.659 \times 10^{10}$$
$$= 109.375\text{dB}$$

Apart from the information sent through the coded message in a signal, some applications also need the signal propagation parameters like the range travelled by the signal, etc. to be derived in situ by the receiver from the received state of the signal itself. In such satellite applications like altimetry, navigation or radar systems, this derivation is typically based upon the time delay between the transmission and reception of the signals and hence in turn depends upon the nominal velocity of the signal. Therefore, any additional delay added by the medium during the propagation causes error to these applications. In such applications, the additional signal delay caused by the propagating medium is of prime importance. This additional delay is caused by the deviation in the nominal refractive index of the medium which in turn is again a function of its constituents.

So, to round the things up, we have found that the performances of the satellite-based system for different applications are governed by the signal parameters like the received signal power or more precisely by the received C/N_0. The time delay of propagation, etc. also influences the performance in certain applications. We have also seen that the received C/N_0 is a function of the atmospheric path loss, while the delay is dependent upon the refractive index (RI) of the medium and hence in turn to the atmospheric constituents. In our later chapters, we shall see in details how each of the propagating factors affects the C/N_0 term. We shall also see, how much is the effect of the degradation of the C/N_0 on the signal acquisition and data retrieval processes and how we can improve or compensate the deterioration of this term. The reasons for the signal delays will also be explored and how to compensate them will be discussed. However, before going to that, the following section will give us an overview of the atmosphere and its role in causing propagation impairments.

1.3 ATMOSPHERE
1.3.1 ATMOSPHERE AND ITS COMPONENTS

We have mentioned in our discussion in the earlier section that, apart from the free space loss, the losses experienced by the propagating signals are mostly atmospheric. Moreover, we have also seen that the propagation delay, which is an important factor, particularly for navigation signals, is also dependent upon the atmospheric constituents. We shall treat each of the impairments caused by the atmospheric elements separately in relevant chapters. But, before that, we need to recognize the salient features of the impairments the atmosphere offers. The most befit phrase in this regard may be derived from Sun Tzu's 'The Art of War', which says 'If you know your enemy, you can win a hundred battles'. So, to understand and win over the threats posed by the atmosphere to the propagating signals, we first resolve to know the atmosphere and its components. Therefore, with absolute reverence to this saying, in the following part of this section, we attempt to get acquainted with some basic features of the earth's atmosphere.

Earth's 'Atmosphere' is the mixture of gases that surround it and are retained by the gravity of the earth. The atmosphere is denser near the Earth's surface and becomes gradually thinner until it fades away into space (Webster, 2010). Particularly near the Earth's surface, the physical matter in the atmosphere attenuates electromagnetic signals due to absorption, scattering and other phenomena.

The earth's atmosphere is mainly constituted by nitrogen and oxygen in addition to the lighter constituents like hydrogen, helium, etc. Water vapour and other trace gases are also found here. However, the constituents vary in density and composition as the altitude increases above the surface.

The earth's atmosphere may be classified on the basis of its temperature profile with height (Fig. 1.4A). On this basis, it can be divided into four distinct primary layers: the troposphere, stratosphere, mesosphere and thermosphere with transition layers in between. We shall start here by explaining these different layers (Earth, 2015).

Troposphere: The troposphere is the layer of the atmosphere that extends from the earth surface up to about 10–14 km above it. The gases in this region are most dense and composed predominantly of molecular nitrogen (N_2) and molecular oxygen (O_2). All weather-related phenomena happen in this lower region. It contains more than 75% of Earth's atmospheric mass and 99% of the water vapour. All the atmospheric elements that influence the EM wave propagation, like the rain, clouds, vapours, gases, etc., are observed here. This region is abundant with neutral particles whose density exponentially decreases with height. The troposphere is also characterized by gradual decrease in temperature with height. However, at the terminal height of the Troposphere, there is a small region of thickness around 3–5 km, called the 'Tropopause'

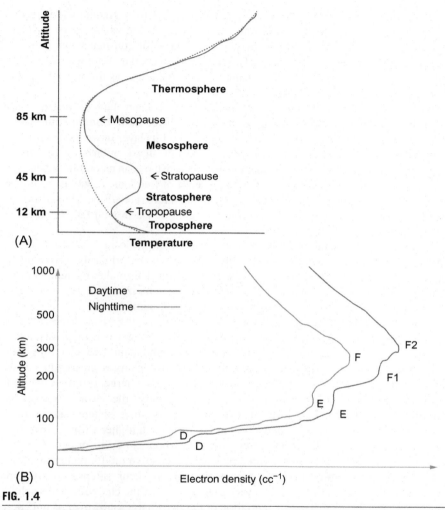

FIG. 1.4

(A) Layers of Earth's atmosphere and (B) vertical profile of Earth's ionosphere.

where the temperature almost remains constant before the inversion in the temperature profile with height takes place in the next layer.

Stratosphere: The stratosphere starts just above the troposphere and extends up to 50 km in height above the earth surface. Within the stratosphere, the incoming solar radiation is able to break up molecular Oxygen (O_2) into individual Oxygen atoms (O), each of which, in turn, may combine with an Oxygen molecule (O_2) to form Ozone, a molecule of Oxygen consisting of three Oxygen atoms (O_3). The concentration of Ozone thus formed is highest near the upper flank of the Stratosphere above 40 km from the earth surface. This concentrated layer of Ozone, called the ozone layer, absorbs ultraviolet radiation from the Sun and consequently gets heated up.

Therefore, the temperature of the Stratosphere is observed to be gradually rising from just above the tropopause to a maximum at around 40–45 km where the ozone concentration is maximum. Above this height, a thin layer of constant temperature is observed in the temperature profile. This layer is called the 'Stratopause' and is observed from 45 to 50 km after which the temperature again inverts in the next layer.

Mesosphere: The temperature above the Stratopause again starts decreasing with height forming a layer called the Mesosphere. This layer starts from around 50 km from the earth surface and extends up to 85 km. From this layer upwards, the neutral particle density gets so rarefied that ions generated by the dissociation of neutral atoms remain in the ionized state for a considerable time before they can recombine again. So, this region onwards is characterized by the presence of free ions. Again at the upper edge of the Mesosphere, a layer of constant temperature, called the Mesopause, is found to exist. The mesosphere is the coldest region on Earth with temperatures as low as $-100°C$

Thermosphere: The thermosphere starts just above the mesosphere and extends to 600 km above the earth surface. The constituents of this layer, primarily consisting of lighter atoms, absorb the energy from the solar radiation. This makes the temperature of this layer very high and most of the particles remain in ionized state. However, due to very low mass density, the total heat content is not significant.

Nevertheless, the atmosphere can also be classified on the basis of the charged particles present. On such a basis, the only major layer that exists is the ionosphere.

Ionosphere: The ionosphere is the layer where abundant free electrons and ionized atoms and molecules are available. These free charges are produced by the ionization of atomic oxygen and nitrogen by highly energetic ultraviolet and X-ray solar radiation. It stretches from about 50 km above the surface to a height of more than 1000 km (Fig. 1.4B). It overlaps with the mesosphere and thermosphere. The atmospheric constituents become so rarefied at higher altitudes that free electrons produced here by photodecomposition method can exist for considerable periods of time before they are captured by a nearby positive ion for recombination. Therefore, although ionization and the recombination occur simultaneously and remain in a dynamic equilibrium, the very long lifetime of the charges creates a definite charge density here. The concentration of the free charges, however, is not constant. This region grows and shrinks based on solar conditions making horizontally stratified layers, viz. D, E and F layers appearing with increasing heights, respectively.

The ionosphere refracts some radio signals transmitted from the ground back towards the earth, enabling communication over great distances as sky waves (Terman, 1955), while passes others to carry out space wave communications. Variability of the ionosphere can interrupt satellite communication, induce errors in satellite-based navigation system and cause severe scintillations in satellite signals. We shall read about the ionosphere in much more details in Chapter 4.

Exosphere: This is the upper limit of our atmosphere. It extends from the top of the thermosphere up to 10,000 km.

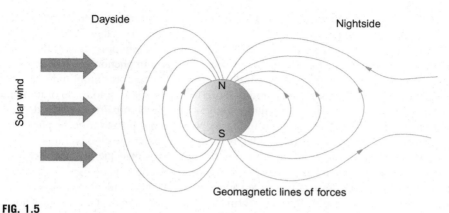

FIG. 1.5

Earth's magnetosphere.

Magnetosphere: It is the region surrounding the earth where the earth's magnetic field is dominant over other fields. To any charged particles moving in this region, the force due to the magnetic field predominantly affects the motion. Earth's magnetic field may be approximated as a field due to magnetic dipole where the dipole is located at the centre of the earth. However, such a field will be symmetric about the axis of the dipole, whereas the earth's magnetic field is distorted due to its interaction with the charged particles of the solar wind. On the dayside, facing the sun, the magnetosphere remains compressed and extends up to about 10 earth radii from the centre of the earth. In the night side, the field is extended to a distance even more than 60 times the earth radius (Hargreaves, 1979; Earth's Magnetosphere, 2013a,b) (Fig. 1.5).

1.4 **ATMOSPHERIC EFFECTS**

That we already have a basic idea about the atmosphere through which the satellite signal propagates, we shall now have an overview of how these layers add impairments to the propagating signals and distort the signal characteristics. A signal is characterized by its amplitude, frequency and phase. An information carrying signal has the information embedded in the systematic temporal variation of any one of these parameters. For this embedded information to be correctly retrieved at the receiver, the received parameters should have values of these characteristic parameters unaltered from those transmitted, save the changes that are expected due to the nominal propagation of the signal over free space. For example, the amplitude is expected to fall due to the free space path loss, the frequency can change due to the Doppler and the phase will have a difference proportional to the propagated path.

However, during the propagation through a medium, the conditions do not remain similar to that of a free space. As the wave passes through any such medium, it interacts with the elements of the medium which have different dielectric and magnetic properties compared to the free space. It consequently either absorbs power from the signal or modifies the refractive index of the medium, or causes the generation of secondary oscillations in the medium resulting in phenomena like scattering, diffraction, etc. Thereby, the medium effectively changes the amplitude, phase, frequency, etc. of the signal compared to those for a free space or to those of normal propagation condition. The signal thus experiences several impairments like depreciation in power, level or variation in the received signal polarization or phase or frequency or added delay or noise, etc. as it passes through the clouds, rains, water vapours or other gases or through the ionized medium. Consequently, the receiver perceives a signal which has the overall deviations from the kind of signal expected. Such deviations make events like failed acquisition, erroneous identification of bits, incorrect derivation of propagation information out of the signal, etc. more probable at the receiver. In other words, it will lead to an overall deterioration of the receiver performance.

Therefore, to ensure a certain quality of service, the impairments that the signal faces over the propagation path are required to be avoided or compensated. We shall study the different impairments and their mitigation techniques in details in subsequent relevant chapters.

1.5 HOW TO BEST UTILIZE THE BOOK?

Before we get more involved with the subject, it is necessary here to briefly talk about the book itself rather than its subject. This section is divided into two parts. In the first part, we shall discuss, how we can utilize this book (or any other similar book) to its best and then, in the second part, we shall discuss the organization of this book.

1.5.1 UTILIZATION OF THIS BOOK

The primary requirement to use this book (or any other similar book) is to know the parts of it. Understanding the sections of a book helps a reader use it more intelligently and enables him to get the maximum benefit out of it. It is customary to divide the printed matter in a book into three broad sections: preliminaries, text matter and subsidiaries (Whittaker, 1972). Out of the preliminaries, the Preface and the Table of contents are the two most important components. Readers, generally, have apathy or a tendency of avoiding it, though it is very much recommended to read this section meticulously. The preface states the author's intention for writing the book. Check, if it matches, at least partly with your intention to read the book. Don't think the contents as just a dumb list of the chapters with page tags. Use this to create a mental map of how the things are logically organized in the book. The following section will tell about the organization of this particular book in a little more details.

The text is the main section of the book that describes all the contents of the book. Undoubtedly, you have to read this section, and that is why you have picked up this book. Read it with thorough details as each sentence that the author has written carries some significance. The most important feature of this book, beyond its text, is the 'Focus' and the MATLAB work outs. You will find such sections attached to every place wherever any important concept has been developed. It is suggestive to the reader to go through these work-out sections and the MATLAB activities, as they come across these sections while reading the text rather than leaving them to be done at the end.

Finally, comes the Bibliography, which is the list of references, books, magazine articles, internet, documents, old notes, verified and authentic online contents and many of such sort that the author has referred to in writing the book. Experienced readers know that every book has its own scope. No author can go beyond that even if he wants to. But what he provides for his readers is the bibliography. Interested readers, can get the answers to their further queries using the details readily available in the references mentioned in the bibliography without much effort.

1.5.2 ORGANIZATION OF THIS BOOK

Having an idea about the organization of the book makes it easy to get a hold over the development of the subject of this book. Actually, in a subconscious way, we make arrangements to keep the information we get in the book organized in our memory. So, when we know about the topics we are going to learn, we create an appropriate space a-priori in our brain even before the information actually comes. Therefore, at the time they really do come, our brain can readily place them at their appropriate prearranged place.

The organization of the book is so made that all the topics are gradually developed with a logical trail from one to the next. This first chapter introduces the reader with the satellite systems and its signals. It gives the overview of the performance metrics of such a system and tells how they are affected by the propagation factors. It then mentions the propagation impairments which affects the signal quality at the receiver and hence deteriorates the receiver performance. It also insinuates the need of compensating the impairment effects.

The second chapter starts with the theoretical foundation of the concerned topics—the Maxwell's equations. In this chapter, the subject of basic electromagnetics is developed from these equations for free space, however, to the extent needed for understanding the formation and the propagation of waves. The following chapter extends the concepts for conditions when the wave is incident on and propagating through a material medium. It is important for our purpose. The different state of affairs arising in the process is explained with emphasis on the electric polarization of the medium and its effect on the electric and magnetic field of the wave, in addition to other features.

The fourth and fifth chapters deal with the impairments related to the ionospheric propagations. Hence, it focuses on the ionosphere and its related features which lead to the physics explaining the impairment it offers to the radio waves propagating through it. In Chapter 4, after describing the ionospheric structures, different physical and empirical models associated with it are described. The characteristic variations in a radio wave arising during the propagation through the ionosphere are explained. Finally, in Chapter 5, the observation of the ionospheric impairments and their measurement techniques have been described. It also addresses the mitigation and applications thereof.

The sixth and the seventh chapter similarly address the impairments pertaining to the tropospheric propagation of radio signals. Chapter 6 describes the tropospheric structure and the physics related to the impairments. It also details the different models used for getting the long-term estimates of the impairments. Chapter 7 contains the description of the mitigation techniques.

Chapter 8 specifically deals with the Noise. It describes the origin of noise and develops the statistical significance of noise. It also establishes the important expressions for the noise terms that get added to the propagating signals. Multipath and interference are also considered and their statistical behaviours and effects on the signal are discussed. Finally, in Chapter 9, the methods of designing a link by establishing the link budget are elaborated.

CONCEPTUAL QUESTIONS

1. What are the factors that may cause degradation of the sky wave signals?
2. Satellite mobile services are more prone to impairments than fixed services—Offer your views on this statement.
3. Which of the transmission parameters can be changed to increase the C/N_0 at the receiver?
4. Other than increased distance, what are the factors which influence the signal quality at low elevation?
5. There are as many free negatively charged electrons as there are positively charged ions in the ionosphere. Which of these are we more concerned about for propagation and why?

REFERENCES

Chakrabarty, B.N., Datta, A.K., 2007. An Introduction to The Principles of Digital Communications. New Age International Pvt. Limited, Chennai.

Chatterjee, B., 1963. Propagation of Radio Waves. Asia Publishing House, Calcutta.

Clarke, A.C., 1945. Extra-terrestrial relays—can rocket stations give worldwide radio coverage? Wirel. World (October), 305–308.

Dolukhanov, M., 1971. Propagation of Radio Waves. Translated from Russian by Boris Kuznetsov, Mir Publishers, Moscow.

Earth, 2015. NASA, http://www.nasa.gov/image-feature/goddard/earths-atmospheric-layers Accessed 14 May 2016.

Earth's Magnetosphere, 2013a. NASA, http://www.nasa.gov/mission_pages/sunearth/science/inner-mag-mos.html Accessed 17 May 2016.

Earth's Magnetosphere, 2013b. NASA—Earth, http://www.nasa.gov/mission_pages/sunearth/science/inner-mag-mos.html Accessed 12 March 2016.

Hargreaves, J.K., 1979. The Upper Atmosphere and Solar-Terrestrial Relations. Van Nostrand Reinhold Co., New York, NY.

Kennedy, F., Davis, B., 1999. Electronic Communication Systems, fourth ed. Tata Mc Graw Hills Publishing Company Limited, New Delhi.

Maral, G., Bousquet, M., 2006. Satellite Communications Systems, fourth ed. John Wiley & Sons Ltd., Chichester.

Mutugi, R.N., 2012. Digital Communication: Theory, Techniques and Applications, second ed. Oxford University Press, Chennai.

Pratt, T., Bostian, C.W., Allnutt, J.E., 1986. Satellite Communications, second ed. John Wiley and Sons Inc., Minneapolis, MN.

Proakis, J.H., Salehi, M., 2008. Digital Communications, fifth ed. Mc Graw-Hill, Columbus, OH.

Relgin, I., Relgin, B., 2001. Telecommunication requirements in telemedicine. Ann. Acad. Studenica 4, 53–61.

Terman, F.E., 1955. Electronic and Radio Engineering. Tata Mac Graw Hills, New Delhi.

Webster's New World Telecom Dictionary, 2010. Wiley Publishing, Inc., Indianapolis, IN.

Whittaker, K., 1972. Using Libraries. Andre Deutsch, London. p. 76.

FURTHER READING

Earth's Atmospheric Layers, 2015. http://www.nasa.gov/image-feature/goddard/earths-atmospheric-layers Accessed 10 March 2016.

Ionosphere, n.d. http://www.swpc.noaa.gov/phenomena/ionosphere Accessed 6 May 2016.

Maxwell's equation and EM waves

2.1 ELECTRIC AND MAGNETIC FIELDS

In this book, we shall deal with the satellite signals which are basically electromagnetic (EM) waves. These waves traverse through space as simultaneous electric and magnetic fields. Each of these fields gets generated from the variations of the other field and they are mutually orthogonal in direction. To understand the signals, it is necessary at the outset to clearly define and explain the origin of the electric and magnetic fields constituting the wave. It is important to realize how these fields are sustained and how they behave, as they propagate through free space or through a medium. In this chapter, we shall learn about the EM wave propagation and its behaviour in free space.

By the term 'Field', in a strict sense, we actually mean any physical quantity which can be specified at all points in the physical space of our interest. It may take different values at different points in this region and may vary with time (Galbis and Maestre, 2012). So, let in a defined region in space, a positive charge exists at a point. A separate independent unit positive electric test charge will then experience a repulsive electric force, at all possible points due to it. The magnitude of the force will be different at different points in this region. So, we can attribute a magnitude and a direction of the force to every point of the space in the region. Therefore, we can say the field of the electric force, or simply the electric field, is present in the mentioned region. The magnitude of the force that will be experienced at any point is called the intensity of the field or the field strength. For the part of the space where the electric force is not perceivable, the intensity is zero and the electric field ceases to exist there. The magnetic field may be defined exactly in the similar manner. Here, both the electric and the magnetic forces are vectors and hence constitute vector fields. Similarly, the fields like the temperature distribution in a room, having no definite direction associated to the value, form a scalar field.

There are different ways to represent a field. To represent a vector field, like the electric field, in a two-dimensional space, we may use arrows. Each arrow will represent the vector physical quantity at the point of its origin. The length of the arrow will represent its magnitude at that point in the field and the head of the arrow will point towards the direction of the vector. Alternatively, we may represent the vector field by 'Lines of Forces'. These are the continuous lines over the whole field such that the tangent on the line at any point represents the direction of the vector at that point. The spatial density of these lines represents its magnitude at any point.

Satellite Signal Propagation, Impairments and Mitigation. http://dx.doi.org/10.1016/B978-0-12-809732-8.00002-8

So, greater the field strength, closer the lines are placed while the lines are dispersed apart for weaker fields. So, the field strength or intensity of a field at any point in a given direction is thus represented by the numbers of lines of forces crossing through a unit area around the point, where the area is normal to that given direction. The two kinds of vector field representations are shown in Fig. 2.1A and B, respectively.

2.2 MAXWELL'S EQUATIONS: ITS MEANING AND IMPORTANCE

The definitions of the fields that we have just now discussed were prerequisites for the rest of this chapter. This will help to explicate the theoretical basis of any propagating EM waves. So, now we are ready to explain the main idea of this chapter—the Maxwell's equations.

An electric charge at rest creates an electric field, imparting electric force on another charge. Similarly, a moving charge or an electric current generates a magnetic field and exerts magnetic force on another current or any other moving charge. When the electric field varies with time, in turn it generates the magnetic field, and in a similar fashion, a time varying magnetic field creates an electric field. Therefore, an oscillating current or accelerating electric charge leads to a time varying magnetic field around it which consequently generates an electric field again. Similarly, a magnetic field, having second-order variation in time, generates the magnetic field around it once again through intermediate generation of varying electric field. In this way, the fields get alternately produced in a coupled manner with the spatial distribution of the fields shifting over space. This forms a wave which, at any location, thus changes with time and also traverses in space. The Maxwell's equations govern the relation between the charge and current with the electric and magnetic fields. They also associate the time and the space variations of these electric and magnetic fields. These relations clearly exhibit the dependence of one field on another and indicate the formation of a wave of combined electric and magnetic fields, resulting what is known as an EM wave. We shall discuss each kind of these fields and their variations once we learn the Maxwell's equations.

Some of you must be wondering why we are set to start with Maxwell's equation. This is because, we want to show here that the Maxwell's equations are the fundamental basis of all the electric and magnetic characteristics of electromagnetic wave and govern all the phenomena that we shall be learning in this book. So, in the following section, we shall first get introduced with these equations and then derive different conditional cases from them.

2.2.1 MAXWELL'S EQUATIONS

Maxwell's equations are the set of equations which define the temporal and spatial characteristics of electric and magnetic field and the interrelations of their variations. If we represent the electric field with **E**, magnetic induction field with **B**, charge

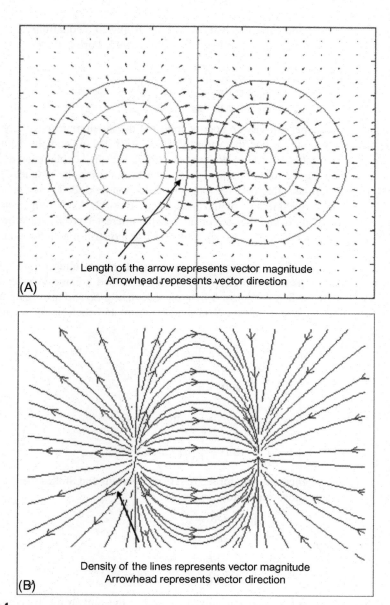

Length of the arrow represents vector magnitude
Arrowhead represents vector direction

(A)

Density of the lines represents vector magnitude
Arrowhead represents vector direction

(B)

FIG. 2.1

Vector field representation (A) using arrows (B) using lines of forces.

volume density with ρ and current density with J, the Maxwell's equations at a point can be written as (Reitz et al., 1990; Feynman et al., 1992; Portis, 1978)

$$
\begin{aligned}
&\text{i.} \quad \nabla \cdot \mathbf{E} = \rho/\varepsilon \\
&\text{ii.} \quad \nabla \cdot \mathbf{B} = 0 \\
&\text{iii.} \quad \nabla \times \mathbf{E} = -\partial \mathbf{B}/\partial t \\
&\text{iv.} \quad \nabla \times \mathbf{B} = \mu \mathbf{J} + \mu\varepsilon \partial \mathbf{E}/\partial t
\end{aligned}
\tag{2.1a}
$$

where ε and μ are the permittivity and permeability of the medium, respectively. For vacuum, the permittivity ε_0 has the value of 8.854×10^{-12} F/m and the value of permeability μ_0 is 1.256×10^{-6} H/m (Reitz et al., 1990). We shall explain and discuss each of these equations in turn. You may find some books using the magnetic field **H** in the above equations rather than the magnetic induction field, **B**. These two terms are related by the permeability μ of the medium as $\mathbf{B} = \mu\mathbf{H}$. As **B** represents the effective magnetic field in a medium, so considering our interest in propagation through medium (including vacuum), here we shall always use the term magnetic induction field **B**.

A better way to visualize these equations is to turn them into integral form. Integrating the equations (i) and (ii) of Eq. (2.1a), over a closed volume V around the point of our interest, which is enclosed by a closed surface S_v and also integrating equations (iii) and (iv) over a surface S, around the point and enclosed by a contour L_s, we get,

$$
\begin{aligned}
&\text{i.} \quad \int_V \nabla \cdot \mathbf{E} \, dV = \int_V (\rho/\varepsilon) \, dV \\[2mm]
&\text{ii.} \quad \int_V \nabla \cdot \mathbf{B} \, dV = 0 \\[2mm]
&\text{iii.} \quad \int_S \nabla \times \mathbf{E} \, dS = -\int_S \partial \mathbf{B}/\partial t \, dS \\[2mm]
&\text{iv.} \quad \int_S \nabla \times \mathbf{B} \, dS = \int_S (\mu \mathbf{J} + \mu\varepsilon \partial \mathbf{E}/\partial t) \, dS
\end{aligned}
\tag{2.1b}
$$

Now using the Divergence theorem for the first two equations and Stokes theorem for the next two, the equations turn into,

$$
\begin{aligned}
&\text{i.} \quad \oint_{S_v} \mathbf{E} \cdot \hat{s} \, dS_v = Q/\varepsilon \\[2mm]
&\text{ii.} \quad \oint_{S_v} \mathbf{B} \cdot \hat{s} \, dS_v = 0 \\[2mm]
&\text{iii.} \quad \oint_{L_s} \mathbf{E} \cdot \hat{w} \, dL_s = -\partial \Phi/\partial t \\[2mm]
&\text{iv.} \quad \oint_{L_s} \mathbf{B} \cdot \hat{w} \, dL_s = \mu I + \mu\varepsilon \, \partial \Omega/\partial t
\end{aligned}
\tag{2.1c}
$$

Here, $\hat{s}$ and $\hat{w}$ represent the normal unit vectors. $\hat{s}$ is the outwardly directed unit normal vector on the surface S_v, enclosing the defined volume V and $\hat{w}$ is the unit vector, tangential along the contour L_s enclosing the area S. Here, $Q = \int \rho \cdot dV$ is the total charge enclosed in the volume considered, $I = \int J \cdot dS$ is the total current flowing over the area S, $\Phi = \int \mathbf{B} \cdot dS$ is the total flux of the magnetic field, while $\Omega = \int \mathbf{E} \cdot dS$ is the total flux for the electric field.

From the concept of the lines of force used in defining the fields and explained earlier, we can understand that the surface integrals of the field, say $\int_{S_v} \mathbf{E} \cdot d\mathbf{S}$, gives the total number of the lines of forces coming out of a volume V enclosed by the surface S_v. Therefore, it represents the effective outward electric field flux from the volume V. Similarly, the line integral, $\oint_{L_s} \mathbf{E} \cdot \hat{w} d\mathbf{L}$, represents the electric work done by the electric fields across a closed loop on unit charge. We shall explain these aspects in details later in this chapter. Here, instead of discussing all these equations at one place, we shall consider two distinct cases—one static, in which the charges and the current do not change with time, and the other dynamic, in which they do vary. Then we shall explain each of these equations with their interpretations and significances under the given assumptions, in turn.

2.3 STATIC CASE: ELECTROSTATICS AND MAGNETOSTATICS

We know by now, only the mathematical form of the Maxwell's equations. Here, we shall show that the theory of static electric and magnetic fields are only the results of enforcing certain conditions on this generalized form. Here, and throughout the rest of this chapter, we shall consider the medium to be vacuum. In Eq. (2.1a) if we assume that the charge and the current are such that they do not change with time, then both $\partial \mathbf{B}/\partial t$ and $\partial \mathbf{E}/\partial t$ terms reduce to zero. Thus, the equations reduce to

$$\text{i.} \quad \mathbf{\nabla} \cdot \mathbf{E} = \frac{\rho}{\varepsilon_0}$$
$$\text{ii.} \quad \mathbf{\nabla} \cdot \mathbf{B} = 0$$
$$\text{iii.} \quad \mathbf{\nabla} \times \mathbf{E} = 0 \quad\quad (2.2a)$$
$$\text{iv.} \quad \mathbf{\nabla} \times \mathbf{B} = \mu_0 \mathbf{J}$$

Notice that, a magnificent thing has happened with this. These assumptions have separated out the terms for the electric and magnetic fields. All the couplings which were there in the original Maxwell's equations between these two types of field were through the first-order derivatives. Now, with our assumptions, those derivatives vanish. As a result, the fields E and B have got beautifully separated from the other's influence and stand independent. In the integral form, the equations turn into

$$\text{i.} \quad \oint_{S_v} \mathbf{E} \cdot \hat{s} d\mathbf{S}_v = \frac{Q}{\varepsilon_0}$$
$$\text{ii.} \quad \oint_{S_v} \mathbf{B} \cdot \hat{s} d\mathbf{S}_v = 0$$
$$\text{iii.} \quad \oint_{L_s} \mathbf{E} \cdot \hat{w} d\mathbf{L}_s = 0 \quad\quad (2.2b)$$
$$\text{iv.} \quad \oint_{L_s} \mathbf{B} . \hat{w} d\mathbf{L}_s = \mu I$$

We shall discuss hereafter, the individual equations for the static case. However, while doing so, we shall treat the two equations for the electric field and the two for the magnetic field separately.

2.3.1 STATIC ELECTRIC FIELD EQUATIONS

Let us consider the equations for the electric fields in Eq. (2.2b) which are

$$\oint_{S_v} \mathbf{E} \cdot \hat{s} dS = \frac{Q}{\varepsilon_0} \qquad \oint_{L_s} \mathbf{E} \cdot \hat{w} dL = 0$$

The first equation is more popularly known as the Gauss' law. It relates the electric field $\mathbf{E}$ on the surface S_v of a closed volume V with the charge Q enclosed in it (Fig. 2.2). It tells us that the net value of the surface integral of the normal component of the electric field, i.e. the net electric outflux from the volume, is proportional to the charge enclosed inside it. The net electric outflux is nothing but the total number of lines of forces emanating from the volume. So, in other words, when there is a finite positive charge ($Q > 0$) enclosed, it acts as a source of electric field, with a total number of Q/ε_0 numbers of lines of forces originating out of it. Similarly, when the enclosed charge is negative ($Q < 0$), equal numbers of lines of forces terminate in the volume. Therefore, if the charge Q becomes zero, the surface integral involving the electric field $\mathbf{E}$ vanishes immediately. The surface integral of the field represents the net electric flux, i.e. numbers of lines coming out from the enclosed volume, minus the numbers of lines going in. So, when Q vanishes, the numbers of lines of forces entering the enclosed volume and those exiting are exactly equal. Therefore, no additional contribution to the electric field lines is made by the enclosed region taken into consideration. Hence, it indicates that the electric charge is the only generator for the electric field.

Since the first equation relates the charge with the field, it should also be equivalent to the Coulomb's law for a point charge. We shall now see that it really is. Consider the equation once more. It states, $\int_{S_v} \mathbf{E} \cdot \hat{s} \, dS = Q/\varepsilon_0$. Now consider a point P at a distance r. A test charge q is placed at this point. The force experienced by this charge q is $\mathbf{F} = q\mathbf{E}$. Again, consider a hollow sphere of radius r centred at the charge Q and hence containing our point P on the surface. Then the surface integral of the field $\mathbf{E}$ over this sphere becomes $\int_{S_v} \mathbf{E} \cdot \hat{s} \, dS$, which is equal to $4\pi r^2 \mathbf{E}$, as per the Gauss Law and due to the symmetry. So, rewriting the equation

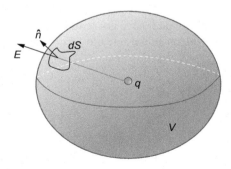

FIG. 2.2

Gauss law.

$$\oint_{S_v} \mathbf{E} \cdot \hat{s}dS_v = \frac{Q}{\varepsilon_0}$$

$$4\pi r^2 \mathbf{E} = \frac{Q}{\varepsilon_0} \qquad (2.3a)$$

$$\text{So,} \quad \mathbf{E} = \frac{Q}{4\pi\varepsilon_0 r^2}$$

This reveals the fact that the field intensity for a point charge follows inverse square law. The force, F, experienced by the charge q can then be written as,

$$F = q\mathbf{E} = \frac{Qq}{4\pi\varepsilon_0 r^2} \qquad (2.3b)$$

This is Coulombs law.

For the electric field, the superposition principle holds good. So, instead of a single charge, if there are a number of distributed charges whose magnitude Q_i and the distances r_i from any reference point are known, then the total field at the point r due to all the charges will just be the vector sum of the individual fields obtained through the above equation.

$$\mathbf{F} = \sum_i \mathbf{F}_i$$

$$= \sum_i \frac{Q_i q}{4\pi\varepsilon_0 (r - r_i)^2} \qquad (2.3c)$$

This is Superposition principle for the electric field. It is emphasized again that the fields being vector, the directions are needed to be taken care of. So, the individual force being in the direction obtained by joining the charge Q_i under consideration to the test charge q, the effective force is along the direction of the vector sum of all the individual components. Two more things are required to be mentioned here. First, the Gauss' law even holds good for cases when the charges change with time. Second, due to the presence of the term ε, the field varies for different medium. We shall study this in the next chapter.

The second equation $\oint_{L_s} \mathbf{E} \cdot \hat{w}dL = \mathbf{0}$, reveals a conservative electric field. The expression, $\oint_{L_s} \mathbf{E} \cdot \hat{w}dL$, represents the work done by the electric field on a unit positive charge over the closed path L_s. As the integral reduces to zero, the work done by the electric field over the closed loop L_s vanishes. So, by definition, the electric field becomes conservative. We shall discuss the matter mathematically in the next subsection. But for now, we can understand its meaning by using the concept of an electric field for a single positive point charge. The field follows an inverse square law and is directed radially outward from this point charge. As the field is symmetric across all radial directions, work is only done when we move the unit test charge radially in or out. The force being perpendicular for all cross radial movement of the test charge, no work is done in such cases. Further, equal work is done for equal radial movement at the same distance from the source charge, irrespective of its angular directions. Work is done in opposite sense for opposite radial movements. So, in the field of a point charge, starting from any arbitrary point, if we reach back to the same point, the whole movement consists of equal outward and inward radial

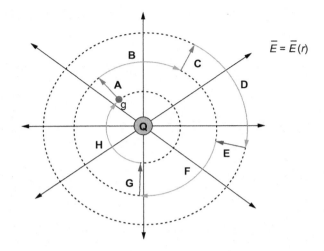

FIG. 2.3

Conservative electric field.

displacement plus some cross radial shift. For example, in Fig. 2.3, the sum of the work done in moving a charge in section A and C is equal and opposite to the work done in section G and E respectively, while no work is done in section B, D, F and H. So, the effective work done is zero. So, when there is a point charge or distribution of point charges, each individually exerts a radial inverse square law force and the field remains conservative.

Mathematically, for a force which varies as $\mathbf{F} = k/r^2 \hat{r}$, where k is a constant, the work done by the force on a unit charge over a closed loop is $W = \oint_{L_s} \mathbf{F} \cdot d\mathbf{L}$, where L_s defines the closed contour of the loop. Using Stokes theorem, this expression of work becomes equal to $\int_S \mathbf{\nabla} \times \mathbf{F} \cdot d\mathbf{S}$, where S is the area enclosed by the contour defined by L_s. It can be shown that $\nabla \times (\hat{r}/r^2) = 0$. Therefore, for such inverse square forces, like the one which originated from an individual point charge, $\nabla \times F = 0$, which implies $\oint_{L_s} \mathbf{F} \cdot d\mathbf{L} = 0$. Hence, the work done on a unit charge, over a closed loop in the field of such a force, is zero. It makes the force conservative. Due to the superposition principle, it is also true for a field by a distribution of charges, too. It also implies that any finite work done in such a force field is also path-independent and depends upon the initial and final position only.

Finally, it is also important to mention here that, since $\mathbf{\nabla} \times \mathbf{E} = 0$, $\mathbf{E}$ can be expressed as the divergence of a scalar function, Φ, i.e. we can represent $\mathbf{E}$ as $\mathbf{E} = -\nabla\Phi$. Again using the relation $\mathbf{\nabla} \cdot \mathbf{E} = \rho/\varepsilon_0$, we get the Poisson's Equation as,

$$\mathbf{\nabla} \cdot \mathbf{E} = \rho/\varepsilon_0$$
$$\text{Or,} \quad -\mathbf{\nabla} \cdot \mathbf{\nabla}\Phi = \rho/\varepsilon_0 \qquad (2.4)$$
$$\nabla^2\Phi = -\rho/\varepsilon_0$$

2.3.2 STATIC MAGNETIC FIELD EQUATIONS

Now, let us consider the equations for the magnetic fields in Eq. (2.2b) which are

$$\oint_{S_v} \mathbf{B} \cdot \hat{s} dS = 0 \qquad \oint_{L_S} \mathbf{B} \cdot \hat{w} dL = \mu_0 I$$

The first equation represents the nonexistence of magnetic monopole. From the equation $\oint_{S_v} \mathbf{B} \cdot \hat{s} dS = 0$, we find that the integral of the normal magnetic induction field $\mathbf{B}$ over any arbitrary closed surface S_v is zero. Using the same explanations as used for the electric fields, we can say that for a magnetic field, for any arbitrary closed volume, the numbers of lines of forces entering will be equal to the number of lines of forces emerging and this is true for any region of choice. No net magnetic lines of forces are added or eliminated in this closed region under any circumstance. Now, the lines of forces in a field can originate from or converge to a single magnetic pole. Therefore, in a magnetic field, the abovementioned fact is possible when there is no separate individual magnetic pole. In other words, equal magnetic poles of opposite kinds always exist together, so there cannot be any magnetic monopole. As it is true for any arbitrarily chosen volume, it can be concluded that magnetic monopoles are universally nonexistent. We have seen that the electric field lines generate from or terminate at the electric monopoles, i.e. the charges. But, unlike electric fields, the magnetic lines of forces are closed without having any discrete point of origin or termination.

The last equation is called the Ampere's law. It says that if you consider any arbitrary closed loop, then the path integral of the magnetic induction field $\mathbf{B}$ along the contour of that loop is proportional to the net current that crosses through the area enclosed by loop. The term net current means the effective sum of all the currents, considering their directions. So, mathematically, it can be written as

$$\oint_{L_S} \mathbf{B} \cdot \hat{w} d\mathbf{l} = \mu_0 \sum_i I_i$$
$$= \mu_0 \int_S J \cdot dS \tag{2.5}$$

where, L_S is the path along a closed contour enclosing an area S and the summation is on all currents crossing through that area. This is true for any shape of the loop over which the path integral has been taken. So, given the net current flowing across a closed area S, we can obtain the path integral of the magnetic induction field, $\mathbf{B}$ along the contour path enclosing the area. Conversely, the spatial distribution of the magnetic field being known, we can find out the net current that flows across this area S by integrating the field $\mathbf{B}$ over the closed path. If the loop we consider does not contain any net current, the path integral of the field remains zero. This triviality also implies,

$$\oint \mathbf{B} \cdot \hat{w} dl = 0$$
$$\text{Or,} \quad \nabla \times \mathbf{B} = 0 \tag{2.6}$$
$$\text{Or,} \quad \mathbf{B} = -\nabla u$$

So, this equation also indicates that, for an region defined by a closed path with no current flowing through it, **B** can be expressed as the gradient of a scalar function.

We have derived the Coulombs law for obtaining the electric field for a point charge from the Maxwell's equation. In a similar way, we can also derive the Biot-Savart's law for magnetic field using it. We shall adopt a comparatively easier way to show the same.

Let us suppose, there is an infinitely extended conductor with current I as shown in the Fig. 2.4. Using Ampere's law, we get that the magnetic field B at a point P at radial distance 'r' from it is

$$B = \mu_0 \frac{I}{2\pi r}$$

$$= \left(\frac{\mu_0}{4\pi}\right) I \frac{2}{r} \tag{2.7a}$$

Since 'r' is constant, we can write $(2/r)$ as a definite integral value of $\int_{\theta=0}^{\pi} (\sin\theta)/r \, d\theta$, where θ is the angle between the direction of the current element dl on the wire and the direction vector from this element to the point, P, and runs from 0 to π.

$$\text{So,} \quad B = \left(\frac{\mu_0}{4\pi}\right) I \int_{\theta=0}^{\pi} \left(\frac{\sin\theta}{r}\right) d\theta$$

$$= \left(\frac{\mu_0}{4\pi}\right) I \int_{\theta=0}^{\pi} \left(\frac{r\sin\theta}{r^2}\right) d\theta \tag{2.7b}$$

But, as $r = s\sin\theta$, where s is the linear distance from the element dl of the wire to the point P, we can write,

$$B = \left(\frac{\mu_0}{4\pi}\right) I \int_{\theta=0}^{\pi} \frac{r\sin\theta}{s^2 \sin^2\theta} d\theta$$

Again, for the element of length dl of the conductor, $-(r/\sin^2\theta) \, d\theta = dl$. Hence,

$$B = -\left(\frac{\mu_0}{4\pi}\right) I \int_{-\infty}^{+\infty} \frac{dl \sin\theta}{s^2}$$

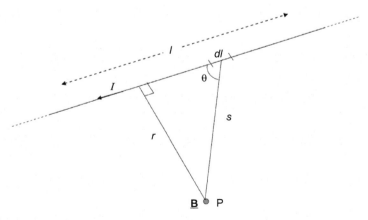

FIG. 2.4

Biot-Savart's law.

So, the contribution to B by the element dl is dB, where

$$dB = -\left(\frac{\mu_0}{4\pi}\right) I \frac{sin\theta \, dl}{s^2}$$
$$= \left(\frac{\mu_0}{4\pi}\right) I \overline{dl} \times \frac{\hat{s}}{s^2}$$

(2.7c)

This is the Biot Savart's Law in differential form, the general relation between the current in a conductor and the consequent magnetic field at a finite distance.

2.3.3 ELECTROSTATIC AND MAGNETOSTATIC ENERGY

2.3.3.1 Electric scalar potential φ and electrostatic energy

Since the electric field exerts force on a charge particle and a magnetic field exerts force on a current carrying conductor, which is nothing but a collection of moving charge, they have the capacity to do work. While potential energy is defined as the capacity of doing work stored in a system of bodies, the term Potential is defined for a point in the field. For electrostatic case, the potential φ at a point is the work done (by external agency) to bring a unit positive charge to that particular point from infinity where the field intensity is zero. In other words, it is the work to be done by a point charge to exist at that point within the field. The potential energy U of a charge q located at a point with potential φ is $U = \varphi q$.

For the static case, the only force experienced by an electric charge at rest is the electric field. So, the electric potential of a point is

$$\varphi(r) = \int_{\infty}^{r} \mathbf{E}(r) \cdot d\mathbf{r}$$

(2.8a)

Taking the gradient on both sides of Eq. (2.8a), we get

$$-\nabla\varphi(r) = -\nabla\left[\int_{\infty}^{r} E(r) \cdot d\mathbf{r}\right] = \mathbf{E}(r)$$

Therefore, this electric potential φ is similar to the Φ we got in Section 2.3.1. However, we still do not say that they have to be equal. We can choose Φ in such a way that Φ and φ may be related as $\Phi = \varphi + g$, where $\nabla g = 0$. So, the gradient of g being zero, its addition does not affect the equality of the gradient of their sum to **E**.

To understand the characteristics of potential, let us take the specific case of a field of a source point charge 'q'. First, consider the hypothetical condition of a unit positive test charge located at infinite distance from the charge q in free space. At this distance, the former will remain out of the influence of the charge with no electrostatic force acting upon it. So, with no mutual force existing, the potential energy of the two charge system under this condition is zero. But, as the test charge comes within a finite range 'r', it feels a repelling force (since the source charge is considered positive here and the test charge is positive by definition). Considering the Coulombs law, the force F on the unit positive point test charge is $\mathbf{F} = +\dfrac{q}{4\pi\varepsilon_0}\dfrac{1}{r^2}\hat{\mathbf{r}}$, where $\hat{\mathbf{r}}$ is the unit vector directed to outward radial direction from the point charge q and 'r' is the magnitude of the distance between the charges. As the test charge is moved towards the source charge q against the repelling Coulomb's force, work is done, by the external agency on the test charge, as a result. Therefore, the work in bringing the unit positive test charge from infinity to a point P by the external agency against the electric force due to 'q' to a distance 'r' from q, amounts to

$$U(r) = \int_\infty^r \frac{q}{4\pi\varepsilon_0}\left(\frac{1}{r^2}\right)\hat{r}\cdot\overline{dr}$$

$$= -\int_\infty^r \frac{q}{4\pi\varepsilon_0}\left(\frac{1}{r^2}\right)dr$$

$$= \frac{q}{4\pi\varepsilon_0}\left(\frac{1}{r}\right)\Big|_\infty^r \quad\quad (2.8b)$$

$$= \frac{q}{4\pi\varepsilon_0}\left(\frac{1}{r}\right)$$

By definition, the potential at this point P, thus becomes

$$\varphi(r) = U$$
$$= \frac{q}{4\pi\varepsilon_0}\left(\frac{1}{r}\right) \quad\quad (2.8c)$$

For a distribution of charge with charge density ρ,

$$\varphi(r) = \int_\infty^r \rho\frac{dV}{4\pi\varepsilon_0 r} \quad\quad (2.8d)$$

So, it gives a definite solution of the Poisson's equation $\nabla^2\varphi = \rho/\varepsilon$, which we obtained in Eq. (2.4). In case of a point charge, since the potential is only dependent upon the radial distance r, irrespective of its direction, the equipotential surfaces form hollow spheres around the charge q. Now, if a charge q_1 is placed at that point P, the potential energy of this two charge system thus becomes,

$$U(r) = q_1\varphi(r)$$

The magnitude of this potential increases as we move the charge q_1 towards the source charge q, since more work is done in bringing the test charge closer to the source against the repulsive force. That is why the direction of the increasing potential is just opposite to the direction of the electrostatic force. So, to get back the vector field $E(r)$ with its direction from the potential $\varphi(r)$, we need to use the negative sign as given in the relation,

$$E(r) = -d\varphi/dx|_{x=r} \quad\quad (2.9a)$$

For a three-dimensional space, the field vector can be obtained as the gradient of the scalar potential φ given as

$$E(r) = -\nabla\varphi \qu\quad (2.9b)$$

Since the potential is a scalar quantity and superimposition principle holds good, the potential at a point due to a number of charges is only the scalar sum of the potentials at that point due to all individual charges. Therefore, to estimate an electric field at a point due to a distributed charge, it is sometimes easier to calculate the total potential at the point due to all the charges and then take its gradient, than directly finding the field from the individual charges using Coulomb's law.

From the above expression, it can be easily derived that for a system of 'n' charges, distributed over space as shown in Fig. 2.5, the total potential energy

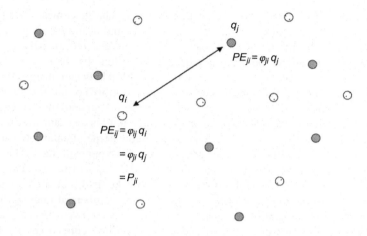

FIG. 2.5

Electrostatic potential.

contained is dependent upon their distribution and also on the magnitude of the charges. The expression for the total potential energy is given by

$$U = {}^1/_2 \sum_i \sum_j \varphi_{ij} q_j \tag{2.10}$$

In the above expression, the potential for the ith charge at the location of the jth charge is φ_{ij} and hence the potential energy is $\varphi_{ij} q_j$. Note that the summation is considered over all the potentials with all the charges. Now, to explain the reason why a factor ½ is introduced, first we need to understand where the potential energy actually resides.

To understand this, we consider a system of two similarly charged metallic balls. When these two balls are brought close enough to each other and hold in that position, a repulsive force is experienced. Thus, in holding the balls in their position, some effort is necessary to put and consequently potential energy get stored therein. But, where is this energy stored? It cannot be in any one of the balls, because if you release the other, it will race out exhibiting kinetic energy converted from this potential energy and hence negating our assumption. It cannot even be shared between the two balls, as each individual ball on release will show kinetic energy exactly equal to the total potential energy. Then it must be in the total system of the two bodies, considered as a whole. The energy is equal to the energy needed to hold these charges in place and thus maintaining the field. Similarly, any system of distributed charge results in an energy contained in the system considered as a whole. This energy is equal to the work done in holding the charges at their positions or equivalently maintaining the resultant field in that volume due to the distribution. This makes it clear that in the above distribution of charges, $\varphi_{12} q_2$ and $\varphi_{21} q_1$ represent the same potential energy between the charges 1 and 2. Therefore, a necessary factor of ½ is introduced to remove the effect of this double count.

From above, it has now become evident that when a charge comes in the field of another charge, the field lines and the associated potential of the region get redistributed. Further, to bring two or more such separate charges and to redistribute and hold the resultant field, some work is required to be done by the external agency on the system. Therefore, there is an essential relationship between the resultant field produced in a given volume and the energy contained in it. To find this relation, at any arbitrary point P, we shall consider a field E and its associated lines of forces in a small unit cubic volume about the point, such that the field is constant within it. This is shown in Fig. 2.6. The lines of force enter the volume perpendicularly through one unit surface and go out from the other unit surface at the opposite face. The energy density at P is the work done to create and hold these lines of forces. The energy density for a given field E is same irrespective of how the field is created. Therefore, it is same as the work done in creating exactly the same numbers of lines of forces in this unit volume by separating necessary charges across the two opposite faces.

From Gauss' law, it can be shown that for such an unit cube, a surface charge density of $\pm\sigma$ on two opposite faces will create a field $E = \sigma/\varepsilon_0$ in the cubic space. So, for creating a field E, the amount of charges required to be separated at each unit surface is $\sigma = \varepsilon_0 E$. The work done in such separation starting from zero field condition untill the field becomes E is the energy contained in the field, E, in this volume. Therefore, at any instant of such separation, if the field existing be $E(\sigma)$, then the incremental amount of work done in separating an infinitesimal charge $d\sigma$ across the unit length is

$$dU = E\,d\sigma$$
$$= E\varepsilon_0 dE \tag{2.11a}$$

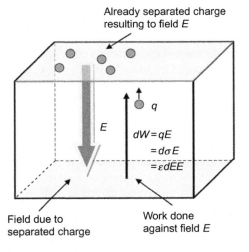

FIG. 2.6

Electrostatic energy density.

Then, the total work done U_1, in separating them across a unit length to finally generate field E is

$$U(E) = \int_0^E \varepsilon E dE \tag{2.11b}$$
$$= \frac{1}{2}\varepsilon E^2$$

The definite integral limits between zero as the initial field, when there is no separated charge to the final value E and the complete field has been developed inside the volume considered. Thus, we get the energy density in an electric field at a point with intensity E as $U = \frac{1}{2}\varepsilon E^2$.

2.3.3.2 Magnetic vector potential A and magnetostatic energy

Unlike the electric field, the magnetic field cannot be defined as a gradient of a scalar potential. This is because, for a field to be expressed as gradient of a scalar, it must be curl free. But **B** is not curl free, in general, i.e. $\nabla \times \mathbf{B} \neq 0$. But, B is divergence free, i.e. $\nabla \cdot \mathbf{B} = 0$. So, considering the fact that $\nabla \cdot \nabla \times \mathbf{A} = 0$, where **A** is any arbitrary vector, we can always represent **B** as the curl of a vector. So, we can always define a vector **A**, such that,

$$\mathbf{B} = \nabla \times \mathbf{A} \tag{2.12}$$

A is called the magnetic vector potential for the field *B*. We shall now go a little bit more deep into this term to understand what exactly this magnetic vector potential means. Whenever a current flows through a conducting loop, it generates a magnetic field and the loop acts like a magnetic dipole. Now, let us consider, there is already a magnetic field present in the region with magnetic induction **B** and an infinitesimally small current carrying loop is brought to a definite point *P* in this field. If the magnetic dipole moment of ths loop is $d\mu$ (remember, it is different from the magnetic permeability), it can be shown that, in terms of the loop parameters, the moment $d\boldsymbol{\mu}$ can be represented as $d\boldsymbol{\mu} = Id\mathbf{S}$. Here, *I* is the current carried by the loop and $d\mathbf{S}$ is its area. $d\boldsymbol{\mu}$ will naturally tend to align along *B* when the loop is set free. Moving the loop needs external work to be done against the torque acting upon it in this condition. So, the maximum potential energy that it can have at this point *P* is $dU = d\boldsymbol{\mu} \cdot \mathbf{B} = Id\mathbf{S} \cdot \mathbf{B}$. In this condition, the moment $\boldsymbol{\mu}$ is directed in direction opposite to the direction of the field, **B**. It will have minimum potential energy and zero torque when the dipole direction is aligned with that of the magnetic field. These are shown in Fig. 2.7.

It can be shown that, the total potential energy, dU_{Total}, that the loop can posses here is quantitatively equal to the mechanical work needed to bring this current carrying loop to this position from infinity (Feynman et al., 1992). So, we can write, the work done to keep this loop in such a position at this point is

$$dU = Id\mathbf{S} \cdot \mathbf{B} \tag{2.13a}$$

Now, any closed contour of arbitrary shape can be imagined to be the sum of such infinitesimal small loops. Therefore, the total work done in bringing a loop of area *S* to that point in this field is

$$U = I \int_S \mathbf{B} \cdot d\mathbf{S}$$

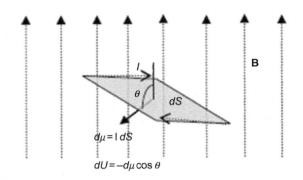

FIG. 2.7

Magnetostatic potential.

As $\nabla \times \mathbf{A} = \mathbf{B}$, replacing the expression for B and then using the Stokes law to replace the surface integral by line integral over the closed contour length L bounding the area S,

$$U = I \int_S \nabla \times A \cdot dS$$
$$= I \oint_L A \cdot dl$$

(2.13b)

Therefore, the work done per unit current is

$$U/I = \oint_L A \cdot dl$$

Therefore, we can consider the magnetic vector potential $\mathbf{A}$ to be a parameter in the magnetic field such that, the total mechanical work done in bringing and holding a magnetic dipole, created by a current loop of unit current in the magnetic field, is equal to the contour integral of A along the same closed loop. Now the value will be different depending upon the orientation of the loop at the same point. This renders A a vector quantity.

We shall now consider some features of A which will be needed in our subsequent discussions. Since $\mathbf{B} = \nabla \times \mathbf{A}$, then adding any curl free variable to A will give rise to the same $\mathbf{B}$. So,

$$\mathbf{B} = \nabla \times \mathbf{A} = \nabla \times (\mathbf{A} + \mathbf{d})$$

(2.14a)

where $\nabla \times \mathbf{d} = 0$. Again, as the condition $\nabla \times \mathbf{d} = 0$ is satisfied, we can write, $\mathbf{d} = \nabla g$, where g is a scalar. So, we can, without changing any physical reality, write

$$\mathbf{B} = \nabla \times (\mathbf{A} + \nabla g)$$

(2.14b)

This gives us the freedom to add divergent terms to A to choose $A' = (A + g)$, such that it eases our mathematical calculations. Therefore, we choose that value of A' as the effective magnetic vector potential such that $B = \nabla \times \mathbf{A'}$ and $\nabla . \nabla \times \mathbf{A'} = 0$. This is called 'Gauge transform' and we shall use this technique again later while considering the dynamic case. Notice that, due to curl free nature of '$\mathbf{d}$', this does not change the significance of $\mathbf{A}$, which is discussed above.

In a manner, similar to the estimation of the electrostatic energy density, magnetostatic energy density may be derived to get

$$U = \frac{1}{2}\mathbf{B}^2/\mu \tag{2.15}$$

This is given as an exercise at the end of this chapter. Similarly, as in the case of electrostatics, the magnetostatic energy may also be derived by considering the work done in maintaining the magnetic field, **B**, by the generating current.

2.4 DYNAMIC CASE

Things really change drastically from what we have seen so far for the static case, when both the charge and the current vary with time. The divergence of the fields representing the spread of the electric and magnetic fields over space remains unchanged for the dynamic cases. However, for the expressions involving the curl of the fields, new terms arise in dynamic cases. To understand the effects of the changes, first consider the Eqs (2.1a)(iii), (2.1c)(iii).

$$\mathbf{\nabla} \times \mathbf{E} = -\partial \mathbf{B}/\partial t$$
$$\text{Or,} \quad \oint_L \mathbf{E} \cdot dl = -\int_S \partial \mathbf{B}/\partial t \cdot d\mathbf{S} = -\partial \Phi/\partial t$$

where **E** is the electric field and Φ is the total magnetic field flux, given by the surface integral of the magnetic induction field, B. It represents the total number of lines of forces in a given area, **S**.

The significance of this equation is that, it says that the electric field may be generated out of the variation of the magnetic field. So, if there is a temporal change in the magnetic field B linked to the area S, it generates an electric field. The spatial distribution of this generated electric field is such that its path integral over the closed loop is proportional to the rate of change of the magnetic flux linked with the area contained by the loop. For a conductor, the path integral of the electric field over a closed loop is nothing but the emf. So, as per the equation, an emf is developed in a closed conductor when a magnetic flux attached to it changes with time. This is Faraday's law. The electric field no longer remains conservative. The conservative character of the field breaks under the dynamic condition, and the Faraday's law is defined.

Before moving to the next section, it is important to mention here a few important points. In the above case of an electric field being generated from the change in magnetic flux, we can write

$$\mathbf{\nabla} \times \mathbf{E} = -\partial/\partial t(\mathbf{\nabla} \times \mathbf{A})$$
$$= -\mathbf{\nabla} \times (\partial \mathbf{A}/\partial t) \tag{2.16a}$$

Therefore, when the total electric field is generated from both electric charges and magnetic flux variations, we can write

$$\mathbf{E} = -\mathbf{\nabla}\varphi - (\partial \mathbf{A}/\partial t) \tag{2.16b}$$

So, when we take the curl of this total electric field E, the gradient term vanishes and only the second term to the right hand side of the equation containing the vector potential, A, exists.

Considering the equation $\nabla \times \mathbf{B} = \mu_0 \mathbf{J} + \mu_0 \varepsilon_0 d\mathbf{E}/dt$, we find that it is an extension over the Ampere's law. The second term to the right hand side of this equation is an addendum over the original Ampere's law. From the first part, it is clear that the path integral of $\mathbf{B}$ over a closed loop is proportional to the current flowing through the loop. However, this equation of Maxwell adds an extra term in addition to the conduction current density $\mathbf{J}$. It states that a change in the electric field with time also contributes to the spatial distribution of the magnetic field. Quantitatively, it means the term $\varepsilon_0 d\mathbf{E}/dt$ also contributes to produce $\mathbf{B}$ similarly as $\mathbf{J}$. Therefore, the spatial distribution of the magnetic field is now also a function of the temporal variation of the electric field. This is the significance of the Maxwell's own contribution over the Ampere's law in the Maxwell's equations.

This equation treats the variations of the electric field, or more precisely, the variation of electric displacement $\mathbf{D} = \varepsilon \mathbf{E}$ (which we shall define in the next chapter), equivalent to a current as far as the generation of the magnetic field is concerned. Now, let us find out the rationale for the equivalence of electric field variation and current density. If in a circuit with a current density $\mathbf{J}$, the constituent charges are allowed to get stored on a parallel plate capacitor, considering the conservation of charge, the variation of the surface charge density σ will lead to the relation $d\sigma/dt = \mathbf{J}$. Now, if this stored charge density σ generates an electric field E, then they are related by the equation $\mathbf{E} = \sigma/\varepsilon$. Therefore, the time differentiation of this field will be,

$$dE/dt = \left(\frac{1}{\varepsilon_0}\right) d\sigma/dt = \left(\frac{1}{\varepsilon_0}\right) \mathbf{J}$$

$$\text{or,} \quad \mathbf{J} = \varepsilon_0 dE/dt$$

(2.17)

Here, the term $\varepsilon_0 d\mathbf{E}/dt = d\mathbf{D}/dt$, which is the time rate of change in the displacement vector, $\mathbf{D}$, is not only quantitatively equal to the conduction current density, $\mathbf{J}$ but also behaves like it. It is called the displacement current. In our next chapter, we shall see why it is so named and how this current is dependent upon the medium and whether or not it varies with frequency.

The above discussion clearly points out the fact that for a changing electric field, the term $\varepsilon d\mathbf{E}/dt$ can be treated equivalently as the conduction current density J. Thus, at places where a part of the conducting charges gets accumulated resulting in the changing electric field with time, a displacement current is generated. The other part that flows freely constitutes the conduction current. This is shown in Fig. 2.8. The total magnetic field is developed by both the current components. Thus, the overall expression is then obtained by adding these two equivalent current density terms in the original Ampere's equation, to get Eq. (2.1a)(iv) as

$$\nabla \times B = \mu_0 J + \mu_0 \varepsilon_0 dE/dt$$

2.5 WAVE EQUATIONS AND THEIR SOLUTION

We have learnt that, according to the Maxwell's equation, whenever there is a temporal change in the magnetic field, an electric field is generated whose nature of spatial variation is governed by Eq. (2.1a)(iii). Similarly, a temporal variation in the electric field leads to the generation of a magnetic field according as Eq. (2.1a)(iv). Thus, if the electric field has a second-order variation, it leads to a time changing

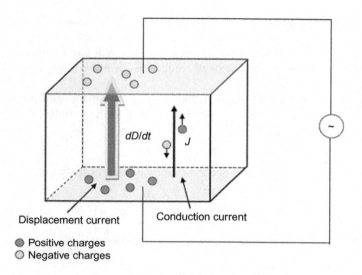

Displacement current   Conduction current

● Positive charges
○ Negative charges

FIG. 2.8

Conduction and displacement current.

magnetic field which in turn again generates an electric field. Now, for the system to sustain unabated, this generated electric field must be equal to the original electric field. Considering this as our criterion, let us see the characteristic features that imply the original electric field in space and time.

Using Eq. (2.1a)(iv) and assuming there is no source of current in the region we are considering, we can write

$$\nabla \times \mathbf{B} = \mu_0 \varepsilon_0 d\mathbf{E}/dt \tag{2.18a}$$

So, the first-order variation in electric field generates a magnetic field. Rearranging the above equation, and differentiating the electric field once again with time, we get

$$d\mathbf{E}/dt = \frac{1}{\mu_0 \varepsilon_0} \nabla \times \mathbf{B}$$
$$\text{or,} \quad d^2\mathbf{E}/dt^2 = \frac{1}{\mu_0 \varepsilon_0} d/dt(\nabla \times \mathbf{B}) \tag{2.18b}$$
$$= \frac{1}{\mu_0 \varepsilon_0} \nabla \times (d\mathbf{B}/dt)$$

But, as $d\mathbf{B}/dt = -\nabla \times \mathbf{E}$, so replacing, we get

$$d^2\mathbf{E}/dt^2 = -\frac{1}{\mu_0 \varepsilon_0} \nabla \times (\nabla \times (\mathbf{E})) \tag{2.18c}$$

Again, using the vector identity, $A \times B \times C = B\,(A \cdot C) - C\,(A \cdot B)$, we get

$$d^2\mathbf{E}/dt^2 = -\frac{1}{\mu_0 \varepsilon_0} \left[\nabla(\nabla \cdot \mathbf{E}) - \nabla^2 \mathbf{E} \right] \tag{2.18d}$$

Assuming there is no free charge in the space considered, we can write $\nabla \cdot \mathbf{E} = \rho/\varepsilon = 0$. Then

$$d^2\mathbf{E}/dt^2 = \frac{1}{\mu_0 \varepsilon_0} \nabla^2 \mathbf{E}$$
$$\text{or,} \quad \nabla^2 \mathbf{E} = \mu_0 \varepsilon_0 \, d^2\mathbf{E}/dt^2 \tag{2.19a}$$

This is a wave equation of the form $d^2y/dx^2 = (1/v^2) \, d^2y/dt^2$, where y is the amplitude of the wave, x is the space coordinate, t is the time and v is the velocity of the wave. This means that whenever there is an electric field which has a nontrivial second-order variation, following the Maxwell's equations, it forms a wave which propagates with velocity $v = 1/\sqrt{(\mu_0\varepsilon_0)}$.

You may wish to see if the magnetic field also behaves in the same manner or not. It will do so and we shall convince ourselves about it from the following derivation. Using Eq. (2.1a)(iii), we get,

$$d\mathbf{B}/dt = -\nabla \times \mathbf{E}$$

Similarly, as before, considering second-order variation of B we get,

$$
\begin{aligned}
d^2\mathbf{B}/dt^2 &= -\nabla \times [d\mathbf{E}/dt] \\
&= -\nabla \times \left[\frac{1}{\mu_0\varepsilon_0} \nabla \times \mathbf{B} \right] \\
&= -\frac{1}{\mu_0\varepsilon_0} \left[\nabla(\nabla \cdot \mathbf{B}) - \nabla^2(\mathbf{B}) \right]
\end{aligned}
$$

Using the fact $\nabla \cdot \mathbf{B} = 0$ from Eq. (2.1b), we get

$$d^2\mathbf{B}/dt^2 = \frac{1}{\mu_0\varepsilon_0} \nabla^2\mathbf{B}$$

$$\text{or,} \quad \nabla^2\mathbf{B} = \mu_0\varepsilon_0 \, d^2\mathbf{B}/dt^2$$

(2.19b)

Hence, we reach to the same wave equation for B.

We need to remember, however, that the wave equation, indicating propagating electric field or magnetic field could only be obtained when there is a resultant intermediate generation and variation of the magnetic or electric field, respectively, following the Maxwell's equation. This is important and this establishes the fact that the wave would not be generated unless there are both electric and magnetic field alternatively generating from each other.

So, one may try imagining the scenario of an EM wave traversing with the electric field strengths changing over space and progressing from the source in a direction away from it. The whole spatial pattern of the field is moving in that definite direction. At any definite snapshot instant, there is a spatial variation of the field along the path of the propagation of the field. Moreover, in the process of this propagation, the field intensity is changing with time at a definite location in the path of the field.

2.5.1 SOLUTION OF WAVE EQUATION AND CONDITIONS FOR EM WAVE

We have seen from the Maxwell's equations that a varying electric and magnetic field form propagating EM waves with formation of the each type of field from the variation of the other. Now, it is the right time to explore the nature of the wave it forms. The general solution comes from Eqs (2.19a), (2.19b) and the solution for such a wave equation is any function of $(x - ct)$, where c is the velocity of the wave. So we can write

$$E = f_1(x - ct) \text{ and } B = f_2(x - ct)$$

(2.20)

One can easily verify this by putting these arbitrary functions f_1 and f_2 into the wave equations. So, consider any two arbitrary functions of $(x-ct)$ as f_1 and f_2 such that, $\mathbf{E}=f_1(x-ct)$ and $\mathbf{B}=f_2(x-ct)$, respectively. Now, taking the function for the electric field and differentiating the function $\mathbf{E}$ with t, we get,

$$d\mathbf{E}/dt = (-c)f_1'$$
$$d^2\mathbf{E}/dt^2 = c^2 f_1'' \tag{2.21a}$$

where f' and f'' are, respectively, the first- and second-order time derivatives of the function f_1. Again, differentiating it with x, we get,

$$d\mathbf{E}/dx = f_1'$$
$$d^2\mathbf{E}/dt^2 = f_1'' \tag{2.21b}$$

So, comparing Eqs (2.21a), (2.21b), we find that $d^2E/dx^2 = 1/c^2 \, d^2E/dt^2$, which thus satisfies our wave equation. The same result can be obtained for the function $B = f_2(x-ct)$, too.

We have found that all of the time-space functions of the form $f = f(x-ct)$ satisfy the wave equation and can be designated to represent $\mathbf{B}$ and $\mathbf{E}$ separately. It is apt to ask at this point that, whether any of these functions, chosen arbitrarily, can represent an EM wave. I believe, you all have the intuitive answer—No. Although the wave equation has been derived from the Maxwell's equations, merely satisfying the wave equation does not guarantee that any arbitrarily designated pair of such function will follow the Maxwell's equations and can represent E and B. In other words, those functions which follow Maxwell's equations are only a subset of all these functions which follow the wave equation. The following focus explains that not all of the functions can represent an EM wave.

Focus 2 .1

Our objective in this Focus is to show that not all functions of the form $f = f(x-ct)$ represent EM waves. For the purpose, consider a magnetic induction field,

$$B = B_0 \sin(\omega t - ky)\hat{j}$$

The field is a function of $(y\text{-}ct)$ and propagating with velocity $c = \omega/k$ along y. Taking derivatives of the field,

$$d^2B/dt^2 = -\omega^2 B \quad \text{and}$$
$$d^2B/dy^2 = -k^2 B = -\omega^2/c^2 B$$

Therefore, $d^2B/dt^2 = 1/c^2 \, d^2B/dy^2$ and hence the wave equation is satisfied. Now, using Eq. (2.1a)(iv), we get

$$\nabla \times B = \mu_0 \varepsilon_0 \, dE/dt$$

But, putting the expression for B, we get

$$\nabla \times B = 0$$

So, E is time invariant

If E is time invariant, no spatial derivative of it can become a function of time.

Now, $dB/dt = \omega B_0 \cos(\omega t - ky)\hat{j}$

Continued

If Maxwell's equation is followed, then the above expression of dB/dt should be equal to $\nabla \times E$. But this expression is a function of time, t. So, if E is time invariant, $\nabla \times E$ cannot be a function of time. Therefore, all the Maxwell's equations do not hold for this field even if it follows the wave equation and hence does not represent an EM wave.

We shall now see what additional conditions are required to be satisfied by the solutions of the wave equation, such that they also satisfy the Maxwell's equations and can rightly represent the E and B pair. For this, let us take the functional representations of $\mathbf{E}$ and $\mathbf{B}$ as

$$\mathbf{E} = E_0 \hat{u}_1 f_1(\omega t - ks) \qquad \mathbf{B} = B_0 \hat{u}_2 f_2(\omega t - ks)$$
$$\text{Or,} \quad \mathbf{E} = E_0 \hat{u}_1 f_1\{-k(s-ct)\} \quad \mathbf{B} = B_0 \hat{u}_2 f_2\{-k(s-ct)\} \tag{2.22}$$

where $\hat{u}_1$ and $\hat{u}_2$ are the two arbitrarily taken unit vectors pointing along the directions of $\mathbf{E}$ and $\mathbf{B}$, respectively. f_1 and f_2 are the two arbitrary functions of $(\omega t - ks)$, where we have considered that the wave moves along an arbitrary direction $\hat{s}$ and $k = \omega/c$. We have also seen above that these equations may be expressed in the generic functional form of $(s - ct)$ which satisfies the wave equation.

Now, to get the conditions for these solutions of the wave equation which ensures that they follow the Maxwell's equations, let us first take the electric field $\mathbf{E}$. $\mathbf{E}$ has components along the three standard axes X, Y and Z and thus can be represented as

$$\mathbf{E} = E_0 \left(u_{1\alpha}\hat{i} + u_{1\beta}\hat{j} + u_{1\gamma}\hat{k} \right) f_1(\omega t - ks) \tag{2.23}$$

where $\hat{u}_{1\alpha}$, $\hat{u}_{1\beta}$ and $\hat{u}_{1\gamma}$ are the components of the unit $\hat{u}_1$ vector along the X, Y and Z axes, with unit vectors $\hat{i}$, $\hat{j}$ and $\hat{k}$, respectively. Using the equation $\nabla \times \mathbf{E} = -dB/dt$ and putting the expression for $\mathbf{E}$ in Eq. (2.23) in this equation, we get,

$$\nabla \times \mathbf{E} = E_0 \nabla \times (\hat{u}_1 f_1(\omega t - ks))$$

$$= E_0 \left[|\hat{u}_{1\gamma}\partial f_1/\partial y - \hat{u}_{1\beta}\partial f_1/\partial z|\,\hat{i} + |\hat{u}_{1\alpha}\partial f_1/\partial z - \hat{u}_{1\gamma}\partial f_1/\partial x|\hat{j} + |\hat{u}_{1\beta}\partial f_1/\partial x - \hat{u}_{1\alpha}\partial f_1/\partial y|\,\hat{k} \right]$$

$$= -E_0 k f_1' \left[|\hat{u}_{1\gamma}\partial s/\partial y - \hat{u}_{1\beta}\partial s/\partial z|\,\hat{i} + |\hat{u}_{1\alpha}\partial s/\partial z - \hat{u}_{1\gamma}\partial s/\partial x|\,\hat{j} + |\hat{u}_{1\beta}\partial s/\partial x - \hat{u}_{1\alpha}\partial s/\partial y|\,\hat{k} \right]$$

$$\tag{2.24a}$$

Now, this equation should be equal to $-dB/dt$ to satisfy the Maxwell's equation in Eq. (2a)(iii). So, considering $u_{2\alpha}$, $u_{2\beta}$ and $u_{2\gamma}$ are the components of the unit vector u_2 along the magnetic field $\mathbf{B}$ with amplitude B_0, we get,

$$-d\mathbf{B}/dt = -B_0 \omega \left[|\hat{u}_{2\alpha}(f_2)|\,\hat{i} + |\hat{u}_{2\beta}(f_2)|\,\hat{j} + |\hat{u}_{2\gamma}(f_2)|\,\hat{k} \right] \tag{2.24b}$$

Similarly, we can take the equation $\nabla \times \mathbf{B} = (1/c^2)\,dE/dt$ and get the relation

$$\nabla \times \mathbf{B} = B_0 \nabla \times \{\hat{u}_2 f_2(\omega t - ks)\}$$

$$= B_0 \left[|\hat{u}_{2\gamma}\partial f_2/\partial y - \hat{u}_{2\beta}\partial f_2/\partial z|\,\hat{i} + |\hat{u}_{2\alpha}\partial f_2/\partial z - \hat{u}_{2\gamma}\partial f_2/\partial x|\,\hat{j} + |\hat{u}_{2\beta}\partial f_2/\partial x - \hat{u}_{2\alpha}\partial f_2/\partial y|\,\hat{k} \right]$$

$$= -B_0 k f_2' \left[|\hat{u}_{2\gamma}\partial s/\partial y - \hat{u}_{2\beta}\partial s/\partial z|\,\hat{i} + |\hat{u}_{2\alpha}\partial s/\partial z - \hat{u}_{2\gamma}\partial s/\partial x|\,\hat{j} + |\hat{u}_{2\beta}\partial s/\partial x - \hat{u}_{2\alpha}\partial s/\partial y|\,\hat{k} \right]$$

$$\tag{2.25a}$$

and

$$\frac{1}{c^2}d\mathbf{E}/dt = \frac{1}{c^2}E_0\omega \left[|\hat{u}_{1a}(f_1)|\,\hat{\boldsymbol{i}} + |\hat{u}_{1b}(f_1)|\,\hat{\boldsymbol{j}} + |\hat{u}_{1c}(f_1)|\,\hat{\boldsymbol{k}} \right] \qquad (2.25b)$$

Therefore, by equating component by component for both the equations, we get

$$A(\mathrm{I})\ \ B_0\omega/k = E_0 \qquad A(\mathrm{II})\ \ (1/c^2)E_0(\omega/k) = B_0$$
$$\mathrm{Or}\ \ B_0 c = E_0 \qquad\qquad\qquad\qquad E_0/c = B_0 \qquad (2.26a)$$

and

$$B(\mathrm{I})\ \ f_1'\left(u_{1\gamma}ds/dy - u_{1\beta}ds/dz\right) = u_{2\alpha}f_2'$$
$$f_1'\left(u_{1\alpha}ds/dz - u_{1\gamma}ds/dx\right) = u_{2\beta}f_2'$$
$$f_1'\left(u_{1\beta}ds/dx - u_{1\alpha}ds/dy\right) = u_{2\gamma}f_2'$$
$$B(\mathrm{II})\ \ f_2'\left(u_{2\gamma}ds/dy - u_{2\beta}ds/dz\right) = -u_{1\alpha}f_1'$$
$$f_2'\left(u_{2\alpha}ds/dz - u_{2\gamma}ds/dx\right) = -u_{1\beta}f_1'$$
$$f_2'\left(u_{2\beta}ds/dx - u_{2\alpha}ds/dy\right) = -u_{1\gamma}f_1' \qquad (2.26b)$$

The equations in $A(\mathrm{I})$ and $A(\mathrm{II})$ are plain and simple stating that there must be a definite relationship between the amplitude of E and B. This relation, as indicated in both the cases, is

$$E_0 = (\omega/k)B_0 \qquad (2.27)$$

For the second set of equation $B(\mathrm{I})$ and $B(\mathrm{II})$ to be valid, two things have to be simultaneously satisfied,

$$f_1' = f_2'\ \ \text{and} \qquad (2.28a)$$

This means that the functional form of both $\mathbf{B}$ and $\mathbf{E}$ must be the same. Additionally

$$u_{1\gamma}ds/dy - u_{1\beta}ds/dz = u_{2\alpha}$$
$$u_{1\alpha}ds/dz - u_{1\gamma}ds/dx = u_{2\beta}$$
$$u_{1\beta}ds/dx - u_{1\alpha}ds/dy = u_{2\gamma}$$
$$u_{2\gamma}ds/dy - u_{2\beta}ds/dz = -u_{1\alpha}$$
$$u_{2\alpha}ds/dz - u_{2\gamma}ds/dx = -u_{1\beta}$$
$$u_{2\beta}ds/dx - u_{2\alpha}ds/dy = -u_{1\gamma} \qquad (2.28b)$$

These equations can also be written as

$$\nabla s \times \hat{u}_1 = \hat{u}_2\ \ \text{and}$$
$$\nabla s \times \hat{u}_2 = -\hat{u}_1$$

These can be simultaneously true, if the direction of increasing s, u_1 and u_2 are mutually orthogonal. This implies that the electric and the magnetic field directions are required to be perpendicular. Further, both of them are perpendicular to the direction of the wave propagation in such a manner that $u_1 \times u_2 = \nabla s$.

To understand the above for a specific case, let us assume that s is measured as distance along the x direction. Then the terms ds/dy and ds/dz vanish while $ds/dx = 1$. This makes $u_{1\alpha} = u_{2\alpha} = 0$ which means that there can neither be an electric field nor a magnetic field in the direction of the propagation of the wave.

Again, as $ds/dx = 1$, it represents, $-u_{1\gamma} = u_{2\beta}$ and $u_{1\beta} = u_{2\gamma}$. Rearranging, we get $u_{1\gamma}u_{2\gamma} = -u_{1\beta}u_{2\beta}$. This implies $\hat{u}_1 \cdot \hat{u}_2 = 0$, which means, the electric and the magnetic field directions are perpendicular. Hence, $\hat{u}_1$ and $\hat{u}_2$ or in turn **E** and **B** are required to be perpendicular to each other.

This suggests that, a function to satisfy both the wave equation and the necessary Maxwell equations so that it can represent a true EM wave, it has to satisfy the above two criteria, viz. (a) space orthogonality of the electric and magnetic fields with both E and B being perpendicular to the direction of the propagation of the wave and (b) magnitudes of the electric and magnetic field must be related as $E_0 = (\omega/k)B_0$.

2.5.2 WAVE EQUATIONS WITH SOURCES

We have already learnt in the previous section about the wave equation and wave characteristics for the traversing EM waves in vacuum. However, we assumed that, there was neither any charge ($\nabla \cdot E = 0$) nor any current ($J = 0$) within the region of our consideration. In other words, both the sources of the electric and the magnetic fields were absent in our region of interest. However, there must be some definite sources present elsewhere, from which these waves got generated. So, we also need to know what happens when there is such a source, like a time varying current or a time varying charge in the region of our interest. In this section, we shall learn about the wave equations and their solutions in presence of such a source.

Let us first reiterate the fact that a static charge generates static electric field. A charge with a velocity results in a first-order variation in the electric field and thus generates a static magnetic field. Therefore, a charge having acceleration results in a second-order variation in the electric field and hence generates a time varying magnetic field which in turn creates an electric field again. In the process, it results in an EM wave. So, wherever there is an accelerating charge, or an equivalent time varying current, an EM wave is generated. Here, we shall consider a conductor carrying a time varying current as the source of an EM wave. Then, we shall derive a relation with which we can relate the changes in the current with the wave characteristics. The simplest approach for doing so is by doing it via the associated potential parameter, viz. the vector magnetic potential A and the scalar electric potential φ. We have already defined these terms before. We may start with the relation

$$\nabla \times E = -\partial B/\partial t$$

writing $B = \nabla \times A$ and substituting in the above equation, we get

$$\nabla \times E = -\partial/\partial t(\nabla \times A)$$
$$= -\nabla \times \partial A/\partial t \qquad (2.29a)$$

or, $\nabla \times \{(E - (-dA/dt)\} = 0$

It says that, now the electric field E is not conservative anymore, as it is not due to a static point charge only. Now, a part of this electric field is also generating out of the magnetic field variation. However, considering the total electric field E, and negating from it the portion $-dA/dt$, that is generated out of the magnetic field variation, the resultant field, representing the static charge field, remains conservative. So, we can write,

$$E + dA/dt = -\nabla\varphi$$
$$\text{or, } E = -dA/dt - \nabla\varphi \tag{2.29b}$$

Recall that, we have derived this equation before while explaining the Maxwell's equations. Now, if we consider Eq. (2.1a)(iv) and replace E with the expression derived above, we get,

$$\nabla \times B = \mu_0 J + \mu_0 \varepsilon_0 \partial E/dt$$
$$\text{so, } \nabla \times \nabla \times A = \mu_0 J - \mu_0 \varepsilon_0 \left(\nabla \partial\varphi/\partial t + \partial^2 A/\partial t^2\right)$$
$$\nabla(\nabla \cdot A) - \nabla^2 A = \mu_0 J - \mu_0 \varepsilon_0 \left(\nabla \partial\varphi/\partial t + \partial^2 A/\partial t^2\right) \tag{2.30}$$
$$\nabla^2 A - \mu_0 \varepsilon_0 \partial^2 A/\partial t^2 = \nabla(\nabla \cdot A) + \mu_0 \varepsilon_0 \nabla \partial\varphi/\partial t - \mu_0 J$$

Now, if we take the term $(\nabla \cdot A) + \mu_0 \varepsilon_0 \, \partial\varphi/\partial t = 0$, then the above equation turns into

$$\nabla^2 A - \mu_0 \varepsilon_0 \partial^2 A/\partial t^2 = -\mu_0 J \tag{2.31}$$

At this point, it is necessary mentioning the rationale of assuming $(\nabla \cdot A) + \mu\varepsilon \, \partial\varphi/\partial t = 0$. This is called gauge fixing. Gauge fixing is a mathematical procedure for coping with the redundant degrees of freedom in a field variable (Gauge fixing, 2016). We have seen previously that adding a gradient term does not alter the significance of A by any sense, other than its absolute value. Therefore, we are free to add any such gradient term to A. However, it should be remembered that the term E in Eq. (2.29b) must also remain invariant under such transform. Therefore, as $E = -\nabla\varphi - dA/dt$, an addition of $\nabla\lambda$ to A must be compensated by a negation of $d\lambda/dt$ from φ to keep the things as fine. But, we have innumerable options of adding $\nabla\lambda$ to A (and equivalently removing $d\lambda/dt$ from φ) and any addition would be equally good. Now, the question is what value should we add to A and why? The answer depends upon what our aim is. Our aim here is to simplify the calculations. We see that, this aim of simplifying things is accomplished when we add such a term to any A such that the relation $(\nabla \cdot A) + \mu\varepsilon \, \partial\varphi/\partial t = 0$ is satisfied. Therefore, for a definite choice of λ, the values of A and φ may be so transformed that the term $\nabla \cdot A + \mu\varepsilon \, \partial\varphi/\partial t$, with the modified values of A and φ, becomes equal to zero. This choice of potentials with the given definite relation between them is called the Lorentz Gauge. This makes Eq. (2.30) as simple and beautiful as we see it in Eq. (2.31).

However, that may not be enough to satisfy you, unless you are told about the implications of this modifications made. To mention the implication of the Laurentz Gauge, we use this relationship in the expression

$$E = -dA/dt - \nabla\varphi$$
$$\text{Or, } \nabla \cdot E = -d(\nabla \cdot A)/dt - \nabla^2\varphi \tag{2.32a}$$

Using the gauge and replacing $\nabla \cdot A$ with $-\mu\varepsilon \, d\varphi/dt$, we get

$$\nabla \cdot E = d(\mu_0 \varepsilon_0 \partial\varphi/\partial t)/dt - \nabla^2\varphi$$

and using Gauss' law $\nabla \cdot E = \rho/\varepsilon_0$,

$$\rho/\varepsilon_0 = \mu_0 \varepsilon_0 \partial^2 \varphi/\partial t^2 - \nabla^2 \varphi$$

Rearranging,

$$\nabla^2 \varphi - \mu_0 \varepsilon_0 \partial^2 \varphi/\partial t^2 = -\rho/\varepsilon_0 \tag{2.32b}$$

Thus, through Eqs (2.31), (2.32b), we get two wave equations where a nonzero charge density ρ or the current density J is present as sources. Rewriting them, we get,

$$\nabla^2 A - \mu_0 \varepsilon_0 \partial^2 A/\partial t^2 = -\mu_0 J$$
$$\nabla^2 \varphi - \mu_0 \varepsilon_0 \partial^2 \varphi/\partial t^2 = -\rho/\varepsilon_0 \tag{2.33a}$$

We also observe that, had there been no source, the above equations would become

$$\nabla^2 A - \mu_0 \varepsilon_0 \partial^2 A/\partial t^2 = 0$$
$$\nabla^2 \varphi - \mu_0 \varepsilon_0 \partial^2 \varphi/\partial t^2 = 0 \tag{2.33b}$$

This means, the parameters A and φ also follow the same wave equations and hence move similarly as E and B in free space. When we consider a region with the source, solving the wave equations for A and φ with sources will get us the time-space variation of A and φ. The solution to the equations for A and φ with their respective sources is the retarded vector and scalar potentials, respectively, given by

$$A(r, t) = \frac{\mu_0}{4\pi} \int \frac{J(t - r/c)}{r} dv \tag{2.34a}$$

$$\varphi(r, t) = \frac{1}{4\pi\varepsilon} \int \frac{\rho(t - r/c)}{r} dv \tag{2.34b}$$

where $c = \sqrt{(1/\mu\varepsilon)}$. Therefore, the solution is exactly like that of the Poisson's equations, $\nabla^2 A = -\mu J$ and $\nabla^2 \varphi = -\rho/\varepsilon$, with only the values being retarded by a time r/c, which is the time taken by moving A and φ to traverse the distance r with velocity c. Using the solution to these equations, we can derive the expression for A and φ at any point due to a portion dv of the space carrying current with density J or charge density ρ. Then, we can find $\nabla \times A$ to obtain B. Similarly, E can also be found from A and φ. Thus, we get both E and B as a function of the generating current. In our next chapter, we shall learn what happens when the EM wave passes through a medium which is not a vacuum.

Box 2.1 Matlab Exercise

In this Box, we shall run a Matlab program that will give us a conceptual idea about how the different fields vary with time. The Matlab program 'spatial_A.m' provides the spatial variation of the vector magnetic field 'A' due to an AC current with time. The focus is on the distribution and effective movement of the magnetic potential. Consider that there is a current carrying conductor at the bottom of the horizontal coordinate, i.e. aligned with the 'X' axis and carrying a sinusoidal current. The A values are distributed over the whole space and changes with time, i.e. on every iteration. One of the snapshots of the same is shown in the figure below. Notice that the regular variation is due to the variation of the current in the conductor and the tilt in the pattern representing delayed response with increasing distance from the source, i.e. A represents a retarded potential. However, the parameters are scaled from their real values to show the effects.

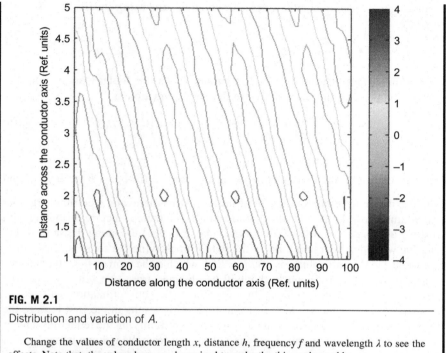

FIG. M 2.1

Distribution and variation of A.

Change the values of conductor length x, distance h, frequency f and wavelength λ to see the effects. Note that, the values here are downsized to make the things observable.

2.6 DIRECTION OF WAVE FLOW AND POYNTING VECTOR

We have seen that in a wave the energy flows outwards from the source. Since the energy is carried by the electric and the magnetic field, it must be a function of E and B. However, rather than showing that some expression of E and B are representing the direction and magnitude of the flow of the energy of the EM wave, we should develop this expression with logical understandings starting from the Maxwell's equations. To achieve this, let us start by considering a unit cubic volume and within it let us consider Eq. (2.1a)(iv),

$$\nabla \times B = \mu_0 J + \mu_0 \varepsilon_0 \partial E / \partial t$$

Now, we know that there must be both magnetic and electric field for the wave to propagate. Hence, dot multiplying the expression with the electric field E, we get

$$E \cdot \nabla \times B = \mu E \cdot J + \mu \varepsilon E \cdot \partial E / \partial t$$
$$= \mu E \cdot J + \tfrac{1}{2} \mu \varepsilon \partial / \partial t (E^2) \tag{2.35a}$$

Using the identity, $\nabla \cdot E \times B = B \cdot \nabla \times E - E \cdot \nabla \times B$, then replacing the term on the left of the Eq. (2.35a) with it and then rearranging, we get

$$E \cdot J = -\frac{1}{\mu_0} \nabla \cdot E \times B + (1/\mu) B \cdot \nabla \times E - \tfrac{1}{2}\varepsilon_0 \partial E^2 / \partial t$$
$$= -\frac{1}{\mu_0} \nabla \cdot E \times B - \frac{1}{\mu_0} B \cdot \partial B / \partial t - \tfrac{1}{2}\varepsilon \partial E^2 / \partial t$$
$$= -\frac{1}{\mu_0} \nabla \cdot E \times B - \tfrac{1}{2} \frac{1}{\mu_0} \partial B^2 / \partial t - \tfrac{1}{2}\varepsilon_0 \partial E^2 / \partial t$$

Integrating the equation on both sides over volume V and using Divergence theorem, we get,

$$\int_V E \cdot J dv = -\frac{1}{\mu_0} \int_S E \times B dS - \int_V \left[\frac{1}{2} \frac{1}{\mu_0} \partial B^2 / \partial t - \frac{1}{2} \varepsilon_0 \partial E^2 / \partial t \right] dv$$

here S is the surface that encloses V. Recognizing $E \cdot J$ as the Joules work, W done per unit volume per unit time, we can write the above equation as

$$dW/dt = -\int \partial / \partial t \left[\frac{1}{2} \frac{1}{\mu_0} B^2 + \frac{1}{2} \varepsilon_0 E^2 \right] dv - \int \frac{1}{\mu_0} E \times B dS \qquad (2.35b)$$

Further, we recognize that $\frac{1}{2} \frac{1}{\mu_0} \partial B^2 / \partial t + \frac{1}{2} \varepsilon_0 \partial E^2 / \partial t = \partial / \partial t \left(\frac{1}{2} \frac{1}{\mu_0} B^2 + \frac{1}{2} \varepsilon_0 E^2 \right)$ and it represents the time rate of change of the stored magnetic and electric energy per unit volume. Now, let us consider a volume in which there is a current flowing with density J across two opposite faces of the cube. E is the field along the same direction. The work done per unit time by the field E in carrying the charges constituting the current J in the unit volume is $dW/dt = E \cdot J$, which is the term on the left hand side of the equation. Energy spent in this will be obtained from two sources. The first source is the stored static energy in the volume. So, this utilization will reduce the stored energy of the volume over unit time. This results in the first term to the right hand side of the equation with the negative sign signifying the reduction. Therefore, the vector that represents energy travelling out from the volume in the form of EM wave per unit area per unit time thus can be expressed as $(1/\mu_0) E \times B$. This is called the Poynting vector. Therefore, the amount of this energy effectively lost in the volume is the energy carried in by the Poynting vector minus the energy carried out, i.e. the negative of the surface integral of the Poynting vector. This is the other source of energy utilized to do the Joules work in the volume and is represented by the second term on the right hand side. This, as a whole, is what is known as the 'Poynting Theorem'. It is the work—energy theorem of thermodynamics. The scenario is illustrated in Fig. 2.9.

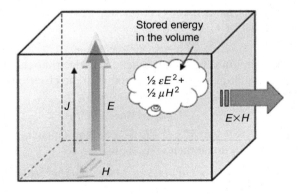

FIG. 2.9

Poynting theorem.

2.7 IMPORTANT OBSERVATIONS

The theoretical derivations done in this chapter will form the basis of all our subsequent understandings. Therefore, it is important to consolidate a few important observations from what we have derived here. We have listed the observations below:

a. Static electric charge creates static electric field and an electric scalar potential.
b. Charges moving in a constant velocity and thus constituting a constant current generates a magnetic field and a magnetic vector potential.
c. Accelerating charges or time varying currents creates EM waves which propagates through the medium.
d. These waves move with velocity $c = 1/\sqrt{(\varepsilon\mu)}$ where the permittivity ε and the permeability μ are parameters of the medium.
e. The motion of the EM wave follows certain definite function with generic form $f(x, t) = f(x - ct)$, while simultaneously following the Maxwell's equations.
f. The electric field E, the magnetic field B and the direction of propagation remain orthogonal
g. The magnitude of E and B related by $|E| = (\omega/k) |B|$. Therefore, a change in magnitude in one implies a change in the other.
h. Both the scalar potential φ and the vector potential A follow the wave equation.
i. In presence of a source, the φ and A solutions at any distance from the source are the retarded potentials derived from the source.
j. The energy carried by the wave moves in the direction and with the magnitude of $(1/\mu_0)E \times B$. This energy is supplied by the source of the wave.

CONCEPTUAL QUESTIONS

1. Derive the Biot Savarts Law from Maxwells equations
2. Derive the expression for the magnetostatic energy density in terms of the magnetic induction filed, B.
3. Does an electron oscillating in response to a changing electric field experience Lorentz force due to the magnetic field developing out of it.
4. Using Poynting theorem, show that for a static electric field suddenly incident upon a dielectric, there is an exchange between the stored magnetic and electric energy

REFERENCES

Feynman, R.P., Leighton, R.B., Sands, M., 1992. Feynman Lectures on Physics. Narosa Publishing House, Chennai.
Galbis, A., Maestre, M., 2012. Vector Analyses Versus Vector Calculus. Springer, New York, NY.

Gauge fixing, 2016. Wikipedia. https://en.wikipedia.org/wiki/Gauge_fixing (Accessed 2 October 2016).

Portis, A.M., 1978. Electromagnetic Fields: Sources and Media. John Wiley & Sons Inc., Hoboken, NJ.

Reitz, J.R., Milford, F.J., Christy, R.W., 1990. Foundations of Electromagnetic Theory, third ed. Narosa Publishing House, New Delhi.

Interaction of waves with medium

3

In our previous chapter, we have learnt about the Maxwell's equations. Using these equations, we have found that, whenever there is an accelerating electric charge, it produces an electromagnetic wave. These waves move in vacuum with orthogonal electric and magnetic fields with a velocity $c = 1/\sqrt{(\mu_0 \varepsilon_0)}$, where μ_0 and ε_0 are, respectively, the permeability and permittivity of vacuum. We have also seen that the energy that these waves carry is supplied by the source generating the wave. Extending this fact, in this chapter we shall see in details the characteristics of the propagating wave and the different conditions which arise when the electromagnetic waves interact with the medium. It is intuitive that, if there be some charges present in a material medium, which are free to move, either fully or partially, and if an EM wave gets incident upon the charges, the latter respond to the wave and start oscillating with it. In the process, these oscillating charges generate some secondary EM wave. Therefore, in such a medium, there coexist two waves—one elementary wave of external origin, which is incident on the medium, and the other which is generated in situ by the oscillation of the charges present there. The two component waves interact and the resultant wave thus formed has some features different from each of the constituent waves.

3.1 ELECTRIC MATERIALS

3.1.1 RESPONSE OF A MEDIUM TO ELECTRIC FIELD: CONDUCTION AND POLARIZATION

Any material medium is made up of molecules, which in turn is constituted by atoms of same or different kinds. Inside the material, the atoms are organized in a definite fashion. In general, these molecules or their higher structural units are geometrically oriented in such a fashion that the centre of the overall negative charges exactly coincides with the centre of the positive charges. This makes the entity electrically unpolarized as a whole.

The material can be either a conductor or an insulator. Conductors have electrons which are not attached to any constituent atom and are free to move. When an electric field is incident upon such a material, these electrons move freely in the direction opposite to that of the field. In the process, they generate some conduction current constituting a finite current density J. Given a field E, the strength of J is dependent upon the character of the material. If 'n' is the density of the free electrons, each of charge 'e', a field 'E' will exert a force $F = Ee$ on each of the electrons which will then tend to accelerate. However, due to the finite inhibiting force offered by the material, the electrons will finally experience a constant mean velocity 'v' where 'v' is proportional to 'E'. The current density thus produced will be $J = N_e e v$. As $v \propto E$, we can write $J = N_e e \gamma E$, where 'γ' is the proportionality constant. 'γ' is large when the inhibiting force is small and vice versa. Therefore, depending upon the value of γ, different conductors will have different current density on exposure to the same electric field, E. The conduction current density generated per unit field, i.e. J/E, is thus an index of the conduction capability and is called the conductivity σ of the material. So,

$$\sigma = \frac{J}{E}$$

$$\text{or,} \quad \sigma = N_e e \gamma \tag{3.1}$$

Thus, we can say that, the conductivity increases when either the number of available charged particles N_e are large or the resistive force inhibiting the flow is small making γ large so that the particles can readily attain large velocity upon incidence of the field.

On the other hand, the charges in an insulator are not free and are attached to their parent atoms. They experience large restoring force on being moved from their positions as it leads to the relative displacements of the charge centres. When such material is exposed to an external electric field E_0, the electric field exerts a force on the constituent charge carrying particles of the medium. The latter respond to it by displacing from their position, the positive charges tend to move in the direction of the external field while the negative charges move in the direction opposite to it. So, for a given basic neutral structural unit of the material, due to this small displacement, the charge centres get separated by an infinitesimal distance creating what is known as 'Polarization'. So, in a simplistic view of the situation, an electric dipole is generated at each such structure and the medium is then said to be polarized. Such materials can hold electric fields across it and are termed as 'Dielectrics'.

The individual dipoles, thus generated within the dielectric medium, get aligned with their axis along the electric field E and are all regularly arranged. The charged particles are displaced within the medium in such a way that everywhere inside the material medium, the positive poles of a dipole remain collocated to the negative poles of the dipole adjacent to it in the direction of polarization and hence get balanced. Thus, for such a regularly polarized dielectric, the only unbalanced charges arise at the two opposite surfaces of the medium in the direction of the field.

The polarized charges produce electric fields, whose direction is opposite to the external field inside the material. Hence, the resultant electric field inside the material E is depreciated from the external polarizing field E_0, i.e. $E < E_0$. The amount of such depreciation depends upon the amount of polarization inside the material.

3.1.2 FIELD INSIDE A DIELECTRIC
3.1.2.1 Electric polarization
We have learnt from the definition of dielectric that, whenever an external polarizing electric field is incident upon it, the effective field inside the dielectric is less than the external field where the amount of reduction in the field is proportional to the amount

of polarization occurring inside. The amount of this polarization is quantified by the total dipole moment 'p' thus created inside the material. The total amount of dipole moment produced in the material is the vector sum of all the individual dipole moments created inside. The collective dipole moment of each of these individual dipole moments of the material is again equal to the dipole moment due to the unbalanced charges only. These unbalanced charges mostly appear at the surfaces. Additionally, any unbalanced charge inside the dielectric also has its contribution. The dipole moment per unit volume is called the electric polarization or simply the polarization and denoted by P.

To understand this term, let us first consider an arbitrary small volume δv in the dielectric medium as shown in the Fig. 3.1. It is polarized by the field E incident on the material in the direction $\hat{r}$. Taking any arbitrary point r_0 as reference, the total dipole moment inside the volume δv is given by (Reitz et al., 1990)

$$\delta p = \int_{\delta v} \rho(r) \times (r - r_0) dv \qquad (3.2)$$

where, $\rho(r)$ is the polarized charge density at any arbitrary point within the volume dv at a distance 'r' from the arbitrary origin, r_0. If the polarization of the ith dipole has been formed by separation of equal and opposite charges δq^+ and δq^- about their charge centres located at distance r_i and if the charges are displaced by $+d/2$ and $-d/2$, respectively, and if the numbers of such charged particles per unit volume be N, then the total integrated polarization value can be written as,

$$\delta p = \left[\sum_{i=1}^{N} \left(r_i + \frac{d}{2} - r_0 \right) \delta q^+ + \sum_{i=1}^{N} \left(r_i - \frac{d}{2} - r_0 \right) \delta q^- \right] \delta v \qquad (3.3a)$$

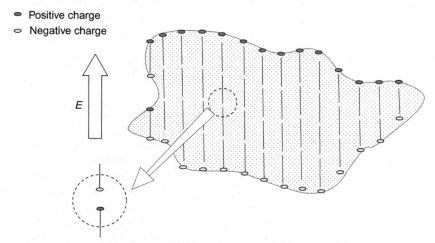

FIG. 3.1

Dipole inside a material of arbitrary shape.

where, the summation indicates consideration of all the displaced pair of charges in the unit volume. A simple multiplication of the volume element δv indicates that the free electron density and other conditions are same over any volume. Now, the positive and negative charges being equal and opposite, i.e. $\delta q_i^- = -\delta q_i^+ = \delta q$ we can write the above as,

$$\delta p = \sum_{i=1}^{N} \delta q \left\{ \left(r_i + \frac{d}{2} - r_0 \right) - \left(r_i - \frac{d}{2} - r_0 \right) \right\} \delta v$$

$$= N \delta q \left(\frac{d}{2} + \frac{d}{2} \right) \delta v \tag{3.3b}$$

$$= N \delta q \, d \, \delta v$$

$$= N p_1 \, \delta v$$

where p_1 is the dipole moment of one individual displaced charge pair. Therefore, we validate our earlier statement that the total dipole moment created is the vector sum of all the individual dipole moments. Extending this idea, we can say that the total dipole moment 'p' is the volume integral of the polarization in a unit volume. So, using our earlier definition of P as the dipole moment per unit volume, we get

$$P = \delta p / \delta v$$

$$= N p_1 \tag{3.3c}$$

$$= N \delta q \, d$$

Since we have assumed that the positive and the negative charges, displaced relative to each other, have the same charge density, the fact that they are displaced do not produce any net charge inside the volume (Feynman et al., 1992). Therefore, the only unbalanced charges δQ_u would be at the two surfaces along the direction of polarization separated by distance 'l'. If σ be the surface density of such polarized charges, and 'A' be the effective surface area, then we can write the polarization p as

$$p = \delta Q_u \, l$$

$$= \sigma A \, l \tag{3.4a}$$

Therefore, the polarization per unit volume P becomes

$$P = \frac{1}{Al} \times \sigma A l \tag{3.4b}$$

$$= \sigma$$

So, P is equal to the surface charge density due to the polarization.

3.1.2.2 Effective electric field

We have already mentioned that the polarized charge in a dielectric, arising out of an external electric field, is proportional to the effective field E on these charges. So, if there were no polarization, the imposed field would have been equal to the external field, E_0. When displacement of the bound charges takes place, the consequent polarization in the material modifies the effective field inside. For dielectric materials, the polarization takes place in such a way that the electric dipole moment developed as a result is in the direction of the field. It causes the polarization field inside the material in direction opposite to the external field. Therefore, the effective field gets depreciated inside. So, if the final field inside the medium be E then the polarization P is proportional to this field and can be written as (Feynman et al., 1992)

$$P = \varepsilon_0 \chi E. \tag{3.5a}$$

The proportionality constant χ is called the electric susceptibility of the dielectric and it indicates how easily the material can be polarized in response to an electric field. Now, the reduction in the field due to the charge separating inside the material due to the polarization is equal to the field that is generated by these charges. This is shown in Fig. 3.2. Now, by Gauss' law, we can write,

$$-\mathrm{d}E = \frac{\sigma}{\varepsilon_0} \tag{3.5b}$$

Since $\sigma = P$, replacing the value of σ by the expression we have obtained for P, we get

$$-\mathrm{d}E = \frac{P}{\varepsilon_0} \tag{3.5c}$$
$$= \chi E$$

where σ is the surface density of the polarized charges. We can write the relationship between the external field E_0 and the effective field inside the medium E using the above as

$$E_0 - \chi E = E$$
$$\text{or} \ \ E_0 = (1 + \chi)E \tag{3.6}$$

For our purposes, χ may be considered to be isotropic and independent of E. When χ is not isotropic, we shall get a vector or a tensor matrix instead of a scalar value for it.

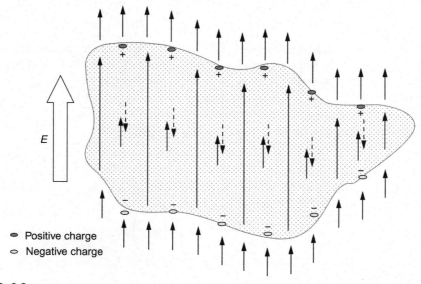

FIG. 3.2

Polarization and field inside a material.

When χ depends upon E, the above Eqs (3.5a), (3.5c) become nonlinear. For Dielectric materials, the electric susceptibility χ is positive which makes $E < E_0$.

So,

$$
\begin{aligned}
\varepsilon_0 E_0 &= \varepsilon_0 (1 + \chi) E \\
&= \varepsilon_0 (K) E \\
&= \varepsilon E
\end{aligned}
\tag{3.7}
$$

where $K = (1 + \chi)$ is called the dielectric constant. Again $\varepsilon = K\varepsilon_0$ is the permittivity of the medium. Therefore, $K = \varepsilon/\varepsilon_0$ is the ratio of the permittivity of the medium to that of the vacuum. Hence, it is also called the relative permittivity. The product εE for any medium represents the displacement vector D, where ε is the permittivity of the medium and E is the normal component of the electric field. From above equation, it is evident that D remains invariant across all material unless there is any free charge at their interface. As from Gauss' law, $\nabla \cdot D = \rho$, the divergence of the displacement vector over any volume is a direct measure of the charge density enclosed in that volume. So, if there is no free charge over a space, there won't be any change in the displacement vector. So, for fields originating from a certain source, the normal component of D is continuous.

We have seen that $P/\varepsilon_0 = -\,dE$ is the polarization field inside a material. This can be also considered as the number of lines of forces per unit area originating due to the polarization occurring inside the material. Further, when will these effective numbers of the lines of forces inside the material change? It is only possible when there is an unbalanced charge inside the material for the lines to terminate. So, we can write, using the Gauss' law,

$$
\nabla \cdot \frac{P}{\varepsilon_0} = -\frac{\rho_m}{\varepsilon_0}
\tag{3.8}
$$

$$
\text{Or, } \nabla \cdot P = -\rho_m
$$

where, ρ_m is the charge density inside the material arising due to the unbalanced polarization. Now, the potential due to such a distribution of dipoles at a distant point r from it may be represented in terms of the dipole moment as

$$
\varphi(r) = \frac{1}{4\pi\varepsilon_0} \int P dv \cdot \frac{\hat{r}}{r^2}
\tag{3.9a}
$$

Now, considering the equality $\nabla(1/r) = \hat{r}/r^2$, we can write,

$$
\varphi(r) = \frac{1}{4\pi\varepsilon_0} \int P \cdot \nabla \left(\frac{1}{r} \right) dv
\tag{3.9b}
$$

Again, $\nabla \cdot (fF) = f(\nabla \cdot F) + F \cdot \nabla f$, where f is a scalar function and F is a vector function. Considering, $f = 1/r$ and $F = P$ in this identity, we get

$$
\varphi(r) = \frac{1}{4\pi\varepsilon_0} \int \left\{ \nabla \cdot \left(\frac{P}{r} \right) - \left(\frac{1}{r} \right) \nabla \cdot P \right\} dv
\tag{3.9c}
$$

We know that $\int \nabla \cdot P dv = \int \nabla \cdot \sigma dv = \int \sigma \, \hat{n} dS$ and $\int \nabla \cdot P dv = -\int \rho dv$, we rewrite the last equation as

$$
\varphi(r) = \frac{1}{4\pi\varepsilon_0} \left\{ \frac{1}{r} \int \sigma \cdot \hat{n} dS + \int \frac{\rho}{r} dv \right\}
\tag{3.9d}
$$

Therefore, it reveals that the total potential has the contribution from both the surface charges and the unbalanced charges inside the material originating due to the polarization.

3.1.3 PRACTICAL DIELECTRIC MATERIALS

Dielectric materials can be defined as those group of electric materials which can be electrically polarized by the application of an external electric field in such a manner that the resultant electric field inside the material is depreciated.

In strict sense, they are practically insulators with no free charges to constitute any conduction current. However, many of the materials found in nature are not perfect insulators, but are partially conducting. These materials simultaneously carry conduction current with density J, as well as the displacement current, with density dD/dt. These are our materials of interest for subsequent chapters, and in the following section, we shall now see how such materials behave for different frequencies and for different conditions.

Amperes law states that the curl of B, or equivalently, the closed path integral of B, has a direct relation with the conduction current density J. From Maxwell's equation, we have seen that the Amperes law is modified by the introduction of the new term called the displacement current. It states that the total curl of B is actually contributed by two different terms, viz. the conduction current density J and the displacement current density dD/dt. We have already seen in Chapter 2 that both these contributory terms represent the time rate of change of charge, the first for the moving charge while the second for the accumulated bound charge. Therefore, considering the Maxwell's equation,

$$\nabla \times B = \mu_0\{J + \varepsilon_0 dE_0/dt\}$$

$$\text{Or,} \quad = \mu_0\{\sigma E + \varepsilon_0 dE_0/dt\} \tag{3.10}$$

$$= \mu_0\{\sigma E + j\omega\varepsilon E\}$$

Here ε_0 is the permittivity of vacuum, and E_0 is the external field, ε is the permittivity of the medium and E is the field inside the medium. We have also used the relation, $D = \varepsilon_0 E_0 = \varepsilon E$, here. Thus, the total current density, irrespective of whether it is conduction current density $J = \sigma E$ or the displacement current density $\varepsilon_0 dE/dt$, or both, constitute the curl of B.

3.1.3.1 Practical conductors and insulators

The relative strengths of conduction and displacement current are represented by the ratio of the two types of currents for a given strength of the electric field, E. Recalling that for a given electric field $E \sim e^{j\omega t}$, the conduction current density is $J_{cond} = \sigma E$ and the displacement current density is $J_{disp} = j\omega\varepsilon E$. The condition is shown in Fig. 3.3. We also find that the conduction current and the displacement current magnitudes are in the ratio of $\sigma : \omega\varepsilon$ and the complex unit vector 'j' represent that they are mutually orthogonal in phase. This ratio, expressed in fractional form, is called the 'Loss Tangent' for a reason we shall see later. For a given value of E, the displacement

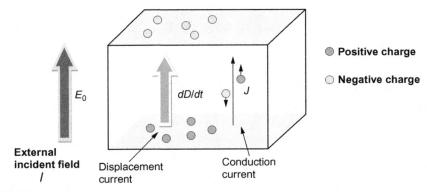

FIG. 3.3

Practical dielectric material.

current increases linearly with frequency, while the conduction current remains invariant with it, considering σ is independent of frequency. However, σ can also be a function of frequency for certain materials.

In this perspective, we can call a material to be a conductor if the conduction current is much stronger than the displacement current, i.e. $\sigma E \gg \omega \varepsilon E$, or $\omega \ll \sigma/\varepsilon$. Similarly, it is called an insulator, if the displacement current is much stronger than the conduction current, the necessary condition being, $\omega \gg \sigma/\varepsilon$. So, considering this fact, it can be concluded that whether such a partially conducting dielectric material will act as a conductor or an insulator depends not only upon the conductivity or permittivity of the material, but also upon the frequency chosen. The following focus area explains the situation taking specific examples of material and comparing the currents at different frequencies.

Focus 3.1

We intend to show here that for a practical dielectric with finite conductivity, the characteristic of the material is dependent upon the frequency. To demonstrate that, we take sea water, which has the following parameter values:

$$\text{Conductivity } \sigma = 4\,\text{S/m}$$

$$\text{Dielectric constant } K = 81 \text{ and}$$

$$\text{Permeability } \mu = 1$$

Taking the value of $\varepsilon_0 = 1/(36\pi) \times 10^{-9}$, we get the absolute permittivity of sea water as

$$\varepsilon = \varepsilon_0 K = 81/(36\pi)10^{-9}$$

Therefore, the loss tangent

$$\tau = (1/\omega) \times (4/81 \times 36\pi \times 10^{\wedge}9)$$

$$= 1/(\omega/2\pi)(8/9 \times 10^{\wedge}9)$$

$$= (/f) \times (0.8889 \times 10^{\wedge}9)$$

So, if $f < {\sim}9\,\text{MHz}$, it makes $\tau > 100$, then the sea water becomes a good conductor. However, if $f > 90\,\text{GHz}$, then $\tau < 1/100$, and it makes the medium a good dielectric

Different situation arises when the conductivity is not a constant for the medium and itself depends upon the frequency. This happens, for example, in the case for the ionosphere where the medium consists of only completely freely moving charges and the conductivity is inversely proportional to the frequency. We shall treat this case when we shall study about the ionosphere in Chapter 4. Therefore, unless the conductivity is varying linearly with ω or higher powers of it, the material exhibits abundance of displacement current over the conduction current at higher frequencies.

3.1.3.2 Loss tangent

We have learnt that an incident electric field on a dielectric results in the movement of the charges. One may question at this point what happens to the energy which is used for moving the charges in the medium. In this section and the following, we shall be answering this question. For a partially conducting material, some of the charges are free, causing flow of conduction current. Rest of the charges are bound causing polarization and will modify the effective internal field. The variation of this field will result in the displacement current. Hence, in such a material, the total admittance of the medium Y is complex, with the real part G_Y representing the conductance (inverse of resistance, R_z) and the imaginary part B_Y representing the susceptance (inverse of reactance, X_z). The corresponding specific values are given by σ and $j\omega\varepsilon$, respectively. Hence, in complex impedance plane, the tangent of the angle that the impedance vector makes with the real axis is $\sigma/\omega\varepsilon$. The total current density J, at any instant t can be expressed as the sum of conduction current and displacement current per unit area and is given by,

$$
\begin{aligned}
J(t) &= J_{\text{cond}}(t) + jJ_{\text{disp}}(t) \\
&= \frac{V(t) \cdot Y}{A}
\end{aligned}
\tag{3.11a}
$$

where, V is the voltage across the length 'l' of the material considered. This voltage V can be expressed in terms of the electric field, E as $V(t) = E(t) \times l$. So, replacing the expression of V, we get,

$$
\begin{aligned}
J(t) &= E(t) l (Y/A) \\
&= E(t) (l/A) \{G_Y + jB_Y\} \\
&= E(t) \frac{G_Y + jB_Y}{A/l} \\
&= \{\sigma + j\omega\varepsilon\} E(t)
\end{aligned}
\tag{3.11b}
$$

Now, the instantaneous power utilized in the real part, i.e. the conduction current per unit volume is given by

$$
\begin{aligned}
P_R &= E \cdot J_{\text{cond}} \\
&= \sigma E(t)^2
\end{aligned}
\tag{3.12a}
$$

Similarly, the instantaneous power utilized per unit volume by the imaginary part, i.e. by the displacement current, is

$$
\begin{aligned}
P_I &= E \cdot J_{\text{disp}} \\
&= \omega\varepsilon E(t)^2
\end{aligned}
\tag{3.12b}
$$

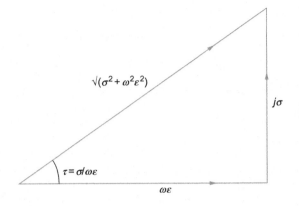

FIG. 3.4

Loss tangent.

The power utilized in the real part is that required to drive the conduction current. This power is utilized to work against the resistance of the medium and is permanently lost. It is well-known that the energy spent here as the electrical energy by the fields gets irreversibly converted to heat. Similarly, the imaginary part arises when there is a time variation in the electric polarization field in the volume. This results in the formation of secondary electromagnetic wave in this volume. The power utilized in the imaginary part is utilized for creating the oscillatory dipole and consequent generation of wave as radiation, e.g. as scattering or any systematic radiation. Refer to Eq. (2.11b) and notice that this term in Eq. (3.12b) is equal to the rate of change in stored energy in unit volume. So, the incident electric field not only does a Joules work but also increases the stored electric energy. In fact, these are components of the Poynting theorem. Further, the ratio of the power lost for Ohmic conduction to the power used in enhancing the stored electric energy per unit volume is

$$|P_R/P_I| = \frac{\sigma}{\omega\varepsilon} \tag{3.12c}$$

Hence, given a frequency ω, the ohmic loss of the electric energy in a given material is proportional to the tangent of the angle that the impedance vector makes with real axis. Hence, the ratio is termed as the loss tangent (Fig. 3.4).

3.2 WAVE EQUATION IN MEDIUM

Until now, we have been using the Maxwell's equations for analysing the different conditions when an electric field is incident on a medium. Now, we shall derive the wave equation for conditions where the medium through which the wave is traversing is partially conducting dielectric. We shall start with the equation

$$\nabla \times B = \mu J + \mu\varepsilon dE/dt$$
$$= \mu\sigma E + \mu\varepsilon dE/dt$$

For the material selected, both the above terms in the right hand side of the equation will prevail. Hence, taking curl on both sides and considering $\nabla \cdot B = 0$, we get,

$$\nabla^2 B = \mu\sigma(dB/dt) + \mu\varepsilon(d^2B/dt^2) \tag{3.13}$$

This is known as the Helmholtz equation and can be used to understand the response of such a material to an incident field. Exactly same equation can be obtained for the electric field E. The only difference here with the wave equation we derived in our previous chapter is that, due to consideration of finite conductivity of the medium, here the first-order derivative term appears to the right side of it.

Using the expression for E or B of the form $\sim e^{j(\omega t - kx)}$, we get,

$$k^2 = \omega^2\mu\varepsilon - \mu\sigma(j\omega)$$
$$k^2 = (\omega/c)^2(1 - j\tau)$$
$$\text{Or,} \quad k = k_0\sqrt{(1 - j\tau)} \tag{3.14}$$
$$= k_0\sqrt{\left\{(1+\tau^2)^{1/2}\angle\theta_\tau\right\}}$$

where k_0 is the propagation constant for free space, $\tau = \sigma/\omega\varepsilon \cdot \sqrt{(1+\tau^2)}$ is the magnitude of the complex number $(1 - j\tau)$ and $\theta_\tau = \tan^{-1}(\tau) = \tan^{-1}(\sigma/\omega\varepsilon)$ is its angular argument. Therefore, from the above expression, we find that, while for free space or pure insulator, the propagation constant k is the only function of frequency ω and velocity components μ and ε, here in a partially conducting medium, the same is complex and also dependent upon the conductivity σ.

From the theory of complex number, we know that the square root of a complex number has amplitude equal to the square root of the amplitude of the original number while the angular argument of the root is half the argument of the original complex number. Using this, the expression for the propagation constant 'k' becomes

$$k = k_0\left\{\sqrt{(1+\tau^2)}\right\}^{\frac{1}{2}}\left\{\cos(\angle\theta_\tau/2) - j\sin(\angle\theta_\tau/2)\right\}$$
$$= k_0\sqrt{sec\theta_\tau}\left[\{\tfrac{1}{2}(1+\cos\theta_\tau)\}^{\frac{1}{2}} - j\{\tfrac{1}{2}(1-\cos\theta_\tau)\}^{\frac{1}{2}}\right]$$
$$= k_0\sqrt{\tfrac{1}{2}}\left[(\sec\theta_\tau + 1)^{\frac{1}{2}} - j(\sec\theta_\tau - 1)^{\frac{1}{2}}\right] \tag{3.15}$$
$$= k_0\sqrt{\tfrac{1}{2}}\left[\left\{\sqrt{(1+\tau^2)} + 1\right\}^{\frac{1}{2}} - j\left\{\sqrt{(1+\tau^2)} - 1\right\}^{\frac{1}{2}}\right]$$
$$= \beta - j\alpha$$

Thus, the wave expression becomes

$$B = B_0 e^{-\alpha x}e^{j(\omega t - \beta x)} \tag{3.16}$$

Therefore, as a result of the finite conductivity, the propagation constant k becomes complex in nature. The real part causes the phase variation, while the imaginary part results in amplitude attenuation as the wave propagates along x. In such cases of finite conductivity, the relation between E and B field amplitude becomes

$$E = \frac{\omega}{k}B$$
$$= \frac{\omega}{(\beta - j\alpha)}B \tag{3.17}$$

Thus, E and B now have a finite phase difference due to the complex nature of 'k' arising due to the conductivity of the material medium. For highly conducting medium, τ is large and the angle θ_τ is $\sim$90 degrees. Hence k, for such material, is given by

$$k = k_0 \sqrt{\tau}(\cos 45° - j\sin 45°)$$
$$= k_0 \sqrt{(\tau/2)(1-j)}$$

(3.18)

Focus 3.2

Our objective of this focus is to learn how to use the properties of a complex number to obtain its square root. Let, at 30 GHz and 20°C, the approximate real and the imaginary part of the permittivity ε of water is $\varepsilon = \varepsilon_r + j\varepsilon_i = 22 + j28$. Thus, the vector ε can be represented as shown in the following figure.

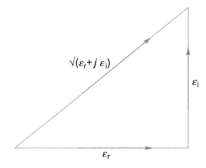

Fig. F3.1. Components of ε.

The magnitude of the ε vector is

$$|\varepsilon| = \sqrt{(22^2 + 28^2)} = 35.6$$

and the angular argument is

$$\theta = \tan^{-1}(28/22) = 51.84°$$

Therefore, the real and the imaginary part of the refractive index n_r and n_i are, respectively,

$$\varepsilon' = \mathrm{Re}\left(\sqrt{\varepsilon}\right) = \sqrt{(35.6)}\cos\left(\tfrac{1}{2} \times 51.84\right) = 5.36 \text{ and}$$

$$\varepsilon'' = \mathrm{Im}\left(\sqrt{\varepsilon}\right) = \sqrt{(35.6)}\sin\left(\tfrac{1}{2} \times 51.84\right) = 2.60$$

In the following section, we shall see how the decay in the amplitude arising as a result of finite σ is accounted for.

Box 3.1 Matlab Exercise

In this box, we wish to demonstrate the effect on an electromagnetic wave when it enters a practical dielectric with finite conductivity. Matlab program 'Pract_dielct.m' was run to obtain the real and imaginary part of the propagation constant, given a medium with permittivity, ε, and a finite conductivity σ. The values of β and α are obtained along with the loss tangent. The wave before and after entering the medium is illustrated.

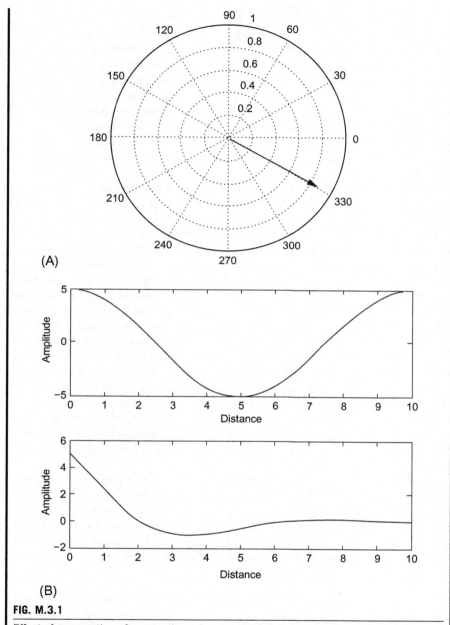

(A)

(B)

FIG. M.3.1

Effect of propagation of a wave through a practical dielectric medium. (A) Vector representation of ε (B) amplitude variations of a signal with real (*upper panel*) and complex (*lower panel*) values of ε.

Change the values of frequency f and observe the results. Also see what happens if the ε of the medium changes.

3.2.1 POWER UTILIZATION IN A DIELECTRIC

Whenever an EM wave is incident on a dielectric, a part of the incident wave power drives the free electrons against the resistive collision forces to generate the conduction current. Some of the incident power is also utilized to move the bound charges. In the first quarter of the phase, when the electric field is increasing from zero, it moves the bound charges from their charge centres. Consequently, an internal electric field is produced that exerts a restoring force on them. As the charges move against these restoring forces, they do a positive work against this force. The necessary energy for the same is obtained from the incident wave since it is the field of this wave which is driving the charges. Therefore, a static electric energy density equal to $\frac{1}{2}\varepsilon E^2$ is developed in the dielectric.

During the next quarter of the phase, the field is such that these charges fall back to make their respective charge centres coincide. In the process, the increased potential energy of the charges is released back. The same process is carried on for the rest half cycle of the incident wave phase with only the charges moving in opposite direction now. As the electric field changes, the magnetic field is also produced and change with time. The magnetic field also stores some static energy into the dielectric equal to $\frac{1}{2}\mu B^2$. Thus, the average stored energy at any point is the sum of the electric and magnetic energy and is dependent upon the magnitude of the field amplitude at that point.

The wave, furthermore, carries its own energy in form of its electrical and magnetic fields, E and B respectively. As the wave move across the dielectric, the fields also move. However, we have seen that due to finite conductivity, the amplitude of the fields reduces with distance. Therefore, a spatially varying density distribution of energy is observed to flow with the wave as a whole. Therefore, if we consider a particular unit volume, some amount of energy effectively enters from one end of this volume with the wave, while some goes out through the other end as the fields of the wave move in and out respectively. This is shown in Fig. 3.5. However, when the electric field E decays with distance and $E_{in} \geq E_{out}$, the energy outflux becomes lesser than the energy influx. In other words, there is a finite divergence of the energy density in the volume.

Now, let us concentrate on what happens to this difference energy density that remains in this volume. This amount is utilized inside the volume. To understand how, recall the Poynting theorem, which we learnt in the last chapter. It states that the amount of net depreciation in the stored energy per unit enclosed volume per unit time is equal to the amount of net power divergence out of the volume plus the power utilized in the volume in Joules work done against the electric resistive force. Expressed mathematically, this can be written as

$$\int E \cdot J\, dv = -\int E \times H\, dS - \int \left[\frac{1}{2}(1/\mu)dB^2/dt + \frac{1}{2}\varepsilon dE^2/dt \right] dv \qquad (3.19a)$$

where $\int E \cdot J\, dv$ is the mechanical work done per unit time in volume dv, $-\int E \times H\, dS$ is the net energy influx per unit time in the volume and $-\int[\frac{1}{2}(1/\mu)\, dB^2/dt + \frac{1}{2}\varepsilon dE^2/dt]dv$ represents the amount of decrement in the stored energy per unit time in the volume. Now when the stored energy considered wholly within the fixed volume does not change on average, the last term in the above equation vanishes. Then,

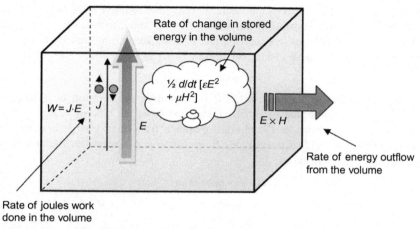

FIG. 3.5

Power utilization in a dielectric.

$$\int E \cdot J dv = -\int E \times H dS \qquad (3.19b)$$

This means, the power lost inside the volume dv by the traversing wave is utilized in the Joules work, where the Joules work is done to carry out the movement of the charges inside the material overcoming the resistance and hence to flow the conduction current with density J. However, here we do not consider scattering.

We shall see in the following section that the dissipation can also happen for non-conducting materials when the mechanical resistance force on the bound charges is considerable. This results in a net depreciation of the energy density with distance.

3.3 DISPERSIVE NATURE OF PARAMETERS

In most of the derivations that we have done in this chapter, it has been assumed that the permittivity and the conductivity of the material are constant over all frequencies. However, if we look deep into the microscopic picture, we shall see that, it is actually not so. To get the dependence of the permittivity on frequency, we recall that as the radio waves traverse through a medium, there is an interaction between the wave and the charges present in the medium. Both the negatively charged electrons and the positively charged ions are affected in the process. However, the positive ions being massive do not move considerably, while the electrons respond to the electric field of the wave by moving in the appropriate direction with respect to the field.

The electric field of the wave, which is oscillatory in nature, forces the electrons in the medium to oscillate with it. If the electrons are free, there is no other force acting on these. However, if the electrons are bound but are able to move with respect to their charge centres, then a restoring force acts on them in the direction opposite to their motion. For bound charges, considering all restoring forces are present, the general equation of motion of these electrons can be written as:

$$m\ddot{x} = eE - g\dot{x} - bx \tag{3.20}$$

Here, m is the mass of the electrons, $\ddot{x}$ is the effective acceleration of it, $g\dot{x}$ is the effective resistive force due to collision, etc. and bx is the electric restoring force acting upon the electrons as the charges displace from their charge centres. E is the incident electric field with amplitude E_0 and angular frequency ω. Then the above equation may be rewritten as,

$$m\ddot{x} + g\dot{x} + bx = eE_0 \exp(j\omega t)$$
$$\text{or,} \quad \ddot{x} + k\dot{x} + \omega_0^2 x = \frac{e}{m} E_0 \exp(j\omega t) \tag{3.21}$$

where $\omega_0^2 = b/m$ and $k = g/m$. Therefore, we get the solution for x and $\dot{x}$ respectively as

$$x = \left(\frac{e}{m}\right) \frac{E}{\omega_0^2 - \omega^2 + j\omega k} \quad \text{and} \tag{3.22a}$$

$$\dot{x} = \left(\frac{e}{m}\right) \frac{j\omega E}{\omega_0^2 - \omega^2 + j\omega k} \tag{3.22b}$$

Now, for perfect dielectrics, i.e. insulators, there are only bound charges. So, there is no conduction current. However, the presence of the significant collision, etc. makes both displacement and velocity a complex term. That is, a part of the displacement as well as velocity occurs in phase with the incident electric field, while there is another component in phase quadrature with it. Now, we have seen in the previous section that the polarization P can be written as $P = Nex$, where N is the charge density, e is the electronic charge and x is the displacement. Replacing the expression for x from the above we get

$$P = \frac{(Ne^2/m)}{\omega_0^2 - \omega^2 + j\omega k} E \tag{3.23a}$$

So, the polarization is not only a linear function of the electric field, E, but is also inversely related to a quadratic function of frequency ω. Therefore, for such materials, the polarization P also has two components, one in phase and the other in quadrature with E. The magnitude of the polarization is

$$|P| = \frac{(Ne^2/m)}{\sqrt{\left(\omega_0^2 - \omega^2\right)^2 + (\omega k)^2}} E \tag{3.23b}$$

If the frequency ω of the incident electric field is equal to the natural frequency ω_0 of the charges, then the polarization is maximum and the condition of 'Resonance' is said to be attained. Under this condition, the polarization becomes

$$P = \frac{Ne^2}{jm\omega k} E \tag{3.23c}$$

Therefore, only the resistive force exists and effectively acts on the charges. The situation is like that of conduction current. We shall see later on that it results in a complex propagation constant as was begot in the case of the conduction current. The polarization falls off as frequency moves away from ω_0 on two sides of it. For frequencies much away from ω_0 and with negligible resistive force, the expression turns into

$$P = \frac{Ne^2}{m\omega^2} E \tag{3.23d}$$

By comparing the above Eqs (3.23a)–(3.23d) for any arbitrary frequency, ω, with the relation, Eq. (3.5c), we get

$$\chi = \frac{P}{E\varepsilon_0} = \frac{(Ne^2/m\varepsilon_0)}{(\omega_0^2 - \omega^2) + j(\omega k)}$$

And hence

$$K = (1 + \chi)$$
$$= 1 + \frac{(Ne^2/m\varepsilon_0)}{(\omega_0^2 - \omega^2) + j(\omega k)} \tag{3.24a}$$

Therefore, dielectric constant K is also complex, made up of real and imaginary component. So, it can be separated as

$$K = K_r + K_i$$
$$= 1 + \frac{(Ne^2/m\varepsilon_0)}{(\omega_0^2 - \omega^2) + j(\omega k)}$$
$$= 1 + \frac{(Ne^2/m\varepsilon_0)}{(\omega_0^2 - \omega^2)^2 + (\omega k)^2} \times [(\omega_0^2 - \omega^2) - j(\omega k)]$$

So,

$$K_r = 1 + \frac{(Ne^2/m\varepsilon_0)}{(\omega_0^2 - \omega^2)^2 + (\omega k)^2} \times (\omega_0^2 - \omega^2) \tag{3.24b}$$

$$K_i = \frac{(Ne^2/m\varepsilon_0)}{(\omega_0^2 - \omega^2)^2 + (\omega k)^2} \times [-j(\omega k)] \tag{3.24c}$$

Therefore,

$$\varepsilon' = \varepsilon_0 K_r = 1 + \frac{(Ne^2/m)}{(\omega_0^2 - \omega^2)^2 + (\omega k)^2} \times (\omega_0^2 - \omega^2) \tag{3.24d}$$

$$\varepsilon'' = \varepsilon_0 K_i = \frac{(Ne^2/m)}{(\omega_0^2 - \omega^2)^2 + (\omega k)^2} \times [-j(\omega k)] \tag{3.24e}$$

Thus, we see that the effective permeability ε ($= \varepsilon' + \varepsilon''$) shows a selective variation with ω. The variations of ε' and ε'' are shown in Fig. 3.6. It also becomes a complex function. It is to be remembered that this complex nature is due to the movement of the bound charges in presence of the resistive force inside the material. Therefore, even

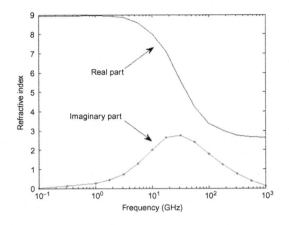

FIG. 3.6

Variation of the real and imaginary part of the RI of water at 20°C (approx.).

Box 3.2 Matlab Exercise

In this box, we work on the real and the imaginary components of dielectric constant, K. Matlab program 'disp_reln.m' was run to obtain the real and imaginary part of the dielectric constant K. The plots show their behaviour near the resonant frequency.

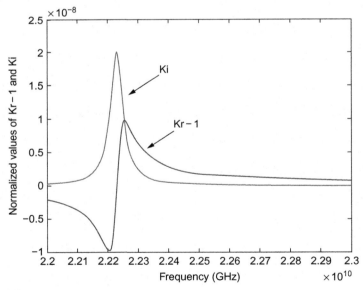

FIG. M.3.1

Effect of propagation of a wave through a practical dielectric medium.

Change the values of frequency f and observe the results. Also see what happens if the ε of the medium changes

though the model was set up for bound charges, the resulting complex ε behaves like that of a conducting medium. Also notice that, the imaginary component vanishes when the corresponding coefficient 'k' becomes zero.

For frequency $\omega \approx \omega_0$, the expressions in Eqs (3.24d), (3.24e) can be approximated as

$$\varepsilon' = \varepsilon_0 K_r = 1 + \frac{1}{2} \frac{(Ne^2/m)}{\omega_0 \left\{ (\omega_0 - \omega)^2 + (k/2)^2 \right\}} \times (\omega_0 - \omega) \qquad (3.24f)$$

$$\varepsilon'' = \varepsilon_0 K_i = \frac{1}{4} \frac{(Ne^2/m)}{\omega_0 \left\{ (\omega_0 - \omega)^2 + (k/2)^2 \right\}} \times (-jk) \qquad (3.24g)$$

The refractive index (*RI*) of the medium is the ratio of the velocity of light in the vacuum compared to that in the medium. Therefore, recalling that the velocity may be defined as $v = 1/\sqrt{(\varepsilon\mu)}$ and considering the permeability μ of the medium is same as that of the vacuum, the *RI* may be expressed as $n = \sqrt{\varepsilon}$. Therefore, from the above discussion, the *RI* may be expressed as $\sqrt{(\varepsilon' + \varepsilon'')}$. Since each of these terms ε' and ε'' is dispersive in nature, so is the *RI*. Further, when a plane wave is incident normally on the surface of such a medium, the reflectivity $|r|$ is given by $|r| = (n_2 - n_1)/(n_2 + n_1)$, where n_1 and n_2 are the *RI* of the first and second medium, respectively. So, if the second medium has purely imaginary value of *RI*, then the $|r|$ becomes 1, i.e. the interface is completely reflective.

3.4 **PROPAGATION IMPAIRMENT FACTORS**

In this section, we shall see in simple terms how exactly the variation of the intensity of an electromagnetic wave occurs due to the propagation medium. We shall learn about the phenomena of absorption and scattering. In addition, we shall understand the basic ideas behind the occurrence of atmospheric diffraction. For each of these phenomenon, here we shall concentrate on the physics behind the occurrence and shall not deal with rigorous mathematics. Each of these topics will be useful to us in subsequent chapters where we shall learn about the different impairments that the satellite signal experience while travelling through the medium containing irregularities. Here, the term irregularity used at different places means that the refractive index that the wave experience has abrupt or random changes.

3.4.1 **ABSORPTION**

We have already come across the principles of absorption earlier in this chapter. There, we have already seen that when a material medium has free electrons to move and electromagnetic wave is incident upon it, the free charges respond to the electric field and start oscillating inside the medium. In the process thay absorb the energy from the wave. We have also seen that for dielectrics, where the mechanical resistance to the oscillation of bound charges are nontrivial, the permittivity becomes complex. The situation becomes similar to a conduction of current. The imaginary part of this permittivity leads to loss of power due to absorption. In other words, whenever there is a mechanical resistance to the motion of the charges leading to the loss of energy in order to overcome it, absorption is said to occur. For dielectrics, this loss shoots up when the frequerncy ω of the wave is near the natural frequency ω_0 of the charges. The work done by the movement of the electrons in an unit volume of the medium leading to absorption is equal to the product of the field and the resultant current due to the motion.

$$
\begin{aligned}
W &= \operatorname{Re}\{E \cdot J_{tot}\} \\
&= \operatorname{Re}\{E \cdot (\sigma E + j\omega\varepsilon E)\} \\
&= \operatorname{Re}\{E \cdot (\sigma E + j\omega(\varepsilon' + j\varepsilon'')E)\} \\
&= E(\sigma - \omega\varepsilon'')E
\end{aligned}
\tag{3.25}
$$

Therefore, even for nonconducting material with $\sigma = 0$, finite absorption occurs due to the imaginary component of the permittivity term. The absorbed power is thus converted into work, or heat in particular. Due to this irreversible loss, the signal power diminishes with consequent reduction in the corresponding electric field.

Focus 3.3

This Focus is to show that the depreciation in the wave amplitude is due to the Joules work done in the volume of a material. For the purpose, let us take a unit volume of the material. Also, let a wave of electric field $|E|$ is incident on one face of it.

Therefore, the power flowing in through this face with the wave is

$$
\begin{aligned}
P_{in} &= \tfrac{1}{2}E \times H \\
&= \tfrac{1}{2}E^2/(c\mu) = \tfrac{1}{2}E^2\sqrt{(\varepsilon/\mu)}
\end{aligned}
$$

After passing through a unit distance, the electric field strength becomes $|E|\exp(-\alpha)$. Then, the power flowing out of the opposite face of the cubic volume is

$$
\begin{aligned}
P_{out} &= \tfrac{1}{2}E^2\sqrt{(\varepsilon/\mu)} \\
&= \tfrac{1}{2}E^2\sqrt{(\varepsilon/\mu)}e^{-2\alpha}
\end{aligned}
$$

So, the power lost in the volume by the signal is

$$
P_{lost} = \tfrac{1}{2}E^2\sqrt{(\varepsilon/\mu)}\left(1 - e^{-2\alpha}\right)
$$

For small α, this can be approximated as

$$
P_{lost} = \tfrac{1}{2}E^2\sqrt{(\varepsilon/\mu)}(2\alpha)
$$

Again, for a good dielectric, when $\tau \ll 1$, it can be shown that $\alpha = \sigma/2\sqrt{(\mu/\varepsilon)}$ (Nasar, 2008). So, the expression $2\alpha\sqrt{(\varepsilon/\mu)} = \sigma$. Therefore, we can write,

$$
\begin{aligned}
P_{lost} &= \tfrac{1}{2}E^2\sigma \\
&= \tfrac{1}{2}E \cdot J
\end{aligned}
$$

But, $\tfrac{1}{2}E \times J$ may be identified as the Ohmic work done in flowing the current per unit volume per unit time. Comparing the loss with the field reduction, we can see that the amount of depreciation in the power occurring as a result of the exponentially decaying electric field over a unit path length in a partially conducting medium is exactly equal to the power used in conducting the current in the volume and thus dissipated by the medium. Therefore, it establishes the required fact.

3.4.2 SCATTERING

A wave, incident upon a medium, will lose power due to the Joules work done inside the medium by the charges, oscillating in response to the incident field, in order to overcome the mechanical forces of resistance. However, a part of the wave power may also be lost due to a phenomenon called Scattering. Scattering of electromagnetic waves may be thought of as the redirection of the waves that takes place when it encounters an obstacle or nonhomogeneity (Yushanov et al., 2013), in course of its propagation. In fact, the process of scattering is a complex interaction between the incident EM wave and the molecular/atomic structure of the scattering object. We shall discuss here the theory of scattering of electromagnetic constituting constitutes the signal. We shall resort to the classical, nonrelativistic approach due to its simplicity. Although quantum-mechanical and fully relativistic treatment provides a better representation of this phenomenon, while explaining we shall take some oversimplified assumptions that will help us to visualize the facts easily.

3.4.2.1 Physics of scattering

When a single-frequency EM plane wave, traversing through a medium, is incident on a neutral dielectric, say a raindrop, the field of the electromagnetic wave interacts with the bound charges within the constituent molecules or other basic neutral structure. These bound charges, experiencing the incident field, get perturbed and hence get displaced from their respective positions responding to the field. Mostly the lighter electrons get displaced rather than the massive positive ions. As a result of this displacement, the constituent charges move from their charge centres leading to unbalanced opposite charges creating polarization of the particles. The incident field, thus, induces electric dipole moment in the neutral material in the direction of the field. The polarity of this dipole gets changes with the change in the incident field's direction and starts oscillating with the incident field. The frequency of this oscillation is same as the frequency of the incident EM wave. However, there can be a phase lag between them. These oscillating charges generate secondary electromagnetic waves and hence radiate energy through it in different directions. These waves may not be coherent with the incident wave actually driving it. This, in most simplistic terms, is how the wave gets scattered, radiating power.

In classical, nonrelativistic physics, the scattering cross section $\sigma_s(\theta)$ of a scattering particle in any direction making angle θ with the direction of propagation (or alternatively angle ψ with the electric field of the incident wave) for a given EM wave is defined as the ratio of the time-averaged scattered power per unit solid angle in that direction, per unit incident power, evaluated at the location of the scattering object. Thus, it is the fraction of the total power incident that is reradiated by the scattering object in that particular direction per unit solid angle (Fig. 3.7).

Extending this concept, the directional scattering cross sectional area of the scattering particle is integrated over all solid angles in all possible directions to get the total scattering cross section. It gives the ratio of the total power scattered to the total power incident on the particle.

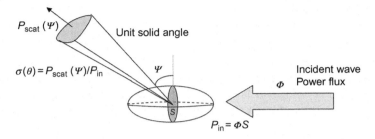

$P_{scat}(\Psi)$

Unit solid angle

$\sigma(\theta) = P_{scat}(\Psi)/P_{in}$

Ψ

Incident wave
Φ Power flux

S

$P_{in} = \Phi S$

FIG. 3.7

Scattering cross section.

If the permittivity ε and hence the refractive index $n = \sqrt{\varepsilon}$ of the material of the scattering object is complex, then the total scattering cross section of the object, which is a function of the refractive index, is also complex in nature. In such cases, the object also absorbs power from the incident wave. Then we can define a cross sectional area, called the total extinction cross section. This is the ratio of the total combined power lost by the wave due to the scattering and absorption to the total power incident on the particle. The difference between the total extinction cross section and the total scattering cross section is called the total absorption cross section. Therefore, to quantify the electromagnetic power absorbed (W_{abs}) and scattered (W_{sca}) by the particle, when P_{in} is the total power incident on it, the total absorption cross section (σ_{abs}) and the total scattering cross section (σ_{sca}) can be defined as

$$\sigma_{abs} = W_{abs}/P_{in}$$
$$\sigma_{sca} = W_{sca}/P_{in}$$

(3.26a)

Thus, the extinction cross section σ_{ext} can be expressed as

$$\sigma_{ext} = (W_{abs} + W_{sca})/P_{in}$$
$$= \sigma_{abs} + \sigma_{sca}$$

(3.26b)

Notice at this point that, the most important variable of the signal we have here is the frequency of the wave, ω, and its corresponding wavelength λ determining the RI, while that of the scattering particle is its diameter, 'd'. The amount and nature of scattering depends upon the relative values of the wavelength to the diameter of the scattering particle. Therefore, we shall first consider these two parameters, and on this basis, shall differentiate the scattering phenomenon into two classes, viz. the Rayleigh scattering and the Mie scattering. It is to be realized at this point that, these two kinds are not any different events occurring, but are the different solutions to the Maxwell's equations for scattering by spherical medium. The former is a conditional solution, while the latter is the generalized form of it.

Rayleigh scattering, named after Lord Rayleigh, is the one which was originally formulated for the small scattering particles and relatively larger wavelengths. The second is the theory of Mie scattering, named after Gustov Mie, and is the one that encompasses the general spherical solution, for both absorbing and nonabsorbing media without any particular bound on particle size. Mie solution being a generalization may be used for describing most of the scattering systems by spherical particles. However, we shall prefer to start with the Rayleigh scattering due to its simplicity and brevity over the Mie scattering formulation.

Rayleigh scattering

The Rayleigh scattering takes place when the dimension of the scattering particles is much smaller than the wavelength of the incident electromagnetic wave. In other words, there is no appreciable change in the phase of the electric field across the dimension of the particle. Therefore, all the individual dipoles created within the particle are forced to oscillate under the influence of a field with uniform phase. Quantitatively, the required condition can be expressed as

$$\left(\frac{2\pi}{\lambda}\right) \times d \ll 1 \qquad (3.27)$$

where 'd' is the diameter of the spherical particle and λ is the wavelength of the wave in the medium of the scattering particle. Physically, this also implies that the time for penetration of the electric field is much less than the period of oscillation of the EM wave. To understand the process of scattering in a spherical particle, recall the fact we have learnt in previous section that an electric field will induce dipole moment inside the material of the sphere. Therefore, the incident wave, having time varying E field, will induce a time varying dipole moment. It can be shown that, for a sphere of diameter 'd', the dipole moment induced due to the incident electric field E_0 will be

$$p = 4\pi\varepsilon(d/2)^3 \frac{(\varepsilon-1)}{(\varepsilon+2)} E_0 \qquad (3.28)$$

Since the forcing electric field upon all the oscillatory dipoles has the uniform phase, we consider that for a distant point, where the distance is much larger than the dimension of the particle, the scattered signal originating from different individual oscillators inside the particle will have the same phase. In other words, the whole particle acts as a discrete unit scattering centre. Therefore, at large distance from such a pole, the electric and the magnetic fields are given by the far field radiation which can be expressed in terms of the polarization p and distance r. Here, it is more convenient to use the direction with respect to that of the incident electric field, given by ψ. So, the expression becomes

$$E(r\psi) = -\frac{k^2}{4\pi\varepsilon_0} p \frac{sin\psi}{r} e^{-j(\omega t - kr)} \qquad (3.29a)$$

where ψ is the angle with the direction of the electric field. Therefore, the magnetic field can be obtained as

$$H(r\psi) = \sqrt{\mu/\varepsilon} E(r\psi)$$
$$= -\sqrt{\mu/\varepsilon} \frac{k^2}{4\pi\varepsilon_0} p \frac{\sin\psi}{r} e^{-j(\omega t - kr)} \qquad (3.29b)$$

Notice that the fields fall off as $1/r$ with distance. The Poynting vector thus produced at distance r and angle ψ from the particle by the oscillating field is

$$S = E \times H$$

$$= \sqrt{\mu/\varepsilon} \frac{k^4}{(4\pi\varepsilon_0)^2} p^2 \sin^2\psi \frac{1}{r^2} \tag{3.30}$$

Since this is the average power radiated by the oscillating dipole in the direction θ per unit area, the average power radiated per unit solid angle is

$$P_{str}(\psi) = \sqrt{\mu/\varepsilon} \frac{k^4}{(4\pi\varepsilon_0)^2} p^2 \sin^2\psi \tag{3.31a}$$

Now, replacing the expression for p_0 in terms of permittivity ε as obtained in Eq. (3.28), we get

$$P_{str}(\psi) = \sqrt{\mu/\varepsilon} \frac{k^4}{(4\pi\varepsilon_0)^2} \sin^2\psi \left[4\pi\varepsilon (d/2)^3 \left(\frac{\varepsilon-1}{\varepsilon+2}\right) E_0 \right]^2$$

$$= 2\sin^2\psi \left(\frac{2\pi}{\lambda}\right)^4 \left(\frac{\varepsilon-1}{\varepsilon+2}\right)^2 (d/2)^6 \left[\frac{1}{2}E_0^2\sqrt{(\mu/\varepsilon)}\right] \tag{3.31b}$$

Identifying $\frac{1}{2}E_0^2\sqrt{(\mu/\varepsilon)}$ as the average incident power, the power radiated per unit solid angle along the direction ψ per unit average incident power I is obtained as

$$P_{str}(\psi)/I = 2\sin^2\psi \left(\frac{2\pi}{\lambda}\right)^4 \left(\frac{\varepsilon-1}{\varepsilon+2}\right)^2 (d/2)^6 \tag{3.31c}$$

Notice that this is the definition of the scattering cross section along direction ψ. Thus

$$\sigma(\psi) = 2\sin^2\psi \left(\frac{2\pi}{\lambda}\right)^4 \left(\frac{\varepsilon-1}{\varepsilon+2}\right)^2 (d/2)^6 \tag{3.32}$$

Integrating the cross section over all directions, i.e. over complete solid angle over a sphere, we get the total scattering cross section as

$$\sigma_{scat} = 2\left(\frac{2\pi}{\lambda}\right)^4 \left(\frac{\varepsilon-1}{\varepsilon+2}\right)^2 (d/2)^6 \int_{\psi=-\pi/2}^{\pi/2} \int_{\varphi=0}^{2\pi} \sin^2\psi \cos\psi \, d\psi \, d\varphi$$

$$= 2\left(\frac{2\pi}{\lambda}\right)^4 \left(\frac{\varepsilon-1}{\varepsilon+2}\right)^2 (d/2)^6 (2/3)(2\pi) \tag{3.33a}$$

Finally, replacing ε as n^2 and rearranging, we get

$$\sigma_{scat} = \frac{2}{3}\pi^5 \frac{d^6}{\lambda^4} \left(\frac{n^2-1}{n^2+2}\right)^2 \tag{3.33b}$$

Therefore, the power radiated by the particles due to the scattering is dependent upon the difference in the refractive index and the power reradiated back is mostly concentrated along the direction of the primary wave (Rana, 2007).

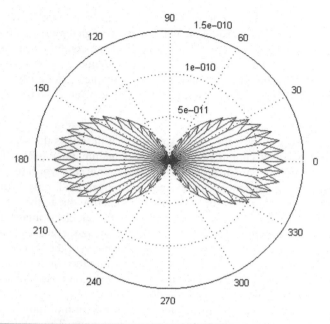

FIG. 3.8

Variation of Rayleigh scattering as a function of θ with respect to the direction of propagation of the wave for a water droplet of diameter 5 mm.

Examination of equations reveals several interesting items. The power radiated by the scattering particles is dependent upon the difference in the *RI* of the material of the scattering medium with respect to its surrounding medium (considered vacuum here with $RI = 1$). Scattering will be more for larger differences. Further, the scattered power is dependent upon direction. Since $\sin \psi$ is maximum when $\psi = \pi/2$, i.e. $\theta = 0$, maximum power is directed along the direction perpendicular to the field, i.e. in the direction of the propagation of the primary wave. Fig. 3.8 shows the radial variation of the Rayleigh scattered power as a function of θ. Functionally, the directional scattered power is proportional to the sixth power of particle size. It is also proportional to the fourth power of frequency or inversely to the fourth power of wavelength λ. Thus, high frequencies are more scattered compared to lower ones. It is the same reason that results in the blue colour of the sunlight to scatter more than other components making the sky Blue in colour. Lastly, it is important to understand that, because the scattering particles are randomly positioned, the scattered light arrives at a particular point with a random collection of phases and makes the scattered wave incoherent.

For a propagating satellite signal, typically in gigahertz range, the atmospheric gaseous particles manifest absorption, but have very small scattering cross section to make any considerable effect. The tiny water droplets in the cloud scatter following the

Rayleigh conditions. For the rain drops, which have diameters from 0.5 up to 5.0 mm, only a few drops of lower dimensions are within the Rayleigh regime for up to Ku band frequencies. But, for Ka and even higher frequencies, the condition for the Rayleigh scattering is not satisfied. Hence, the signals of these frequencies are scattered where the solution to the scattered power follows a more complex feature described by the Mie scattering solution. We shall learn about it in the next subsection.

Mie scattering

Mie scattering theory is the generalized solution that describes the scattering of an electromagnetic wave by a homogeneous spherical medium having RI different from that of the medium through which the wave is traversing. It is worth reiterating that Mie scattering is not any independent physical phenomenon. It is only a definite solution to the Maxwell's equations for the multipole radiation due to the electric polarization of the molecules in the scattering particles when an electromagnetic wave is incident upon it. It gives the solutions for scattering where the phase of the incident signal can even change considerably within the dimension of the scattering particle. Therefore, unlike Rayleigh scattering, it does not require the condition of $2\pi d/\lambda \ll 1$ to be fulfilled. Dust, smoke, rain drop particles are common causes of scattering to propagating waves following the Mie solution.

The Mie solution takes the form of an infinite series of multipole expansion of polarization in a spherical medium due to an incident wave. The electric and magnetic field of such polarization can be expressed also in the form of an infinite series with each element representing the contribution of one particular order of multipole expansion. Now, recall that, the power scattered in any direction can be estimated from the Poynting vector in that direction. If the expression for the fields E and H are completely known, the Poynting vector can also be derived, as we have done it for Rayleigh scattering. So, knowing the expression for E and H, the cross section can also be derived by utilizing these expressions for any particular direction. Nowadays, a number of efficient algorithms are available for getting the Mie solution (Hergert and Wriedt, 2012). However, solving for the Mie solutions is beyond the scope of the book. So, we shall only mention here the expressions for the scattering cross sections obtained and will discuss the features of the solution.

It is interesting to note that the Mie scattering cross section is directive in nature and may be expressed as a combination of different orthogonal modes. The function has maxima and minima for different values of θ, where θ is the angle the scattering direction makes with the direction of propagation of the incident wave. The functions have nulls at certain θ values, thus creating directive lobes in between the nulls. For larger dimensions of the scattering particle, the number of the nontrivial modes increases. This results in an increasing number of lobes in the spatial profile of the cross section with resultant narrower widths. The backward lobes vanish for odd values of the modal order, while the forward lobe is always present. The forward lobe intensity enhances for conditions when the backward lobe diminishes. Thus, with increasing diameter of the sphere, the forward lobe becomes stronger and narrower.

The total scattering and absorption cross section are readily calculated from the directional cross section by integrating the directive values over all directions. The total scattering cross section is indicated as σ_s, the total absorption cross section as σ_a and the total extinction cross section as σ_e. These values can be obtained as a function of some a_n and b_n coefficients derived from the boundary conditions of the solutions to Maxwell's equations under the scattering condition. According to the Mie solution, the scattering and extinction cross section, each normalized to the particle geometric cross section α ($\alpha = \pi d^2/4$), can be expressed as

$$Q_s = \sigma_s/\alpha = \frac{2}{x^2}\sum(2n_p + 1)\left[|a_n|^2 + |b_n|^2\right]$$
$$Q_e = \sigma_e/\alpha = \frac{2}{x^2}\sum(2n_p + 1)[\text{Re}(a_n) + \text{Re}(b_n)] \tag{3.34}$$

where $x = (2\pi/\lambda)\,(d/2)$ and 'd' is the particle diameter and λ is the wavelength of the incident wave and 'n_p' is the order of the multipole expansion of the polarization due to charge oscillation inside the particle. The coefficients 'a_n' and 'b_n' represent the contribution of the multipoles of order 'n_p'. Note that, the absorption cross section is the difference of the extinction and scattering cross sections. The variations of the extinction cross-section σ_e for different diameters of water drop acting as the scattering particle, for frequencies 20 and 30 GHz is shown in Fig. 3.9.

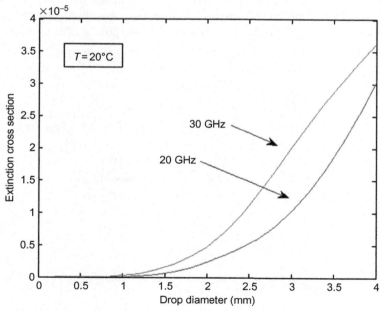

FIG. 3.9

Variation of Mie scattering intensity with frequency and dimension.

3.4.3 **DIFFRACTION**

An electromagnetic wave moves through a medium forming wave fronts. A wave front is the plane obtained by joining all points having the same phase. Thus, for a point source, all points of a wave, travelling through a uniform medium, at a definite radial distance from the source are in the same phase of oscillation and they constitute a wave front. Whenever a wave front is incident upon a partially blocked screen with narrow openings, the blocked portion inhibits a portion of the wave front to propagate. But, each point on the wave front at the openings acts as a secondary source. As a result, these portions of the wave spread out forming secondary waves and thus advancing the wave front. At the receiver, such secondary waves, originating from different such portions, arrive with different phase and amplitudes and combine to form the resultant wave with arbitrarily different amplitude and phase. The temporal and spatial variation of the resultant amplitude and phase depends upon the relative strength and phase of the contributory secondary sources. Thus, they make some particular spatial intensity distribution. This is called diffraction (Ghatak, 1992).

The geometrical attributes of the distribution also depend upon the dimension of the different openings forming the secondary sources. To understand the formation of the diffraction pattern and the contribution of different portions of the wave front to it, let us consider a wave front WW', propagating in the $\hat{z}$ direction, as shown in Fig. 3.10. In order to determine the field at any arbitrary point P at a perpendicular distance 'b' from wave front due to the disturbances reaching from different parts of the front, we consider different concentric annular areas about the centre O, where O is the point on the wave front which is radially above the point P. The distances from

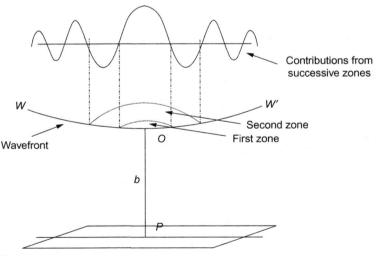

FIG. 3.10

Diffraction.

two adjacent annular regions to the receiver are separated by $\lambda/2$, which corresponds to a phase difference of π.

It can be shown that the radius of the nth annular region is $r_n = +\sqrt{(n\lambda b)}$ and also the area of each annular region is approximately the same and equals to $\Delta A = \pi\lambda b$. Therefore, the resultant amplitude reaching P from all points of the wave front is

$$\mathbf{u}(\mathbf{P}) = \sum u_i = u_1 - u_2 + u_3 - u_4 + \ldots$$

where u_n represents the net amplitude of the secondary wavelets emanating from the nth zone, i.e. the nth annular region. The alternate positive and negative signs show that the resultant disturbances produced by two consecutive zones are π out of phase with respect to each other. Fig. 3.10 also shows the contribution variation from different parts of the wave front to a point at the centre.

Considering the incident wave to be uniform, the distance increases with the increase in zonal number and hence the amplitude monotonically decreases. Thus, we may write

$$|\mathbf{u1}| > |\mathbf{u2}| > |\mathbf{u3}| > |\mathbf{u4}| > \ldots$$

Therefore, the contributions from adjacent zonal regions not only differ in phase by π but also gradually decrease in magnitude with increasing zonal numbers. The decrement in the magnitude is approximately linear and it can be shown that the total resultant amplitude produced by the entire wave front is only half the amplitude produced by the first half period zone, i.e. $u(P) = \frac{1}{2}u_1$. This is because, the arithmetic mean of the contribution of two successive odd zones is exactly equal and opposite to that of the even zone they have in between. Thus, half the sum of contribution of the first zone and third zone is exactly cancelled out by the equal and opposite contribution of the second zone. Similarly, the rest half of the third zone and half of the fifth zone contribution cancel out the contribution of the fourth zone. This goes on for increasing zonal numbers. Therefore, ultimately the half of the first zone and the half of the last zone effectively contribute. However, as the latter is much smaller than the former, the effective intensity at P is $\frac{1}{2}u_1$. Similarly, a regular pattern of intensity is observed across the plane around P.

Now, if different portions of the wave front are either blocked or their phase gets changed from other parts of the wave front, the front develops aberration in terms of phase. Then, the contribution of the entire front to the point P will differ significantly. For example, if a zone with positive contribution is blocked, the effective amplitude will decrease at P, while if that of a zone with negative contribution is blocked, the amplitude will increase. This modification will be different at different points around P, since for each individual point, the centre and the zonal contribution from different parts of the wave front will be different. Further, if the blockage or the equivalent phase change varies with time, then the effective signal strength at P will also vary with time. Therefore, effectively, the intensity pattern around P will become irregular.

CONCEPTUAL QUESTIONS

1. Show that for good practical dielectric, i.e. when $\tau \ll 1$, $\alpha = \sigma/2\sqrt{(\mu/\varepsilon)}$.
2. Comment on the statement—Near the resonant frequency, the dielectric behaves just like a conductor.
3. Rain drop sizes vary from 0.5 to 5.0 mm. Cloud droplets are even smaller and <0.1 mm. What are the features of the scattered wave you expect from these elements for different bands of signal frequencies?
4. What should be the order of the size of the irregularities which may cause considerable variation in the diffraction pattern for the satellite signals?

REFERENCES

Feynman, R.P., Leighton, R.B., Sands, M., 1992. Feynman Lectures on Physics. Narosa Publishing House, New Delhi.

Ghatak, A., 1992. Optics, second ed. Tata Mc Graw Hills, New Delhi.

Hergert, W., Wriedt, T. (Eds.), 2012. The Mie Theory. In: Springer Series in Optical Sciences. http://dx.doi.org/10.1007/978-3-642-28738-1_2.

Nasar, S.A., 2008. 2008+ Solved Problems in Electromagnetics. SciTech Publishing Inc., Raleigh, NC.

Rana, F., 2007. Electromagnetic Scattering. Lecture 34, Cornell University. https://courses.cit.cornell.edu/ece303/Lectures/lecture34.pdf Accessed 16 October 2016.

Reitz, J.R., Milford, F.J., Christy, R.W., 1990. Foundations of Electromagnetic Theory, third ed. Narosa Publishing House, New Delhi.

Yushanov, S., Crompton, J.S., Koppenhoefer, K.C., 2013. Mie scattering of electromagnetic waves. In: Proceedings of COMSOL 2013, Boston, MA.

Ionosphere and ionospheric impairments

4

4.1 IONOSPHERE

In Chapter 1, we had a basic picture of our atmosphere including ionosphere. There, we defined ionosphere as the region of the upper atmosphere extending from 50 km to more than 1000 km above the earth's surface, where the free electrons are available in abundance in addition to of neutral particles.

There are many formal definitions for the Earth's ionosphere. But, as our prime focus is on the satellite signal propagation through ionospheric plasma, we adopt here the definition for the ionosphere which is particularly defined for radio wave propagation and is relevant to telecommunications. This states that—'the ionosphere of the Earth is that portion of the upper atmosphere where ions and electrons of thermal energy are present in quantities sufficient to affect the propagation of radio waves' (IEEE, 1998).

This ionosphere is the site for the major impairment to the propagating waves, especially for satellite signals in L band and below. The effects diminish rapidly with increasing frequency. Therefore, it causes more harm to the navigation signals which operate at L band than for the pure satellite communication signals which operate at C or even higher bands.

The ionospheric impairments pose a threat to the navigation applications. This may even prove detrimental, especially for those meant for critical uses. Therefore, to understand the nature of impairment, it is important first to understand its constitutional features in a comprehensive fashion. So, first, we shall read in details about the structure of the ionosphere and about its prominent characteristics.

4.1.1 BASIC STRUCTURE OF IONOSPHERE

The neutral molecular density in the atmosphere decreases with height and at large heights from the earth surface, it is extremely rare. Due to this, the sunlight coming from above can carry large energy up to a considerable depth into the earth's atmosphere without decay. Below these heights, the neutral particles of the earth's atmosphere absorb large quantities of this radiant energy from the sun. This energy is enough to dissociate these molecules into charged ions and free electrons. The ionization, thus resulting from the photo-decomposition of the thin upper atmospheric gases by the radiation from the Sun, produces electron-ion pairs.

Satellite Signal Propagation, Impairments and Mitigation. http://dx.doi.org/10.1016/B978-0-12-809732-8.00004-1

There also exists a parallel recombination process of the charged pairs. This recombination, however, is dependent upon the availability of neutral molecules. Due to the sparse particle density at these heights, the ions thus created spend a considerable lifetime before they recombine and cease to exist as charged entities. A region is therefore created at heights between 50 km up to 1000 km above the earth surface, where the ionization is appreciable with a considerable availability of free charged particles. This region is called the ionosphere. However, the total number of electrons always remains equal to the total number of positive ions and hence the ionosphere's overall charge is neutral in nature. Further, the ionization process remains in a dynamic equilibrium with the recombination process to create a definite stable electron density.

The density of ionizable neutral atoms decreases exponentially with altitude, whereas the incoming solar radiation intensity, necessary for ionization, decreases exponentially as it penetrates down. This gives the ionosphere a finite range of extent. The lower limit is defined by the penetration power of the high energy solar radiation, while the upper limit is restricted by the availability of the neutral atoms for dissociation. Inside the two flanking limits of the ionosphere, the ionized charge densities are horizontally stratified forming layers of different densities. The vertical density gradient occurs due to difference in the ionization and the recombination rate. The lower layers among these strata are ionized by highly penetrating components of the solar radiation like hard X-rays, while in the upper layers, the soft X-rays or extreme UV component of the sun rays does the ionization.

The vertical profile of the electron density in the ionosphere can be distinctly separated into few regions. Out of these different regions of the ionosphere, the lowermost is called the D region. It starts from around 50 km to about 90 km in height. The peak is observed around 72 km (Rishbeth and Garriot, 1969). This region is basically solar-controlled with electron density of $\sim 10^8$ electrons/m^3.

E region extends from 90 km up to around 160 km from the earth surface (IEEE, 1998). The ionized layer within this region has a daytime peak, at a height called h_mE, which is at around 110 km. The ionization within the E region is mainly contributed by the extreme ultra violet rays (EUV) and soft X-rays to some extent. Hence, it is highly correlated to the incident solar flux. Therefore, the normal E layer is present only during the daytime with the typical densities between 10^{10} and 10^{11} electrons/m^3 (Hargreaves, 1979). After sunset, the peak E layer density reduces to around 5×10^9 electrons/m^3 (Rishbeth and Garriot, 1969).

The F region of the ionosphere extends from about 160 km to more than 1000 km in altitude. It has two different ionized layers in it. The lower of the two is called the F1 layer and exists only during the daytime. It extends from above 180 km to about 250 km in height with the peak somewhere around 220 km. The electron density observed here are about $3–5 \times 10^{11}$/m^3 during moderate solar activity period.

The higher of the two layers, called the F2 layer, has the maximum electron density in the total ionosphere. The typical electron density of this layer is 10^{12}/m^3. The peak height of F2 layer, called h_mF2, is highly variable and created somewhere between 300 and 450 km (Hargreaves, 1979; Rishbeth and Garriot, 1969). This range is considering the day and night variation and also the variation during the high and low solar activity

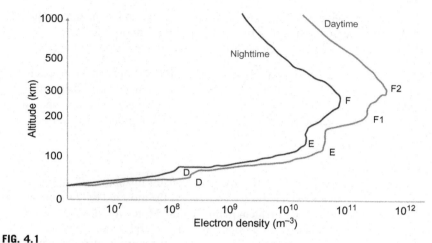

FIG. 4.1

Vertical profile of Ionosphere (not to scale).

period. This layer exists even in the nighttime, but with a reduced electron density. Both the peak height and the peak density of F2 show considerable diurnal variations. It also varies with seasons, solar activities and other factors as well. At large altitudes above the F2 peak, the ionosphere decays monotonically with increasing altitude with a typical exponential decay rate (Rishbeth and Garriot, 1969).

The peak layer densities and the heights are typically measured from the ground-based instruments like ionosonde and are based upon the maximum frequency that gets reflected back from the layers and the time delay of arrival. We shall derive the relationship between the reflection and the frequency later in Chapter 5, whereas in this chapter, our emphasis will be to see how these layered structures of the ionosphere may be explained with simple physics model.

The variation of the ionospheric profile with height is shown in Fig. 4.1. The peak production height and the peak ionization value vary with solar zenith angle, χ. However, as the solar energy flux varies with space and time, the electron density exhibits both spatial and temporal variations. Further, the temporal variation is considering the rotation of the earth causing diurnal variations and its revolution around the sun causing the changes over the seasons. The distribution and dynamics of this ionospheric plasma are both dependent upon a complex coupling between the neutral atmosphere, solar heating, photo ionization, electrical conductivity and neutral winds, all interacting with the magnetic field of the Earth.

4.1.2 SPATIAL AND TEMPORAL DISTRIBUTION OF IONOSPHERE
4.1.2.1 Global distribution
The ionosphere varies significantly with geographic location in terms of its character including electron density. The global distribution of the ionosphere may be divided into three distinct regions (Acharya, 2014):

(a) Equatorial region: extending up to ±25° on both sides of the magnetic equator.
(b) Mid-latitude region: extending from latitudes ±25° up to ±65°.
(c) High-latitude region: This region extends from above 65° up to the poles.

Each of these regions has their own characteristics and has effects on the propagating radio signals. However, bearing in mind the brief scope that we have to describe the ionosphere, we restrict our current discussion only to few very important and interesting characteristic phenomena occurring in these three regions, which affects the satellite signals.

Equatorial ionosphere

The equatorial region extends from the magnetic equator to about ±25 of magnetic latitude and is very dynamic in nature. This region has the most ion production rate as most of the solar flux is directly incident in this region. Moreover, it is very sensitive to variations in the geomagnetic conditions and the neutral winds that control the movements of these charged particles. The equatorial region is also characterized by the occurrence of a prominent phenomenon called equatorial ionospheric anomaly. Effactually, these results in large delay and sometimes rapid power fluctuations of the radio signals propagating through it. These impairments, viz. delay and power fluctuations, affect the satellite signals in L band and below, especially those used for navigational purposes. The effects will be discussed later in this chapter, while now we consider the details of the different phenomena associated with this region in the following few subsections.

Equatorial ionospheric anomaly. The ionized particles, i.e. electrons and ions, are generated as a result of photoionization. Therefore, these particles are expected in larger densities at locations where the intensity of the solar flux is more. The ionospheric electron density at the equator during the equinoctial period is thus expected to be high. This is because, at these locations the ionosphere is illuminated with vertical incidence of the solar radiation at the solstices, and hence, receives maximum available power. But for the equatorial ionosphere, an anomalous behaviour is observed during this period. It rather shows a trough in electron density at the equator with a consequent dynamic enhancement in the density at a certain low-latitude location. This happens because the ionized particles that exist in the ionosphere in the form of plasma are moved from the region of the magnetic equator to either side of it up to around 15° North and South magnetic latitude. This phenomenon is called the equatorial ionization anomaly (EIA). This was first observed by Appleton (Appleton, 1946). He showed this anomalous nature of the equatorial density distribution with the F region electron density depreciating at noon to form what he called as 'Noon Bite Out'. The consequent peaks appear at around ±15° dip latitude. Mitra (1946) was the pioneer to propose an explanation of the phenomenon attributing the anomaly to the diffusion of the plasma from the equator towards the poles along magnetic lines of forces. However, the shortcoming of Mitra's theory was corrected later by Martyn (1947) by introducing the idea of equatorial vertical drift. He explained the anomaly in the light of the combined action of the vertical updrift of the plasma at the magnetic equator followed by diffusion of these charged particles to higher latitudes along the geomagnetic lines of forces.

The development of the EIA is a result of the plasma transport, which in turn depends on a complex interaction of a number of ionospheric processes. It shows considerable variability with local time, longitude and season. It sets off at the equator due to the vertical drift of the plasma occurring as a result of the $E \times B$ force acting upon the charges in the F region. The electric field E is directed eastward and is generated at the E region due to a dynamo action of the geophysical forces like the neutral wind, etc. This horizontal eastward electric field gets linked to the F region via the magnetic lines of forces due to highly conductive nature. The geomagnetic field B is also horizontal and northwards at the equator. The $E \times B$ force is thus created at the F region and is directed vertically up and acts equally on both positive and negative charges. Therefore, both ions and electrons move in the upward direction resulting in no net current. This drift thus lifts the charged plasma vertically from the site of its abundance in F region and puts them to a greater altitude. The neutral particles at these extended heights being very scanty, the raised plasma do not readily recombine and go on accumulating. The consequent rise in plasma density results in the plasma particles to diffuse along the geomagnetic lines of forces under the influence of gravity. Thus, the lifted plasma, controlled by gravity-aided diffusion, now moves along the magnetic field lines, sliding towards the higher latitudes. In this way, the combination of vertical lift due to the drift and subsequent diffusion gives the plasma a 'fountain-like' pattern and hence called the 'Fountain effect'. It redistributes the electron density by transporting them from the equatorial region resulting in a local trough and move them towards higher latitude where they are dumped forming the crest around $\pm 15°$ geomagnetic latitude. This observation is what is known as the Equatorial Anomaly. This is shown in Fig. 4.2.

The EIA exhibits day-to-day variability due to the variation in the wind, effective geomagnetic field and also due to various other geophysical reasons. EIA is sensitive to the variations of the seasons and solar activities, as well. The daily variability affects both the extent of the anomaly, i.e. the latitudinal distance from the geomagnetic equator at which the anomaly crest is formed, and the strength of the anomaly. The magnitude of this variability depends upon the strength of the controlling geomagnetic and solar parameters and the extent by which they affect the transport process of the plasma.

Prereversal enhancement

At the equatorial region, just before the sunset, the eastward electric field in the F region of the ionosphere gets enhanced before it becomes westward during night-times. This is called the prereversal enhancement (PRE).

This enhanced eastward electric field, in conjunction with the northward magnetic field, causes an extended $E \times B$ upward drift in the dusk sector. This causes

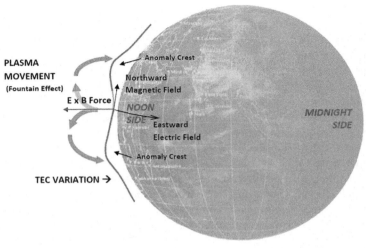

FIG. 4.2

Plasma fountain effect and equatorial anomaly.

a large volume of plasma to rise suddenly to greater heights leading to slower recombination, and hence, showing enhanced vertical TEC. This elevated plasma leads to the initiation of Rayleigh-Taylor instability, which in turn, affects the irregularity formation in the ionosphere significantly, including formation of the plasma bubbles. The magnitude of PRE depends upon season, solar cycle and longitude. The mechanism by which this enhancement is established depends upon the F region dynamo and is briefly mentioned below.

Few significant models have been proposed for explaining the PRE. According to the Farley-Bonelli model (Farley et al., 1986), during the daytime, at the heights of the F layers, there prevails a zonal wind which flows from the subsolar point to both the directions away from it. Thus, at the locations near the sunset terminator, the wind flows from the subsolar location eastwards towards the dusk. As the wind, carrying the charges, flows through the terminator, due to the gradient in the conductivity, there is a divergence in the spatial transport of the charges. This results in an eastward electric field at the location. This field is sustained when the conjugate points of the field lines also simultaneously cross the terminator. This leads to an enhanced eastward field causing PRE.

Another mechanism suggested by in the Haerendel–Eccles model (Haerendel and Eccles, 1992) states that the equatorial electro jet (EEJ), which flows as an eastward current at the peak heights of the E region driven by the Eastward electric field, *partially* closes after sunset through a poorly conducting F region valley, thus creating an effective enhanced eastward electric field. This is suggested to be the dominating mechanism for PRE (Kelley et al., 2009).

Due to the effect of the interplanetary electric field, carried by the solar wind, sometimes strong westward field may penetrate up to the F region and overcompensate the enhanced eastward field. In the next subsection, we shall see how the enhanced electron density, forming as a result of the PRE, causes plasma instability leading to the formation of irregularities in the ionosphere.

Equatorial spread-F

Equatorial spread-F (ESF) is a geonatural phenomenon in which the vertical profiles of the equatorial ionosphere are redistributed after sunset. They occur on time scales of up to hours and across length scaling from up from few to hundreds of kilometres and is caused by the plasma instabilities.

It is a phenomenon, occurring from the postsunset period up to a time around midnight, except for magnetically disturbed conditions. It occurs in a belt near the magnetic equator. The first detection of an ESF event was reported when 'diffuse echoes' of ionosonde from the F region were observed using an ionosonde. This gave the impression of the fact that the F region densities have penetrated downwards as if the F region has spread vertically (Farley et al., 1970). One of the distinguishing features of this phenomenon is the occurrence of local and small-scale plasma depletions, known as Ionospheric Bubbles. These are small confined region, with abrupt low electron density surrounded by relatively higher densities.

With these observations, the origin of the spread-F has been explained in the light of plasma instability. We have already learnt in the previous section that the PRE raises a large volume of plasma to greater heights, just near the sunset, increasing their effective height. However, after the sunset, this driving mechanism ceases to continue. But it has already resulted in the formation of a heavy dense layer of plasma on the top of relatively rarer ones below it. The layer below also gets rapidly rarefied due to the chemical recombination process, at a rate much faster than those at the top due to the abundance of the neutral molecules at lower heights. Therefore, a top-heavy condition is created with a steep positive upward gradient.

Now the heavy dense plasma at the top starts moving down under the action of gravity. This motion initiates the Gravitational Rayleigh Taylor (GRT) instability (Dungey, 1956). As they pass down perpendicularly across the Northward geomagnetic lines, they experience a Lorentz force. So, the positive and the negative charges experience a relative displacement in two opposite sides horizontally. This results in horizontal electric field E, which further enhances the downward movement of the plasma by generating the $E \times B$ force. This causes the instability. This instability results in random and abrupt penetration of high F region density into lower density regions adjacently below it resulting in 'spread-F'. Finger-like projections of high density plasma is observed amid lower densities surrounding it, as shown in Fig. 4.3.

The large size perturbations in the unstable bottom side of the F region essentially start growing nonlinearly. The projections of high density can ultimately surround a region of low density around it on all sides. This forms low-density regions enclosed by high-density regions, called Bubbles (Woodman and Hoz, 1976). The bubble will

FIG. 4.3

Equatorial spread-F.

rise by the effect of buoyancy and will continue to do so even in regions with stable gradients, provided the density entrapped in the bubble is lower than that of the surroundings. Similar irregularities are observed at the topside of the F region. Bubbles or plumes have been observed which extend from the bottomside through to the topside.

The seasonal behaviour of the Spread-F depends on longitude. To be more precise, it is very much dependent upon the declination of the location. The spread-F maximizes at a place for that particular season when the sunset terminator line is aligned with the magnetic meridians. This ensures the simultaneity of sunset at the conjugate points on the magnetic lines of forces joining the lower F and E region of the ionosphere, thus both locations being transferred simultaneously across the sunset terminator. It is only when this condition is fulfilled that the enhanced eastward electric field generated at the dusk sector is not electrically shorted at the conjugate points and thus maintain the field to initiate plasma rise leading to GRT instability. Therefore, due to the westward declination, it is greatest during the December solstice in the American sector, but minimum during the summer. Conversely, it is maximum in summer in the pacific sector, where the declination is Eastward. The same reason results in maximizing of the spread-F activity during the equinoctial months of

September and March in the Indian region where the declination here is close to zero. (Abdu et al., 1981; Iyer et al., 2003). Spread-F shows considerable day-to-day variability within a particular season as well. The plasma irregularities that occur in the F region thus influence the performance of the satellite applications including navigation and communication. As the waves propagate through the medium, they experience a random local but rapid spatial variation of the electron density, resulting in high scintillation, i.e. a rapid fluctuations in its amplitude and phase.

Mid-latitude ionosphere. The region of the ionosphere extending between 25° and 65° latitudes on both the Northern and Southern hemisphere constitutes the mid-latitude. Compared to the equatorial and polar counterparts, the ionosphere in this region is largely benign. In this region, the motion of ionospheric plasma is dominated by winds in the neutral atmosphere and the Earth's magnetic field. Following are the two important characteristics observable in the mid-latitude ionosphere (Scali, 1992).

Mid-latitude trough

One significant feature of the mid-latitudes is the mid-latitude trough. This major feature of the F region ionosphere is formed at the boundary between the mid-latitude and auroral ionosphere around 65° latitudes and consists of a region of exceedingly low density of electrons and hence called the trough. This region is primarily formed on the nightside of the globe. The poleward edge of the trough is located adjacent to the auroral ovals with a width of few degrees of latitude. It can stretch longitudinally across the nightside.

The mid-latitude trough is formed as a result when the plasma enters a region that is convecting very slowly. This plasma, as a result, remains confined in the nightside for long resulting in a situation referred to as 'Stagnation'. Due to this, the plasma continues to recombine without any further generation of the ions. So, the effective electron density reduces largely forming a trough.

Winter anomaly

At mid-latitudes, in the F2 layer, the daytime electron density is observed to be lower in summer than in winter. This phenomenon is known as winter anomaly.

The production of free electrons and ions through photoionization is higher in the summer, as expected, due to the large solar flux leading to dissociation of neutral atoms. However, during this time the seasonal changes in the molecular-to-atomic ratio of the neutral atmosphere are such that it causes the summer ion loss rate due to recombination to be overwhelmingly higher than normal. Consequently, the excess loss during the summertime overcompensates the increase in summertime production. Therefore, the effective F2 electron density gets lower in the local summer months.

The major reason behind the occurrences of winter anomalys, as mentioned above, is associated with the seasonal and hemispheric change in the neutral composition of the thermosphere (King, 1961; Rishbeth and Setty, 1961; Rishbeth et al., 2000, 2004). The summer-to-winter interhemispheric wind is suggested to be the source of the neutral composition change (King, 1964). Although this seasonal change in neutral composition is the prime reason of such anomaly, other factors also contribute to the effect. The seasonal variations of the vibrational temperature of the constituent molecules also affect the winter anomaly in the F layer. However, the contribution of the vibrational-excited molecular gases to the F layer winter anomaly is less significant (Torr et al., 1980; Lee et al, 2011).

The winter anomaly is more pronounced during the solar maximum period and in the Northern Hemisphere than it is during the solar minimum period and in the Southern Hemisphere (Torr and Torr, 1973). This distinction has been validated even for magnetically quiet conditions. Although the phenomenon is conspicuous near the F2 heights, it gradually disappears with an increase in altitude above the F-peak (Balan et al., 1998, 2000; Kawamura et al., 2002).

High latitude/polar ionosphere. At high latitude, ionosphere is very dynamic. However, due to the high latitudinal location, the solar rays reach here with very low elevation, and hence, with much lesser energy. So, the ion production by photoionization is relatively less significant here. Therefore, polar ionosphere has less photo-ionized charges. This region, on the other hand, is rich with energetic charged particles. These particles reach here mostly from the extra-terrestrial sources. The solar wind thrashes these highly energized charges onto the magnetosphere and the latter are conveyed to this region along the lines of force which terminate at these locations. The ionosphere in this zone, therefore, is rich in precipitating electrons and ions, causing ionization through collision. These charged particles contribute to the production rate. With less photoionization, the polar ionosphere behaves rather differently than the other ionospheric regions. The behaviour of the high-latitude ionosphere is rather closely related to the physics of magnetosphere.

The major difference in polar ionospheric characteristics is in the variability of its electron density. This is primarily because, the influx of the solar charge particle, causing ionization, is much more variable than solar irradiance. In addition to the irregular variations in the D and E regions, the F region, too, is also different from that in the lower latitudes. Further, it maintains its existence during the long winter night.

It has also been observed that, at very high latitudes, the critical frequency and height of the ionospheric F2 layer are strongly dependent on Universal Time. The variations in the ionospheric density are described better by universal time rather than local time. Unlike the equatorial and mid-latitude ionospheres, here the maximum density in F2 peak occurs near 0600 UT for Antarctic stations close to the magnetic pole, while the Arctic maximizes at 1880 UT. These are just the times when the respective geomagnetic poles are closest to the sun to earth line (Challinor, 1969; Hargreaves, 1979). Further, the neutral wind also plays an important role in exhibiting the anomalous behaviour. The daytime neutral wind blows over the magnetic pole toward the Geographic pole at these times. Therefore, the polar ionization, which is carried by the wind along the magnetic field lines, is lifted to greater heights. The extended height results in the reduction of the loss process of the plasma. Because of the reduction of loss rate with altitude, the peak electron density is maintained at a higher level even at night, than would otherwise be the case.

There is also a marked latitudinal variation which in essence is controlled by the shape of the geomagnetic field. The F region is enhanced in the Auroral oval which is usually to be found between magnetic latitudes 60° and 80° and forms an eccentric loop around the magnetic pole. On the equator side of the Auroral oval, the F region is depleted over a region forming the mid-latitude trough. We have already treated this in the previous section. Within the polar caps, i.e. poleward side of the Auroral oval, the magnetic field lines are connected to the regions of outer magnetosphere. These lines do not close back to the earth, but remain open in space. Thus, they provide paths for the lighter ions from the plasma to escape the ionosphere. The flow is known as the polar ionospheric wind.

Polar irregularities

Two of the most significant mechanisms which are responsible for the polar irregularities are Precipitation and Convection. Precipitation refers to the process of energetic particles arriving from space in the Earth's atmosphere. The solar wind that continuously strikes upon the earth's dayside magnetosphere carries with it the charged particles. These charged particles get frozen with the magnetic lines of force and start moving along it towards the pole. These geomagnetic field lines, forming the outer layers of the magnetosphere on the dayside, terminate at the poles. Due to very high conductivity, there is no resistance to the motion of charged particles parallel to a magnetic field. So, this plasma readily flows along the magnetic field lines and, due to the strong solar wind pressure, precipitate on the polar ionosphere. It occurs throughout the polar regions, but most intensely in the auroral ovals, as the geomagnetic lines reach the ionospheric heights at this region. These charged precipitated particles create patches of high electron density. The variations in the precipitation cause patches to disperse into smaller irregular structures and results in irregularities.

The magnetosphere–solar wind interaction also generates zonal electric fields. This electric fields are carried along magnetic field lines so that the electric field distribution at the latter region is an image of that of the former. This is called 'electric field mapping'. Now, at the polar locations, the Earth's magnetic field is nearly vertical, while the mapped electric field have a significant component perpendicular to it. Thus, the direction of the $E \times B$ drift is in the direction, mutually perpendicular to both. This situation creates plasma instability, called the plasma gradient drift instability (GDI).

The gradient drift instability (Simon, 1963) is also known as the $E \times B$ or cross-field instability. If an enhancement in the plasma density is present, the charges in the patches move in response to the pressure gradient, etc. This horizontal movement of the charges, in presence of vertical magnetic field B, results in Lorentz force that leads to an effective charge separation. The separation occurs in such a manner that the resultant electric field, E. The polarization electric field thus generated produces an $E \times B$ force which aids further movement of the charges in the same direction which initiated the process. This cycle continues, producing the instability. This instability is a common mechanism by which the patch structures break down into smaller regions of irregular plasma density (Hargreaves, 1992). Electron precipitation and GDI are believed to be the main sources of scintillation at high latitude (Kelley, 1989).

4.1.3 BROAD FACTORS AFFECTING IONOSPHERIC CHARACTERISTICS

It was apparent from the previous discussions that the ionosphere varies both spatially and temporally depending upon various factors. In this section, we explicitly spell out the most important factors out of them which are responsible for its variations.

4.1.3.1 Solar flux variation

The temporal variations of the ionosphere are mainly determined by the solar flux variation which leads to the variation of the photoionization. These variations may be divided into three categories depending upon the interval of variation. These are diurnal variation, seasonal variation and long-term variation. The diurnal variation is mainly governed by the amount of solar flux being received by the ionosphere over the time of a solar day. In other words, the duration of the reception of the solar flux determines the ionization amount of the region. As a result, there is a demarcated distinction between the dayside and the nightside of the ionosphere. Further, the regions in the lower geographical latitudes obtain stronger solar flux than that of the regions near the poles. Therefore, they have higher values of photoionized electron densities than the latter. The ionosphere also varies with season. The two main factors that determine the variation are the longer duration of the available solar radiation over a day and also the solar zenith angle. For example, during summer, the solar zenith angle is such that the solar flux intensity is high and simultaneously the duration of the day is large. Both the factors aid larger photoionization, leading to higher electron density formation for longer periods. Consequently, the winter experiences lower photoionization leading to a slim ionosphere. The solar magnetic activities determine the solar flux, and hence, ionospheric electron density values. The 11-year solar cycle is, therefore, a major decisive factor for the ionospheric variation. During the solar high activity periods, when the sun spot numbers and hence the mean solar intensity is higher, the ionospheric electron density values surge compared to the same during the solar quiet periods. Nevertheless, the influence of the solar flux on polar ionosphere is less significant.

4.1.3.2 Wind

The thermospheric neutral wind plays a very important role in shaping the ionosphere and generating its variations. The movement of the plasma in the ionosphere is both mechanically and electrodynamically driven.

The mechanical movement of the charged particles results in the transport of the ionospheric plasma. It occurs as a result of the drag impressed upon them by the neutral wind. Unlike the neutral air, the ionized particles are not free to move horizontally, as their movement is limited by the Lorenz force generating due to the Earth's magnetic field. As a result, any movement of the neutral wind along the geomagnetic lines of forces will readily blow the ionized plasma along with it. The field-aligned movement, occurring as a result, will vary the altitude of these charges which in turn will influence the recombination rate, and hence, modify their average lifetime. Movement across the lines are also determined by various factors including the magnetic field strength, collision rate, etc., defining the effective conductivity.

The electric fields are also generated by dynamo action in the ionosphere when there is a mechanical movement of the unbalanced charged particles across the geomagnetic field. This electric field, in turn, drives the plasma and thus redistributes the charge density and eventually shapes the ionosphere.

4.1.3.3 Space weather

Space weather is the concept of changing environmental conditions in near-Earth space. It deals with the interactions of the solar plasma and fields with Earth's magnetic field. It also includes the study of interplanetary (and occasionally interstellar) space. It is important here to understand that space weather is distinct from the concept of tropospheric weather in the earth's atmosphere.

The space weather is influenced by the speed and density of the solar wind and also by the interplanetary magnetic field (IMF) carried by the solar wind plasma. Coronal mass ejections occurring at the sun and their associated shock waves are also important drivers of space weather as they determine the content and the velocity of the solar wind, when occur. The solar wind modifies the earth's magnetic field which in turn affect Earth's local space environment. Moreover, space weather exerts a profound influence to the ionosphere, particularly to the polar region and to the equatorial region to some extent. Accordingly, an understanding of space environmental conditions is also important in appreciating the variations in the polar, and to some extent, the lower latitudinal ionosphere.

A variety of physical phenomena are associated with space weather which influence the nominal ionosphere. These include geomagnetic storms and substorms, ring currents, ionospheric disturbances and scintillation, aurora and geomagnetically induced currents at Earth's surface and many more.

One major facet of space weather is that of the prompt penetration electric fields (PPEFs) (Sastri, 1988; Abdu et al., 1995; Abdu, 1997; Sobral et al., 1997, 2001; Basu et al., 2001a,b; Sastri et al., 2002). The term prompt penetration electric field refers to the interplanetary electric fields that appear almost immediately in the Earth's equatorial ionosphere after these electric fields have been convected by the solar wind to the magnetosphere.

4.1.4 MODELS OF IONOSPHERE

There are mainly two broad types of models for ionosphere, viz. the theoretical models and the empirical or semiempirical models. Theoretical models are physics-based and thus needs understanding of all the processes happening in the ionosphere to precisely describe it. Empirical models, on the other hand, are data-driven and thus avoid the uncertainties of the physical understandings (Acharya, 2014). We need to reiterate at this point that, our main purpose in this chapter is to understand the constitution of the ionosphere and its effect on the propagating satellite signals. So, to focus on our objectives, we shall only concentrate on those models that will help understanding the formation and the constitutional features of the ionosphere and will also unleash the features of the ionosphere that have direct bearing on the satellite signals. So, considering our objective here, we shall discuss only the simplest of the models which serve our purpose. We shall learn about other models used for mitigating ionospheric impairments in the next chapter.

4.1.4.1 Chapmans physical model

Sydney Chapman (Chapman, 1931) was the pioneer to provide a very simplistic model of the ionosphere in terms of its production rate and electron density. This model is able to describe the behaviour of an ionospheric layer and its variations during the day. The mathematical model is based on a list of simple assumptions:

i. The solar radiation is monochromatic
ii. The atmosphere is composed of only one element which has an exponential vertical distribution
iii. The atmosphere is horizontally stratified
iv. The temperature is constant so the scale height $\left(H = \dfrac{kT}{gm} \right)$ is constant too;
v. Ionization is caused only by photoionization through absorption and electron loss only by recombination.

According to Chapman, the ionization takes place when the neutral atoms absorb sufficient energy from the solar radiation to get their electrons freed from their orbits. Therefore, the numbers of electron-ion pair produced in an infinitesimal volume of the atmosphere are proportional to the amount of solar power absorbed in that volume. Now, let us consider a vertical layer of the atmosphere at a height h and of vertical width dh through which the solar radiation is incident at an angle, χ. This angle χ is called the solar zenith angle. It is the angle between the vertical direction to the point on the plane of the Earth's surface and the line of direction of the Sun. Then, the path traversed through the layer by the radiation is

$$dl = dh \sec \chi \tag{4.1}$$

Again, let the intensity of the solar radiation at this height h be $S(h)$. Then, if the number density of neutral particles in this layer be $n(h)$, the amount of solar radiation power absorbed by this portion of the layer is proportional to $S(h)$, $n(h)$ and the total traversed length. Therefore, we can write the loss in the solar radiation power $-dS$ as,

$$-dS = \sigma S(h) n(h) dh \sec \chi \tag{4.2}$$

where σ is the proportionality constant and is called the ionization efficiency. It is the (statistical) number representing the amount of energy absorbed per unit energy incident on an ionizable particle. It is like the absorption cross-section that we have read in Chapter 3. A constant value of σ implicitly assumes that the atmosphere is composed of a single neutral species and hence the absorption cross-section is constant. Constant cross-section is also equivalent to assuming monochromatic radiation. We can rearrange the above equation to write

$$dS/S = -\sigma n(h) \sec \chi dh$$

$$\text{or,} \quad S(h) = S_0 \exp \left\{ -\sigma \int_h^\infty n(h) dh \sec \chi \right\} \tag{4.3}$$

where S_0 may be considered as the intensity of radiation just above the atmosphere or before any absorption takes place. Recalling that we have taken a pragmatic assumption that the neutral particles decrease exponentially with height, the expression for the neutral particles as a function of height becomes, $n = n_0 \exp(-h/H)$, where H is the scale height.

Now that we have already got the expression for the solar radiation intensity as a function of height and the same for the neutral particles, we can use them to get our value of ionization production at the height of h. As the numbers of photoionized charges produced are proportional to the amount of photo energy absorbed, we can write the production function of ionization per unit volume as,

$$q(h) = k\sigma n(h) S(h) \tag{4.4}$$

where, k is the proportionality constant. Again, as σ, n_0 and S_0 are constants,

$$
\begin{aligned}
q(h) &= k\sigma n_0 \exp\left(-\frac{h}{H}\right) S_0 \exp\left\{-\sigma \int_h^\infty n_0 \exp\left(-\frac{h}{H}\right) dh \sec\chi\right\} \\
&= k\sigma n_0 S_0 \exp\left(-\frac{h}{H}\right) \exp\left\{-\sigma n_0 H \exp\left(-\frac{h}{H}\right) \sec\chi\right\} \\
&= k\sigma n_0 S_0 \exp\left\{-\frac{h}{H} - \sigma n_0 H \exp\left(-\frac{h}{H}\right) \sec\chi\right\}
\end{aligned}
\tag{4.5a}
$$

Chapman's theory also assumes that the atmosphere is isothermal so that the scale height H is independent of altitude. This function shows how q depends on the solar zenith angle χ and the height h. Now, first we shall find the height, h_m, at which this production is maximum, when the zenith angle is χ. Rewriting the expression for the production, we get

$$q(h\chi) = k\sigma n_0 S_0 \exp\left\{-\frac{h}{H} - \sigma n_0 H \exp\left(-\frac{h}{H}\right) \sec\chi\right\} \tag{4.5b}$$

To find the height h_m, at which q maximizes, we first get the derivate dq/dh, and then equate it to zero. When this is achieved, we get,

$$
\begin{aligned}
&d/dh\left[k\sigma n_0 S_0 \exp\left\{-\frac{h}{H} - \sigma n_0 H \exp\left(-\frac{h}{H}\right)\right\} \sec\chi\right] = 0 \\
&\text{or,}\ \ k\sigma n_0 S_0 \exp\left\{-\frac{h}{H} - \sigma n_0 H \exp\left(-\frac{h}{H}\right) \sec\chi\right\} \\
&\quad \left\{-\frac{1}{H} - \sigma n_0 H \exp\left(-\frac{h}{H}\right)\left(-\frac{1}{H}\right) \sec\chi\right\} = 0 \\
&\text{or,}\ \ \sigma n_0 H \sec\chi = \exp\left(\frac{h_m}{H}\right)
\end{aligned}
\tag{4.6}
$$

It is clear from this equation that the height of maximum production increases with increasing χ, and hence, is lowest for overhead sun. Now, at this location, where the zenith angle is zero, i.e. the location placed vertically under the sun, $\chi = 0$, so sec $\chi = 1$. Hence, the corresponding peak height h_{m0} satisfies the relation $\sigma n_0 H = \exp(h_{m0}/H)$. Therefore, using this relation of h_{m0} in the expression for q, we get

$$
\begin{aligned}
q(h, \chi = 0) &= k\sigma n_0 S_0 \exp\left[-\frac{h}{H} - \sigma n_0 H \exp\left\{\left(-\frac{h}{H}\right)\right\}\right] \\
&= \left(\frac{k}{H}\right) S_0 \exp\left(\frac{h_{m0}}{H}\right) \exp\left[-\frac{h}{H} - \exp\left(-\frac{h}{H}\right) \exp\left(\frac{h_{m0}}{H}\right)\right] \\
&= \left(\frac{k}{H}\right) S_0 \exp\left[-\frac{(h - h_{m0})}{H} - \exp\left\{-\frac{(h - h_{m0})}{H}\right\}\right]
\end{aligned}
\tag{4.7}
$$

Again the production function q at $\chi = 0$ and $h = h_m$ becomes

$$q(h_{m0}, \chi = 0) = \frac{kS_0}{H} \exp(-1)$$

$$= q_0$$

(4.8)

Therefore, $q(h, \chi = 0)$ can also be represented as

$$q(h, \chi = 0) = q_0 \exp(1) \exp\left[-\frac{(h - h_m)}{H} - \exp\left\{ -\frac{(h - h_{m0})}{H} \right\} \right]$$

$$= q_0 \exp\left[1 - \frac{(h - h_{m0})}{H} - \exp\left\{ -\frac{(h - h_{m0})}{H} \right\} \right]$$

(4.9)

Taking this height of maximum ionization production under the vertical sun as reference and calling the term $(h - h_{m0})/H$ as the reduced height 'z', we can express the generic production function as

$$q(z, \chi) = q_0 \exp\{ 1 - z - \exp(-z) \sec \chi \}$$

(4.10)

This expression suggests that the maximum production is encountered at $\chi = 0$, i.e. when the sun is shining overhead. So, with increasing zenith angles, the maximum height h_m rises to higher altitudes, while the maximum production value decreases. When the zenith angle is high enough, i.e. for nearly grazing elevation angle of the sun, the curvature of the Earth and the atmosphere should be taken into account and the theory should be modified since the model assumes flat horizontal atmospheric structure.

For heights above the maximum ionization, the production profile decays with height in the same way as the atmospheric density. This relation arises because at those heights the intensity of the radiation is only weakly reduced, and the rate of production is essentially proportional to the neutral particle density. The production function under this condition may be represented by making z very large in Eq. 4.10. The parameters of the Chapman's model are illustrated in Fig. 4.4.

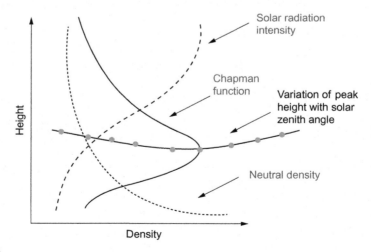

FIG. 4.4

Parameters of Chapman's function.

However, the generation of the electrons is continuously being negated by the recombination process. Therefore, the effective electron density present in the ionosphere is the value that is attained at the equilibrium due to this generation and loss process. For lower heights, where the neutral particles are found in abundance, the loss factor is proportional to the square of the electron density, N_e (Rishbeth and Garriot, 1969). Hence, if α is the constant of proportionality at equilibrium when the production and the losses are equal,

$$q(z) = \alpha N_e{}^2$$

$$\text{or,} \quad N_e(z) = \sqrt{\frac{q(z)}{\alpha}} \tag{4.11a}$$

For higher altitudes, the loss due to recombination is linear with the electron density. Hence, with β as the proportionality constant present here

$$q(z) = \beta N_e$$

$$\text{or,} \quad N_e(z) = \frac{q(z)}{\beta} \tag{4.11b}$$

Focus 4.1 Recombination Processes in the Ionosphere

Our focus in this section will be to understand the recombination process happening in the ionosphere. For this, we shall consider three participating particles, viz. the electrons with concentration N_e, atomic ions with concentration N_A^+ and molecular ions with concentration N_M^+. We also assume that the ion-atom interchange that occurs through the exchange of an electron involves a molecular gas, whose concentration is $n[M]$. Then, we can write, the rate of change in electron numbers as

$$dN_e/dt = q - \alpha N_e N_M{}^+$$

As atomic ions and electrons are simultaneously generated, so,

$$dN_{A^+}/dt = q - \gamma n[M]N_{A^+}$$
$$dN_{M^+}/dt = \gamma n[M]N_{A^+} - \alpha N_e N_{M^+}$$

we assume here that direct generation of molecular ions by photoionization does not occur.

Now under equilibrium, none of the concentration changes with time. So, we can write

$$\gamma n[M]N_{A^+} = \alpha N_e N_{M^+}$$
$$\text{or,} \quad N_{A^+}/N_{M^+} = \alpha N_e/\gamma n[M]$$
$$= \alpha N_e/\beta \quad \text{where} \quad \beta = \gamma n[M]$$

Now, charge neutrality demands,

$$N_e = N_{M^+} + N_{A^+}$$

Combining the above two, we get,

$$\alpha \beta N_e^2 - \alpha q N_e + \beta q = 0$$
$$\text{or,} \quad 1/q = 1/(\beta N_e) + 1/(\alpha N_e^2)$$

So, this equation reduces to,

$$q = \alpha N_e^2 \quad \text{when} \quad \beta \gg \alpha N_e, \text{ i.e. } N_{M^+} \gg N_{A^+}$$
$$\text{and} \quad q = \beta N_e \quad \text{when} \quad \beta \ll \alpha N_e, \text{ i.e. } N_{M^+} \ll N_{A^+}$$

Therefore, at lower heights, where $N_M^+ \gg N_A^+$, the recombination process follows square law formula, while at higher heights where $N_M^+ \ll N_A^+$, the recombination is linear with the existing electron density.

Box 4.1 Chapman's Model

In this box section, we shall run the MATLAB programme chapman.m to generate the figure representing the Chapman's ion production function for different values of reduced height 'z'. Notice that the peak appears at $z = 0$ and the gradients are different on two sides of the peak.

   Also, observe the second figure where the production function is converted to electron density. We have assumed abrupt change in loss process from square law to linear. Find the formation of a secondary peak at the point of transition.

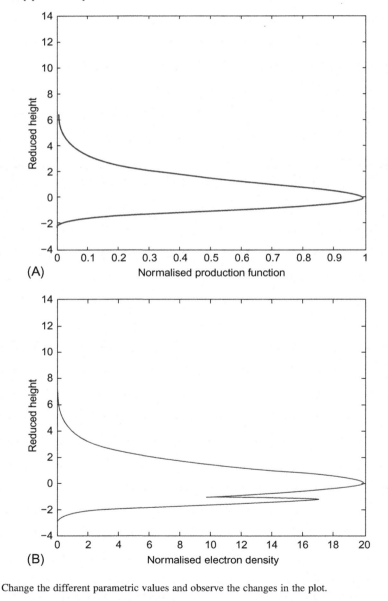

(A)

(B)

Change the different parametric values and observe the changes in the plot.

4.2 IONOSPHERIC PROPAGATION

4.2.1 INCIDENCE OF EM WAVE ON FREE CHARGES

Whenever a radio wave passes through a medium, there is an interaction between the wave and the charges present in the medium. As the wave traverse through the ionosphere, its electric field exerts a force on the ionospheric electrons. The electric field of the wave, which is oscillatory in nature, forces the free electrons in the medium to oscillate with it. To the extent that the earth's magnetic field does not complicate the situation, the electrons vibrate in a sinusoidal manner along the direction of the electric field. Although both the negatively charged electrons and the positively charged ions are affected in the process, the positive ions being massive do not move considerably. The lighter electrons, however, respond readily to the electric field by moving in the appropriate direction with respect to it. We have already seen this for a conductor in Chapter 3.

In the ionosphere, however, the electrons are free and there is no other force acting on these. For such completely free charges, the equation of motion can be formed in the same manner as we have formed for the bound charges in Chapter 3, only with the condition that the resistive and the restoring forces are zero here. Hence, we get

$$m\ddot{x} = eE \exp(j\omega t) \tag{4.12}$$

where, m is the mass of the electron, e is the charge and $\ddot{x}$ is the acceleration of the electrons produced as a result. E is the incident electric field with the angular frequency ω. Therefore, we get the solution as,

$$x = -\left(\frac{e}{m\omega^2}\right)E \quad \text{and} \tag{4.13a}$$

$$\dot{x} = -j\left(\frac{e}{m\omega}\right)E \tag{4.13b}$$

So, using the above expression of $\dot{x}$, the expression for the resultant current density J, generated due to the motion of the electrons, where the motion is a result of the imposed electric field is given by:

$$J = N_e e\dot{x}$$
$$= -j\left(\frac{N_e e^2}{m\omega}\right) \cdot E \tag{4.14}$$

where N_e is the electron density in the region. So, the resultant conductivity σ can be expressed as

$$\sigma = \frac{J}{E}$$
$$= -j\left(\frac{N_e e^2}{m\omega}\right) \tag{4.15}$$

Therefore, we find that, for a given electric field E, the current density J and hence the conductivity σ is directly proportional to the electron density N_e and to the square of the electronic charge. This is evident, as for more numbers of charge bearers and for larger charge carried by each, the amount of total flux of charges through a unit area increases, thus increasing J. Again, due to the negative charge of the electrons, the direction of its displacement is opposite to that of the field. Since the current density is proportional to the electron velocity, there is a quadrature phase difference between the electric field strength and the current. Further, the finite inertial mass of the electrons hinders the motion leading to inverse variation with m. Again, with increase in frequency, as the velocity of the vibrating electrons diminishes, it makes the conductivity vary inversely to the frequency.

4.2.2 WAVE PROPAGATION IN PLASMA

The net effect that the vibrating electrons in the ionosphere have on the propagating wave is that these individual charges act as a secondary source of radiation and interact with the primary propagating wave to modify the effective velocity of the wave. This modified phase velocity of the wave in the medium can also be described as due to a difference in refractive index (RI) of the medium than free space. The change in the RI is also due to the motion of the charges arising from the finite conductivity σ, which in turn influences the effective relative permittivity of the medium. The fact that finite conductivity of a medium modifies the refractive index of a medium can also be derived (Feynman et al., 1992; Reitz et al., 1990) from the Maxwell's equation. So, considering that the ionosphere has a conductivity σ, the corresponding Maxwell's equation can be written as

$$\nabla \times B = \mu_0 J + \mu_0 \varepsilon_0 dE/dt$$
$$= \mu_0 \sigma E + \mu_0 \varepsilon_0 dE/dt \tag{4.16}$$

Here, μ_0 and ε_0 are the permeability and permittivity of free space in absence of conducting electrons, respectively. We have also used here the general relation that, for a medium with propagating wave with field strength given by E, the current density J can be expressed as $J = \sigma E$. B is the resultant magnetic field arising due to the motion of the charges. Since the field of the propagating signal is sinusoidal with angular frequency ω, we can write the electric field E as

$$E(t) = E_0 \exp(j\omega t - k_0 x)$$
$$\text{or,} \quad dE/dt = j\omega E(t)$$
$$\text{So,} \quad E(t) = -\left(\frac{j}{\omega}\right) dE/dt \tag{4.17}$$

Replacing this expression in Eq. (4.16), we get

$$\nabla \times B = \mu \varepsilon_0 \left(1 - j\frac{\sigma}{\omega \varepsilon_0}\right) dE/dt \tag{4.18}$$

Thus, due to the partial conductivity of the medium, an imaginary term appears in the expression and the effective relative permittivity gets transformed into a complex parameter

$$\varepsilon' = \varepsilon_0 \left(1 - j\frac{\sigma}{\omega \varepsilon_0}\right) \tag{4.19}$$

Using the relation between permittivity of a medium and its refractive index, the complex refractive index 'n' of the medium can be obtained as

$$n^2 = \frac{\varepsilon'}{\varepsilon_0}$$
$$= 1 - j\frac{\sigma}{\omega\varepsilon_0} \tag{4.20a}$$

Replacing the expression of σ that we have derived as Eq. (4.15) in the previous section, we get the refractive index as

$$n^2 = 1 - \frac{N_e e^2}{m\varepsilon_0\omega^2} \tag{4.20b}$$

$$= 1 - \left(\frac{\omega_p^2}{\omega^2}\right) \tag{4.20c}$$

$\omega_p = N_e\, e^2/(m\varepsilon_0)$ is called the plasma frequency. This is the frequency at which the plasma in the ionosphere would oscillate if allowed to vibrate freely after an initial force is applied to it (Reitz et al., 1990). Compare Eq. (4.20c) with Eq. (3.24a). It is interesting to note that, we have derived the same equation as Eq. (3.24a) but with the condition $\omega_0 = 0$ and $k = 0$, which is the condition for the ionosphere.

The expression for the ionospheric refractive index 'n' is obtained by putting values of the constants e, m and ε_0 in Eq. (4.20b), and replacing ω as $2\pi f$, we get

$$n = \sqrt{1 - \frac{80.6N_e}{f^2}} \tag{4.21a}$$

where N_e is the density of free electrons responsible for conductivity σ and $f = \omega/2\pi$ is the frequency of the wave in Hertz. For the propagating satellite signals, the frequency being very high compared to the plasma frequency, Eqs (4.21a) can be approximated to

$$n = 1 - \frac{40.3N_e}{f^2} \tag{4.21b}$$

We shall learn about the effects of such variation of 'n' with Ne and f in the next section. But, before discussing that topic, it is important here to explain the significance of the Plasma frequency, ω_p. From the above expression, we can see that the refractive index n is imaginary when $\omega < \omega_p$. For frequencies $\omega > \omega_p$, the index n becomes real. Now, the reflection coefficient at the interface of two media may be expressed as

$$R = \frac{n_2 - n_1}{n_2 + n_1} \tag{4.22a}$$

where n_1 is the refractive index of the first medium from which the wave is incident on the second, whose refractive index is n_2. Therefore, when the signal is incident from free space to ionosphere and the latter has imaginary refractive index as $-jn_2$, then r becomes

$$R = \frac{-jn_2 - 1}{-jn_2 + 1} \tag{4.22b}$$

So, the magnitude of R becomes $|R| = 1$ and the whole of the incident wave gets reflected back to the first medium with a phase shift of π. That is, it bounces back to the first medium from the ionospheric interface, when the refractive index turns imaginary. Then, how does the equation of the propagating wave get changed? With imaginary refractive index, the equation of the electric field E of the transmitted wave becomes

$$
\begin{aligned}
E &= E_0 \exp(j\omega t) \exp\{-jk_0(-jn_2)x\} \\
&= E_0 \exp(j\omega t) \exp(-k_0|n_2|x)
\end{aligned}
\tag{4.23}
$$

Thus, the wave is nonprogressive and exponentially decays with distance, thus forming what is known as Evanescent wave. Therefore, for conditions, $\omega_p > \omega$, the signals get reflected completely back from the ionospheric layer where the above condition is satisfied.

Now, putting the expressions for ω_p, we get,

$$
\omega_p{}^2 = \frac{N_e e^2}{m\varepsilon_0}
\tag{4.24a}
$$

This indicates that the plasma frequency is directly dependent upon the electron density N_e. So, for a particular region of the ionosphere, the maximum frequency that will be reflected back is proportional to the density of its peak layer. This threshold frequency for normal incidence which is reflected back from the peak of a layer is called the Critical frequency, f_0, for that layer. Also, for a given ionospheric region, we can write the expression for the maximum peak density N_m in terms of its corresponding critical frequency as

$$
\begin{aligned}
N_{m|layer} &= \left(4\pi^2 \frac{m\varepsilon_0}{e^2}\right) f_0{}^2{}_{|layer} \\
&= \left(\frac{1}{80.6}\right) f_0{}^2{}_{|layer}
\end{aligned}
\tag{4.24b}
$$

Using this basic relation, we measure the peak densities $N_{m|layer}$ of the E, F1 and F2 layers of the ionosphere knowing their critical frequencies, $f_{0|layer}$. We shall learn about this in more details in our next chapter.

Finally, when the frequency is above the ω_p, then the RI is real and the wave passes through the second medium, i.e. the ionosphere with a bending, such that the laws of reflection are followed. Hence,

$$
\frac{\sin i}{\sin r} = \frac{n_2}{n_1}
\tag{4.25a}
$$

Where 'i' is the angle of incidence and 'r' is the angle of refraction. Therefore, higher the deviation of this ratio from 1, larger the bending of the wave. For oblique incidence, the critical angle is defined as that angle of incidence for which the refracted wave grazes through the interface of the two medium, i.e. $r = 90°$. So, if θ_c be the critical angle when the signal is incident on the ionosphere, then

$$
\sin \theta_c = \sqrt{1 - \frac{\omega_p{}^2}{\omega^2}}
\tag{4.25b}
$$

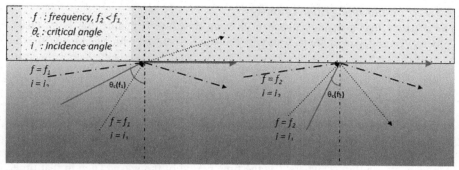

FIG. 4.5

Different scenarios for reflection of radio waves.

So, for a given plasma frequency,

$$\cos \theta_c = \frac{\omega_p}{\omega} \qquad (4.25c)$$

So, for increasing electron density, ω_p increases, increasing the ratio in the right hand side of Eq. (4.25c). Hence, θ_c decreases. While, with increasing frequency ω, the cosine of the critical angle decreases. As a result, the critical angle value θ_c increases. This means, for lower frequencies, or higher electron densities, the critical angles being smaller, waves are reflected back even for lower angles of incidences. In other words, for every angle of incidence, there is an upper threshold for frequency, which is reflected back and above which the wave passes through the ionosphere. This threshold decreases with lower incidence angle and is minimum for normal incidence. This means, to get any higher frequency back through reflection, we need an oblique incidence. For normal incidence,

$$\cos \theta c = \cos 0 = 1 = \frac{\omega_p}{\omega} \quad \text{or,} \quad \omega_p = \omega \qquad (4.26)$$

This corresponds to our previous consideration. The different scenarios are shown in Fig. 4.5.

4.3 EFFECTS

4.3.1 GROUP DELAY AND PHASE ADVANCEMENT

We have already seen the effect of the ionospheric electron density on an incident radio wave which is near the critical frequency in terms of reflection and transmission. In this section, we shall only focus on those signals which are much higher than the plasma frequency and are always transmitted through the layers. These signals are qualified to be used as satellite signals.

4.3.1.1 Phase and group velocity

The distance that a definite phase of a wave traverses per unit time is its phase velocity. To understand, let us start with the wave equation, which can be written as

$$S(t) = A \exp\{j(\omega t - kx)\}$$
$$= A \exp\left\{j\left(\omega t - \frac{2\pi}{\lambda}x\right)\right\} \tag{4.27}$$

$2\pi/\lambda = k$ is the phase change occurring per unit distance and called the phase constant and $\omega = 2\pi/T$, which is the phase change per unit time, is the angular frequency.

Now, let us take a definite point x in space. At this point, a particular phase φ reappears after a time T, where T is the time period of the wave. Then in that time, that particular phase which was initially located at distance of the wavelength λ has traversed the distance λ to reach that point. Therefore, the phase velocity is given in terms λ and period T as

$$v_p = \frac{\lambda}{T}$$
$$= \frac{2\pi/T}{2\pi/\lambda}$$

This is the velocity at which the phase of any signal wave travels. Again, since $v_p = c/n$, where 'c' is the velocity of the wave in vacuum and is equal to the speed of light, we get $2\pi/\lambda = (\omega/c)n$, we can write,

$$S(t) = A \exp\left\{j\omega t - j\left(\frac{\omega}{c}\right)nx\right\} \tag{4.28}$$

As n is now a function of density N_e and frequency 'f', v_p also varies with these factors. With increase in density N_e, the refractive index n reduces towards zero, and hence, the phase velocity increases. Again, notice that, for ionosphere, as $v_p = c/n$ and as $n = (1 - 40.3\, N_e/f^2) < 1$, $v_p > c$. That is, the phase velocity is greater than the velocity of light. This is possible as it is only the phase of the wave that moves and not any mass or energy.

For a modulated signal, containing multiple frequency components, since the phase velocity of the individual wave components is different from that of the vacuum, the relative phase change of the wave in traversing a fixed distance differs due to the presence of the dispersive medium, from what it would be with vacuum alone. Therefore, the difference in mutual relative phase change, which is actually observed, for traversing a distance l_m in the medium and that would be expected for traversing the same distance in vacuum would be

$$\delta\varphi = \frac{2\pi}{\lambda}(n-1)l_m$$
$$= k_0(n-1)l_m \tag{4.29}$$

Due to the frequency dependence of n, this change will therefore be different between different frequency component of the waves. This will also be dependent upon the electron density N_e. Now, replacing n by the expression derived, we get,

$$|\delta\varphi| = \frac{2\pi}{\lambda}\left(40.3\frac{N_e}{f^2}\right)l_m \tag{4.30a}$$

When N_e is a function of the path length l_m through the medium, $N_e = N_e(l_m)$, the total phase change over the path traversed through the ionosphere can be written as

$$|\delta\varphi| = \frac{2\pi}{\lambda}\left(\frac{40.3}{f^2}\right)\int N_e(l_m)dl_m \tag{4.30b}$$

So whenever electromagnetic waves pass through this region, its phase moves faster compared to that in vacuum resulting in an effective phase advancement at the receiver.

The group velocity of a wave is the velocity with which the overall shape of the waves' amplitudes, known as the envelope of the wave, propagates. In the wave group, individual wavelets of differing wavelengths travel at different speeds. Now, it is known that, in a wave, the components of different frequencies combine with different relative strengths and phases to create the actual wave envelope. However, as these components start travelling with different phase velocities, this definite phase combination breaks up as the wave moves on.

Now, considering that the wave envelope traverses a distance l in time t, then $v_g = l/t$. This means in time t, the same set of phase in the definite combination, which was at any initial point, has moved by distance l. For this to happen, for any component with frequency $\omega_0 + d\omega$, the change in its phase, relative to that of frequency ω_0, on traversing a distance l, must be counterbalanced by the relative phase change between them occurring in the time, in the interval needed to travel distance l. That is,

$$d\omega t = -dk l$$
$$l/t = -d\omega/dk$$

Therefore,

$$v_g = -d\omega/dk \tag{4.31}$$

As we have previously mentioned that the velocity with which the phase of the wave travel is

$$v_p = \frac{c}{1 - 40.3\dfrac{N_e}{f^2}} > c \tag{4.32a}$$

Further, it can be shown that, $v_p \times v_g = c^2$. Consequently, the group velocity with which the codes move becomes

$$v_g = \frac{c^2}{c/\left(1 - 40.3\dfrac{N_e}{f^2}\right)} \tag{4.32b}$$
$$= c\left(1 - 40.3\frac{N_e}{f^2}\right) < c$$

As the energy contained in a wave at any part of it is determined by its amplitude and therefore, by its shape, it is the group velocity with which the energy moves in a wave. Thus, from the above, it is evident that the energy contained in the signal traverses slower in the ionosphere than in vacuum causing an additional code delay. Sometimes, a term similar 'Group Refractive Index' n_g is used to represent the ratio of the velocity of wave in vacuum to the group velocity of the wave in the medium. So, this term can be represented as (Hargreaves, 1979)

$$n_g = c/v_g$$
$$= c(dk/d\omega)$$
$$= d/d\omega(ck) \quad\quad (4.33)$$
$$= d/d\omega(n\omega)$$

Where n is the phase refractive index of the medium. So, it implies that, when n is independent of ω, the group refractive index is same as the phase refractive index term, n.

The delay has its conspicuous effects where the time of propagation of the wave is a major parameter in the system. For example, in Radar reflectometry and Satellite-based Radio Navigation system, the total time of propagation is used for calculating the distance. Therefore, in such applications any additional delay unaccounted for will lead to erroneous estimation of the range and will eventually cause errors in the positioning result.

4.3.2 SCINTILLATION AND FADING

Ionospheric scintillation is a rapid fluctuation of the phase and/or amplitude of a radio-frequency signal received at a receiver while the signal passes through the ionosphere. Scintillation is caused by small-scale irregularity structure in the ionospheric electron density along the signal path. It occurs as a result of interference of different components of the signal that reach the receiver after being refracted and/or diffracted through these ionospheric irregularities. Scintillation is usually quantified by two indices, viz. S4 for amplitude scintillation and σ_ϕ for phase scintillation. S4 is the metric for measuring amplitude scintillation, while σ_ϕ is for the phase, both determined from the measured signals taken over a predefined interval, typically taken as 1 min. We shall define these terms as functions of measurable quantities in our next chapter. These indices reflect the statistical variability of the signal over the period of time, and sometimes used to represent, but imprecisely, the ionospheric condition at the definite location. Scintillation is mostly observed in the low and high latitudes, while it is minimum at the mid-latitudes. Apart from this geographical location, scintillation is a strong function of local time, season, geomagnetic activity, solar cycle, etc. Some geophysical forces like the ionospheric dynamos and gravity waves also affect the same. In short, it is influenced by anything that can create irregularities in the ionospheric electron density distribution. With this basic introduction, we shall, from the following section, have a detailed treatise on the subject. Here, we shall study the physics behind the occurrence of the scintillation. This will help us in understanding the morphology and the behaviour of scintillation. We shall also learn about the details of its measurements and related aspects in the next chapter.

4.3.2.1 Physics of scintillation

The basic reason for the origin of the ionospheric scintillation is the combination of signal components reaching the receiver through different paths of the ionosphere with random and irregular distribution of densities. These components have random time varying amplitude and phase as the condition of the propagation paths through which these components traverse vary with time. Therefore, when they combine at the receiver, the resultant signal thus formed has rapidly fluctuating resultant amplitude and phase, leading to the scintillation. We shall later mathematically see how the different components combine to generate a scintillating signal. Here, we shall rather concentrate on the reasons that create the multipath components.

Scintillation theory relates the observed signal statistics to the statistics of ionospheric electron density fluctuations. The general problem of propagation of wave in a random medium is difficult to treat numerically. However, it can be greatly simplified for the conditions when the wavelength is much smaller than the characteristic scale size of irregularities. Further simplifications are possible if the irregular layer is so thin that $\sqrt{(\lambda L)}$ is much smaller than the extent of the irregularity contributing to scintillation. Here, λ is the wavelength of the signal and L is the thickness of the irregularity, i.e. the irregularity layer can be considered to be a thin sheet. This assumption leads to the so-called phase screen theory of scintillation. According to this, the irregular layer is replaced by a thin screen, which changes only the wave's phase. The screen is located in the ionosphere at the height of the maximum electron density (Wernik et al., 2004).

The plasma irregularities are located at the ionosphere and are mostly concentrated in the F region and below the F2 peak. The conspicuous part of the ionosphere has a vertical extent of about 400 km. Considering the height of the navigation satellites, the signals of which are most susceptible to scintillation, at around 20,000 kms from the surface or even more, the vertical width of this irregular part of the ionosphere constitutes only 2% of the total height or even less than that. Moreover, for the satellite signals in the L band and lower S band with wavelengths around 0.2 m, $\sqrt{(\lambda L)}$ becomes about 300 m, which is much smaller than the largest irregularity size of the electron densities in the ionosphere, the latter being of the order of kilometres. Therefore, it may be justifiably considered as a thin screen in the path of the signal.

The phase screen model is probably the most common and simple model based on diffraction theory that describes the occurrence of scintillation. This model assumes that the plasma density irregularities are concentrated within a thin layer at a height that is typical of the F2 layer peak height and demonstrates the effects of this layer of irregularity on the wave front of a propagating plane wave incident normally on the screen. A wave front is the plane obtained by joining the points having the same phase of a signal. Now, as the wave front gets incident and subsequently propagate through this medium with irregularities, the processes of scattering, diffraction or both may occur. The wave that finally emerges out of the screen has random variations in phase across it. The variation in the optical path length within this thin layer of irregularity is

$$\delta\varphi = -2\pi r_e \frac{\delta N}{k_0{}^2} \tag{4.34}$$

where, $\delta\varphi$ is the change in phase, δN is the thickness of the layer and r_e is the classical radius of electron and $k_0 = 2\pi/\lambda$. (Wernik et al., 2004). The resulting patterns of amplitude and phase variations on the ground are then derived by combining signals reaching a receiver and interfering after being diffracted from different parts of the emerged wave. This model provides an insight into the types of irregularities that are likely to produce scintillations, as well as the characteristics of the resulting signals. This approach is valid provided $\delta\varphi$ is not too large. If the phase variation is significant, the phase on the wavefront becomes incoherent and hence no interference occurs. The scenario is depicted in Fig. 4.6.

To provide a mathematical substantiation, we take the situation of a wave travelling through a thin screen of irregularity. We assume that, first the plane wave, traversing vertically down along the z axis, is incident normally on a plane-parallel irregular slab. If we consider the situation that the dielectric fluctuations are very small such that the scale size of the refractive index variation is many times larger than the wavelength, then the wave equation can be written using Maxwell's Equation, as

$$\nabla^2 E + k^2 \{1 + \delta_\varepsilon(r,t)\}E = 0$$
$$\text{Or,} \quad \nabla^2 E + k'^2 E = 0 \tag{4.35}$$

where $\delta_\varepsilon(r,t)$ is the fluctuating part of the dielectric permittivity caused by electron density irregularities, i.e. $\varepsilon = \varepsilon_0 + \varepsilon_0\,\delta_\varepsilon(r,t)$. So, $k'^2 = k^2\{1 + \delta_\varepsilon(r,t)\}$. Notice that, the effective wave number k' in the above differential equation is made up of randomly fluctuating parameters. This k' varies randomly with time and space, i.e. $k' = k'(r,z,t)$, and hence, forms the basis of the scintillation theory. Here, r is the distance in transverse direction, i.e. in the horizontal plane. Now, let $u(r,z,t)$ be the complex solution for the amplitude of the electric field of the wave propagating along z. u accommodates all the variational components due to the irregularities. Then, the electric field of the wave may be represented as

$$E(t) = u(r,z,t)\exp(-jkz) \tag{4.36}$$

So, the above differential equation becomes

$$-2jk'\,du/dz + \nabla^2 u - k^2 u = -k'^2 u \tag{4.37a}$$

If now we consider that the wavelength is small compared to the irregularity scale length, and if the condition $\sqrt{(\lambda L)} \ll r_0$ is satisfied, where L is the thickness of the

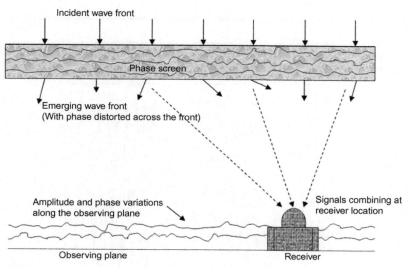

FIG. 4.6

Scintillation due to diffraction from a phase screen.

Box 4.2 Scintillation

In this box section, we shall run the MATLAB programme scinti.m to understand the basis of occurrence of the scintillation in the received signal.

This programme generates a scintillating signal from combination of two component signals s1 and s2. Each of the components has its own random phase variations which can be considered to be the result of its passage through the phase screen. The component signals are shown in below figure.

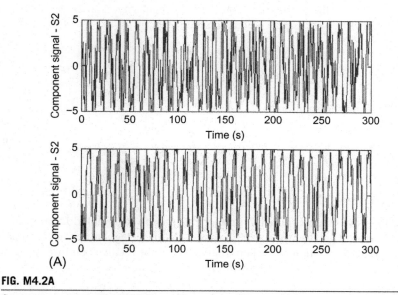

FIG. M4.2A

Component signals of scintillation with random phase variations.   *Continued*

Box 4.2

The combined scintillating signal is in below figure. Notice that, this signal varies in phase as well as in its amplitude.

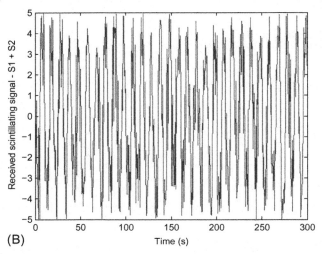

(B)

FIG. M4.2B

Resultant signal with scintillation.

Vary the strengths of phase variation in the components (default 3 and 2) and see the results.

irregularity and r_0 is the outer scale length of the irregularity, then $2k'\partial u/\partial z \gg \nabla^2 u$ (Priyadarshi, 2015). So, the above equation reduces to

$$-2jk\,du/dz = -k^2\delta_\varepsilon(rt)u$$

$$\text{Or,}\quad du/dz = -j\frac{k}{2}\delta_\varepsilon(rt)u \tag{4.37b}$$

Therefore, we get the solution as

$$u(r,t) = A\exp\{-j\varphi(\delta_\varepsilon, t)\}$$
$$= A\exp\{-j\varphi(r, t)\} \tag{4.38}$$

where $\varphi(z,t) = k/2\displaystyle\int_0^z \delta_\varepsilon(z,t)dz$

So, as the wave emerges out of this region, the phase and amplitude homogeneity of the planar front of the wave gets distorted. As the irregularities have random spatio-temporal distribution of the perturbation $\varepsilon\delta_\varepsilon$ of the relative permittivity ε, this distortion is also random, changing with time and location.

Recall from Chapter 3 that for a plane wave incident on a screen or other obstacle, each point acts as a secondary source. Hence, this resultant distorted wavefront acts as an extended source as a whole. Therefore, the signal component from each point of this front diffracts to reach the receiver. Thus, an equivalent multipath is formed with varying phase for each individual path. This is illustrated in Fig. 4.6. The signals on each path will combine in a phase-wise sense, causing a random fluctuation in the amplitude and phase in the resultant signal at the receiver. Since each of these component signals varies with time due to the temporal variations in the irregularities, the resultant signal also shows fluctuations in amplitude and phase along the time scale, causing Scintillation. (Kintner et al, 2000). The resultant wave received at the ground will contain the contribution, however with different magnitudes, from all diffractive elements of the emerged wave front and at all radial distances on the screen from the point of passage of the direct ray on the phase screen (Priyadarshi, 2015; Basu, 2003).

If the size of the irregularities is of the order of the first Fresnel zone, significant scintillations are produced. The fluctuating amplitude and phase diffraction patterns are a function of both the cross-radial distance of the irregularity from the direct ray and the radial range to the irregularities. Thus, for a fixed source (satellite) and a fixed irregularity distribution over space, a fixed spatial pattern of diffraction is generated. Scintillations are produced at a receiver when these spatial diffraction patterns are transformed into temporal ones, either through relative motion of the irregularity patterns through the line of sight of the satellite from receiver, or by changes in the structure of the irregularities with time.

4.3.2.2 Morphology of scintillation

The basic understanding we had from the previous section is the fact that, the primary requirement to have scintillation is the formation of the irregularities. Therefore, it is the formation of numbers of very small localized regions of relatively higher or lower electron densities against the nominal background is what causes the scintillation.

Scintillations occur predominantly in the equatorial region over a band across the magnetic equator and in the polar regions. It may also occur sometimes in the mid-latitudes, although the probability of occurrence here is rare. While the basic ingredient for the scintillation being the irregularity of the plasma density and physics behind the occurrences of scintillation remains the same, the processes that produce the ionospheric electron density irregularity leading to this phenomenon in these two regions are quite different. This also leads to significant differences in the characteristics of the resulting scintillations.

Equatorial scintillations are produced by irregularities in the layers, formed mostly in the F layer of the equatorial ionosphere. This irregularity starts to occur just after sunset and tends to disappear soon after midnight. In these regions, the most severe scintillations are observed near the locations of the Equatorial anomaly peak where the vertical density gradient of electron density remains the highest. Equatorial irregularities of the electron density are initiated by instabilities in the F layer of the ionosphere during the post-sunset period. This instability leads to the formation of the Equatorial spread-F, in which higher densities of electrons from top penetrate down into the comparatively lower density background at the bottom. This has been discussed in a previous section. This also results in the formation of plasma bubbles, which are enclosed lower density regions surrounded by high density plasma. These structures tend to produce very strong scintillation effects. Smaller irregular patches are created out of them and they typically spread upto few 100 km along the magnetic meridians. They tend to move with velocities of the order of 50–200 m/s. Consequently, scintillations are often experienced in patches that can last for an hour or so with periods of little or no activity in between. Eventually, in the absence of solar radiation, the background density falls. In turn, the irregularities begin to fade along with the associated scintillation activity. This usually occurs around local midnight, although at times scintillations can persist until early morning.

Equatorial scintillations show a strong seasonal dependence. This is in concurrence to the seasonal dependence of the spread-F. It has been established in Section 4.1.2 that spread-F is very much dependent upon the declination of the location. The spread-F maximizes at a place for that particular season when the sunset terminator line and the magnetic meridians are more or less aligned. This satisfies the prerequisite of conjugate points in F and E regions being transferred simultaneously across the sunset terminator. Due to the westward declination, it is greatest during the December solstice in the American sector, but minimum during the summer. Conversely, it is maximum in summer in the pacific sector, where the declination is Eastward. For the same reason, the scintillation activity is highest in the Indian sector, during the equinoctial months of September and March as the declination here is close to zero. (Abdu et al., 1981; Iyer et al., 2003). The equatorial scintillations in the Indian region also occur in conjunction with the anomaly, as both are most probable during equinoctial months (Acharya et al., 2007). Further, they tend to be worse during the years of solar maximum when the anomaly is at its greatest.

The occurrence and variation of electron density at the polar region due to either the photochemical effect or the transport effect of the plasma are much less compared to the equatorial region. Here, scintillation is mainly caused by the irregularities occurring as a result of the space weather activities including contributions from the precipitating charges in continuous solar wind and by the occurrences of solar flares, CME, etc. We have discussed the polar irregularities in previous sections. The irregularities necessary for scintillation are caused by the process of precipitation and Gradient Drift Instability (GDI), which are already described in earlier sections.

Being linked to solar activity, polar scintillations also show a strong dependence on the 11-year solar cycle. Therefore, it is most intense during solar maximum periods and almost nonexistent during minima. Unlike equatorial scintillations, they show little diurnal variation in their rate of occurrence and can continue for larger intervals.

Although typically remain confined to the polar region, it should be mentioned that during intense magnetic storms, ionospheric disturbances can extend well into the mid-latitudes, disrupting the navigation and communication signals in relevant bands through scintillation activity.

4.3.2.3 Characteristics of scintillations

The effects of ionospheric scintillations may be modelled as a complex modulation of the unperturbed signal, where the variations are produced by a random distribution of ionospheric irregularities of different sizes. The phase and amplitude variations of this resultant signal exhibiting scintillation may be modelled as wide sense stationary (WSS) stochastic processes. For any WSS, the autocorrelation and the power spectral density (PSD) form the Fourier pairs. Thus, representing the random process by its PSD also indicates its temporal characteristics. Hence, they are defined statistically, in terms of their power spectral densities, probability density functions and variances.

Spectral characteristics

The PSD of a fully uncorrelated random noise-like signal is white. However, the scintillations, occurring as a result of the vector combination of different components of the same signal with phase and amplitude variations, are not fully random but have a temporal correlation existing over a small finite time. Since for any WSS, the autocorrelation and the PSD form the Fourier pair, larger the autocorrelation time, shorter is the extent of the power spectrum. This alters the PSD of the process from its white nature to a bandlimited spectral function. The power spectral density of phase scintillations follows an inverse power law relationship of the form

$$S_{\varphi p}(f) = \frac{T}{\left(f_o^2 + f^2\right)^{p/2}} \text{rad}^2/\text{Hz} \tag{4.39}$$

where f_o is a frequency that corresponds to the ionospheric outer scale size, f is the frequency of phase fluctuations and T is a constant. T determines the strength of the scintillation and is dependent upon the relative direction and effective velocity of propagation of the radio wave and the irregularities. The effective velocity is a function of the velocity of the ionospheric pierce point through the irregularity layer due to the relative motion of the satellite and the receiver and also of the drift velocity of the irregularities, the latter being typically 50–200 m/s. T is also dependent upon the geometry and orientation of the irregularities, as well as on the thickness of the irregularity layer and the effective path traversed by the signal through it. Since $f_o \ll 1$, therefore when $f = 1$, $S_{\phi p} \approx T$. Thus, T becomes the magnitude of the power spectral density at a frequency of 1 Hz. p is termed the spectral index and usually lies in the range 1–4 (Fig. 4.7).

The power spectral density of amplitude scintillations follows a similar power law relationship for high fluctuation frequencies, but is heavily attenuated at low frequencies. It means the scintillation signal can have a very slow variation in phase, but not a very slow variation in amplitude. This is called Fresnel filtering. Fresnel filtering occurs because amplitude scintillations are mainly produced by diffraction effects which are only significant when the outer scale size of the irregularity is of the order of the Fresnel zone radius. Larger the irregularity size, slower the variation in amplitude. However, as the irregularities are limited in size and very large

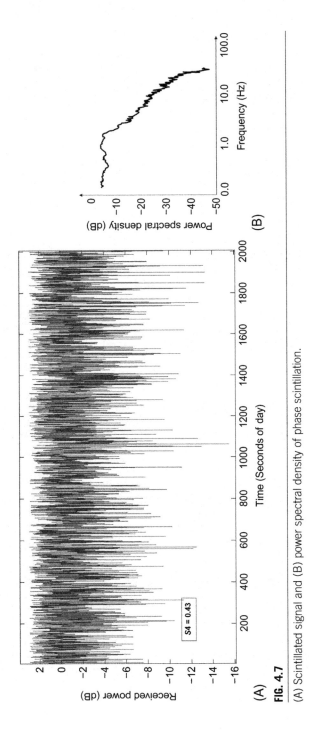

FIG. 4.7

(A) Scintillated signal and (B) power spectral density of phase scintillation.

irregularities are unlikely to occur, very slow variations are not observed. The cut-off frequency of the amplitude scintillation power spectrum (the Fresnel cut-off frequency) is thus determined by the Fresnel radius and is given by

$$f_c = \frac{1}{\sqrt{2}} \frac{v_e}{z_F} \text{Hz} \tag{4.40}$$

where v_e is the relative drift velocity of the irregularities and $z_F = \sqrt{\{\lambda z_1 z_2 / (z_1 + z_2)\}}$ is the Fresnel zone radius. z_1 and z_2 are the distances between the mean ionospheric irregularity layer and the satellite and receiver, respectively. λ is the wavelength of the signal. Since f_c is the lowest frequency in the amplitude power spectrum and the spectrum height monotonically decreases with frequency, f_c is also the frequency that corresponds to the peak of the amplitude scintillation power spectrum.

At typical ionospheric heights ($\sim$400 km for the F2 layer peak) and assuming vertical propagation, z_F is of the order of 276 m at the GPS L1 frequency. For an irregularity drift velocity of 100 m/s (a typical equatorial value) and assuming ionospheric pierce point variation is zero, i.e. $v_i = 0$, the Fresnel cut-off frequency is $\sim$0.26 Hz.

Probability density function

Measurements of the probability density functions (PDF's) of scintillations have shown that phase scintillations follow a zero-mean Gaussian PDF, while amplitude scintillations follow the Rician or Nakagami-m PDF. Although other distribution functions have been proposed for scintillations, the Gaussian/Nakagami-m distribution functions were found to provide the best fit.

Phase scintillations follow a zero-mean Gaussian PDF and are therefore defined completely by the variance, viz.

$$p(\varphi) = \frac{1}{\sqrt{2\pi\sigma_\varphi^2}} \exp\left(-\frac{\varphi^2}{2\sigma_\varphi^2}\right) \tag{4.41}$$

The temporal variation of the phase scintillation, therefore, can be completely represented by the RMS value of the phase deviation which is also equal to the standard deviation of the phase deviation by virtue of its zero-mean Gaussian nature and given by σ_φ. The RMS value can be obtained from the integral of the power spectral density of phase scintillations, considering zero-mean distribution of the same which is given by the relation

$$\sigma_{\varphi p} = \sqrt{<\varphi^2>} = \sqrt{\int S_\varphi(f)df} \tag{4.42}$$

where φ is the carrier phase (assumed to be zero-mean) and S_φ is the power spectral density of phase scintillation. Consequently, σ_φ is a function of T, p and the outer scale size parameter, f_0.

Effects of scintillation

Among various phenomena of ionosphere, ionospheric scintillation due to electron density irregularities causes deep fading of the propagating signals. Signal to noise ratio or more precisely carrier to noise density ratio (C/N_0) of a certain satellite channel fluctuates rapidly and sometimes drops very deeply if strong scintillation occurs. The scintillation and fading can create serious problems for satellite communications and navigation services. Reduction of the C/N_0 of a of a satellite signal deteriorates the signal quality which directly affects the performance of the system.

A lower C/N_0 not only results in unreliable communication with enhanced BER, but also can hinder the receiver's capacity to detect, acquire and track the signal. The phase scintillation causes the acquisition and tracking of the signal very difficult. Since these processes involve the chasing of the received signal phase, the abrupt deviations resulting from the scintillation hinder the process.

One of the effects of scintillation, which is most typically responsible for complete loss of lock of the phase tracking loops, is a canonical fade. Canonical fades refer to a power fade of sufficient depth and rapidity with associated abrupt phase deviation of near-half-cycle. In other words, canonical fades occur whenever deep fades (amplitude scintillation) and rapid, near-half-cycle phase changes (phase scintillation) occur simultaneously. These fades are particularly detrimental to carrier tracking loops because the receiver is not able to differentiate between a phase variation due to modulation and a phase variation caused by scintillation. In an extreme case, repeated cycle slips can cause complete loss of lock of the signal. In this case, reacquisition may be necessary in order to continue tracking and demodulation of the navigation data.

Scintillation primarily affects signals in the lower latitudes near the magnetic equator between $\pm 20°$ latitude as well as in higher latitudes near the poles. Navigation signals are particularly vulnerable to ionospheric scintillation primarily because they are typically in lower L band and have low-power spread spectrum signals transmitted below the noise floor.

CONCEPTUAL QUESTIONS

1. Using Eq. (4.6) for any arbitrary zenith angle χ, derive an equation like 4.10 where the height z is with reference to $h_{m\chi}$, the peak height obtained at location with zenith χ, and not h_{m0}. In Fig. 4.4, find the location of the peak for minimum and maximum zenith angle.
2. Using Eqs (4.33), (4.21b), show the relation between v_p and v_g.
3. Why GDI is important at the poles and GRT at the equator?
4. Lower frequency signals have predominant slower components in their amplitude scintillation than higher frequencies—Comment of the statement.

REFERENCES

Abdu, M.A., 1997. Major phenomena of the equatorial ionospherethermosphere system under disturbed conditions. J. Atmos. Sol. Terr. Phys. 59 (13), 1505.

Abdu, M.A., Bitteacourt, J.A., Batista, I.S., 1981. magnetic declination control of the equatorial f region dynamo electric field development and spread F irregularities. J. Geophys. Res. 86, 11443.

Abdu, M.A., Batista, I.S., Walker, G.O., Sobral, J.H.A., Trivedi, N.B., de Paula, E.R., 1995. Equatorial ionospheric electric field during magnetospheric disturbances: Local time/longitude dependences from recent EITS campaigns. J. Atmos. Terr. Phys. 57, 1065.

Acharya, R., 2014. Understanding Satellite Navigation. Academic Press, Elsevier, USA.

Acharya, R., Nagori, N., Jain, N., Sunda, S., Regar, S., Sivaraman, M.R., Bandyopadhyay, K., 2007. Ionosphric studies for the implementation of GAGAN. Indian J. Radio Space Phys. 36 (5), 394–404.

Appleton, E.V., 1946. Two anomalies in the ionosphere. Nature 157, 691.

Balan, N., Otsuka, Y., Bailey, G., Fukao, S., 1998. Equinoctial asymmetries in the ionosphere and thermosphere observed by the MU radar. J. Geophys. Res. 103 (A5), 9481–9495.

Balan, N., Otsuka, Y., Fukao, S., Abdu, M.A., Bailey, G.J., 2000. Annual variations of the ionosphere: a review based on MU radar observations. Adv. Space Res. 25 (1), 153–162.

Basu, S., 2003. Ionospheric Scintillation: A Tutorial. In: Cedar Workshop, Colorado.

Basu, S., Basu, Su, Groves, K.M., Yeh, H.-C., Su, S.-Y., Rich, F.J., Sultan, P.J., Keskinen, M.J., 2001a. Response of the equatorial ionosphere in the South Atlantic region to the great magnetic storm of July 15, 2000. Geophys. Res. Lett. 28, 3577.

Basu, Su, et al., 2001b. Ionospheric effects of major magnetic storms during the international space weather period of September and October 1999: GPS observations, VHF/UHF scintillations, and in situ density structures at middle and equatorial latitudes. J. Geophys. Res. 106, 30,389.

Challinor, R.A., 1969. Universal time control of polar ionosphere. Nature 221, 941–943. http://dx.doi.org/10.1038/221941b0.

Chapman, S., 1931. The absorption and dissociative or ionizing effect of monochromatic radiation in an atmosphere on a rotating earth. Proc. Phys. Soc. Lond. 43 (26), 484.

Dungey, J.W., 1956. Convective diffusion in the equatorial F region. Atmos. Terr. Phys. 9, 304.

Farley, D.T., Balsley, B.B., Woodman, R.F., McClure, J.P., 1970. Equatorial spread F: implications of VHF radar observations. J. Geophys. Res. 75, 7199.

Farley, D.T., Bonelli, E., Fejer, B.G., Larsen, M.F., 1986. The prereversal enhancement of the zonal electric field in the equatorial ionosphere. J. Geophys. Res. 91 (A12), 13723–13728. http://dx.doi.org/10.1029/JA091iA12p13723.

Feynman, R.P., Leighton, R.B., Sands, M., 1992. Feynman Lectures on Physics. Narosa Publishing House, India.

Haerendel, G., Eccles, J.V., 1992. The role of equatorial electrojet in the evening ionosphere. J. Geophys. Res. 97, 1181–1192.

Hargreaves, J.K., 1979. The Upper Atmosphere and Solar-Terrestrial Relations. Van Nostrand Reinhold, New York.

Hargreaves, J.K., 1992. The Solar-Terrestrial Environment. Cambridge University Press, Cambridge, UK.

IEEE, 1998. IEEE Standard Definitions of Terms for Radio Wave Propagation, 211-1997. IEEE. http://dx.doi.org/10.1109/IEEESTD.1998.87897.

Iyer, K.N., Jivani, M.N., Patban, B.M., Sharma, S., Chandra, H., Abdu, M.A., 2003. Equatorial Spread-F: statistical comparison between ionosonde and scintillation observation and longitude dependence. Adv. Space Res. 31 (3), 735–740.

Kawamura, S., Balan, N., Otsuka, Y., Fukao, S., 2002. Annual and semiannual variations of the midlatitude ionosphere under low solar activity. J. Geophys. Res. 107 (A8), 1166. http://dx.doi.org/10.1029/2001JA000267.

Kelley, M.C., 1989. The Earth's Ionosphere. Academic Press, San Diego, CA.

Kelley, M.C., Ilma, R.R., Crowley, G., 2009. On the origin of pre-reversal enhancement of the zonal equatorial electric field. Ann. Geophys. 27, 2053–2056.

King, G., 1961. The seasonal anomalies in the F region. J. Geophys. Res. 66 (12), 4149–4154.

King, G.A.M., 1964. The dissociation of oxygen and high level circulation in the atmosphere. J. Atmos. Sci. 21 (3), 201–237.

Kintner, P., Humphreys, T., Hinks, J., 2000. GNSS and Ionospheric Scintillation—How to Survive the Next Solar Maximum. Technical Article, July/August.

Lee, W.K., Kil, H., Kwak, Y.S., Wu, Q., Cho, S., Park, J.U., 2011. The winter anomaly in the middle-latitude F region during the solar minimum period observed by the constellation observing system for meteorology, ionosphere and climate. J. Geophys. Res. 116, A02302. http://dx.doi.org/10.1029/2010JA015815.

Martyn, D.F., 1947. Atmospheric tides in the ionosphere – I: solar tides in the F2 region. Proc. Phys. Soc. Lond. A189, 241–260.

Mitra, S.K., 1946. Geomagnetic control of region F2 of the ionosphere. Nature 158, 668–669.

Priyadarshi, S., 2015. A review of ionospheric scintillation models. Surv. Geophys. 36, 295–324. http://dx.doi.org/10.1007/s10712-015-9319-1.

Reitz, J.R., Milford, F.J., Christy, R.W., 1990. Foundations of Electromagnetic Theory, third ed. Narosa Publishing House, India.

Rishbeth, H., Garriot, O.K., 1969. Introduction to Ionospheric Physics. Academic Press, New York.

Rishbeth, H., Setty, C.S.G.K., 1961. The F-layer at sunrise. J. Atmos. Terr. Phys. 21, 263–276.

Rishbeth, H., Müller-Wodarg, I.C.F., Zou, L., Fuller-Rowell, T.J., Millward, G.H., Moffett, R.J., Idenden, D.W., Aylward, A.D., 2000. Annual and semiannual variations in the ionospheric F2 layer: II. Physical discussion. Ann. Geophys. 18, 945–956.

Rishbeth, H., Heelis, R.A., Müller-Wodarg, I.C.F., 2004. Variations of thermospheric composition according to AE-C data and CTIP modeling. Ann. Geophys. 22, 441–452.

Sastri, J.H., 1988. Equatorial electric fields of the disturbance dynamo origin. Ann. Geophys. 6, 635.

Sastri, J.H., Niranjan, K., Subbarao, K.S.V., 2002. Response of the equatorial ionosphere in the Indian (midnight) sector to the severe magnetic storm of July 15, 2000. Geophys. Res. Lett. 9 (13), 1651. http://dx.doi.org/10.1029/2002GL015133.

Scali, J.L., 1992. The mid latitude trough—a review. Scientific report-2, Phillips Laboratory. Report number: PL-TR-92-2100.

Simon, A., 1963. Instability of partially ionized plasma in crossed electric and magnetic fields. Phys. Fluids 6, 382.

Sobral, J.H.A., Abdu, M.A., Gonzalez, W.D., Batista, I., 1997. Lowlatitude ionospheric response during intense magnetic storms at solar maximum. J. Geophys. Res. 102, 14,305.

Sobral, J.H.A., Abdu, M.A., Gonzalez, W.D., Yamashita, C.S., Clua de Gonzalez, A.L., Batista, I., Zamlutti, C.J., 2001. Responses of the low latitude ionosphere to very intense geomagnetic storms. J. Atmos. Sol. Terr. Phys. 63, 965.

Torr, M.R., Torr, D.G., 1973. The seasonal behaviour of the F2-layer of the ionosphere. J. Atmos. Terr. Phys. 35, 2237–2251.

Torr, G.D., Torr, M.R., Richards, P.G., 1980. Causes of the F region winter Anomaly. Geophys. Res. Lett. 7 (5), 301–304.

Wernik, A.W., Alfonsi, L., Materassi, M., 2004. Ionospheric Irregularities, scintillations and its effect on systems. Acta Geophys. Pol. 52(2).

Woodman, R.F., La Hoz, C., 1976. Radar observations of F region equatorial irregularities. J. Geophys. Res. 81, 5447.

FURTHER READING

Basu, S., Kelley, M.C., 1979. A review of recent observations of equatorial scintillations and their relationship to current theories of F region irregularity generation. Radio Sci. 14, 471.

Dungey, J.W., 1961. Interplanetary magnetic field and the auroral zones. Phys. Rev. Lett. 6, 47.

Sultan, P.J., 1996. Linear theory and modelling of the Rayleigh-Taylor instability leading to the occurrence of equatorial spread F. J. Geophys. Res. 101, 26875.

Ionospheric impairments: Measurement and mitigation

5

We have already learnt about the ionosphere and its impairing effects on the propagating satellite signals in Chapter 4. The origin of these impairments, including the related physics, proviso for occurrence and salient features, have been described there. In this chapter, we shall deal with more practical aspects of these impairments. We shall learn about the methods of measuring them, including the related instruments which measure the relevant parameters. Further, we shall also study the different models of their predictions and learn the standard methods of mitigating them. Finally, we shall also discuss how these impairing factors are advantageously utilized in different applications.

5.1 MEASUREMENTS

In this section, we shall learn about the measurements of two parameters—the ionospheric group delay or phase advancement and the scintillation. We shall first introduce different measuring instruments and discuss the methods of measurements. The characteristic variations of the measured parameters and the factors affecting them have already been explored in Chapter 4. So, here we shall briefly discuss the models and the pertinent issues in the measurement processes.

5.1.1 MEASUREMENT OF TEC/ELECTRON DENSITY

5.1.1.1 Direct measurements

Ionosonde

Ionosondes are instruments for sounding the ionosphere. An ionosonde consists of a high-frequency (HF) transmitter which emits radio frequency directed to the ionosphere. It also consists of a receiver which can automatically track and receive the transmitted signal reflected back by it.

Ionosonde automatically transmits from the ground in upward direction, a range of frequencies. The frequency coverage is from 1 to more than 30 MHz, though normally sweeps are confined to approximately between around 2–12 MHz. The total sweep of the frequency is digitally controlled. The transmitter sweeps the frequency range by transmitting short pulses of unmodulated waves. These pulses are reflected by various layers of the ionosphere, at heights of around 100–400 km, and their echoes are received by the receiver and analyzed by the control system.

Satellite Signal Propagation, Impairments and Mitigation. http://dx.doi.org/10.1016/B978-0-12-809732-8.00005-3

To understand the working of the ionosonde, we recall from Chapter 3 that the refractive index of the ionosphere can be given as

$$n = \sqrt{1 - \frac{\omega_p^2}{\omega^2}} \qquad (5.1)$$

$\omega_p^2 = N_e e^2/m\varepsilon_0$ is the plasma frequency of the ionosphere, where N_e is the electron density of the ionosphere, e is the electronic charge, m is the mass of the electron and ε_0 is the permittivity of free space. The wave gets reflected back from the layer of the ionosphere only when n turns out to be imaginary, i.e. $\omega < \omega_p$, creating evanescent wave in the forward direction. Signals of frequency higher than ω_p get passed through it. Every layer has its own ω_p depending upon the corresponding N_e. So, the maximum frequency that gets reflected, or in other words, the minimum frequency that can get passed through a layer is dependent upon the electron density of the layer, N_e (Hargreaves, 1979).

From the structure of the ionosphere, we know that, starting from the height of about 50 km from the surface of the earth with minimum electron density, the ionosphere has gradual increase in electron density N_e with height up to the peak. Therefore, if the transmitted signal starts to sweep up from a very low frequency, then for all frequencies below the ω_p for the lowest level of the ionosphere, the signal will be reflected back. As the signal frequency exceeds this value, it will penetrate through the lowest layer, but will eventually rise to a height where it will encounter the density whose plasma frequency just equals the frequency of the wave. Therefore, it will get reflected from that level. Similarly, as the frequency of the wave ω is increased, it gets reflected from higher levels of larger electron density. So, for any signal with frequency $\omega_1 {\scriptstyle(= 2\pi f_1)}$, reflecting from a layer with density N_{e1}, the following relation is satisfied.

$$\omega_1^2 = \frac{N_{e1} e^2}{m\varepsilon_0} \qquad (5.2a)$$

$$\text{Or,} \quad N_e = \frac{4\pi^2 m\varepsilon_0}{e^2} f_1^2 \qquad (5.2b)$$

As the frequency increases, each pulse penetrates further up to a denser region before it is reflected. Therefore, with the increase in frequency over the sweep, at a certain maximum frequency, the signal reflects back from the peak layer of the ionospheric density. This highest frequency is called the 'critical frequency' for the layer, and for all frequencies above, this will penetrate through the peak. If they do not find any higher density at any greater height, these signals will never return back. For a layer with critical frequency f_c, using Eqs (5.2a), (5.2b), the corresponding peak density N_m of the layer is given by

$$N_m = \frac{4\pi^2 m\varepsilon_0}{e^2} f_c \qquad (5.3a)$$

Putting the standard values of the constants in the above equation, we get

$$N_m = 0.124 \times 10^{11} f_c^2 \qquad (5.3b)$$

Here f_c is the critical frequency of a layer in MHz, N_m is the peak electron concentration of the layer in m^{-3}, e is the charge on an electron and m is the mass of an electron. Eventually, as the critical frequency is exceeded, the wave penetrates even the peak layer without being reflected. For the transmitted waves, the corresponding density for reflection is never encountered in the ionosphere.

As the transmitter sweeps from lower to higher frequencies, the receiver records the return signal reflected from the different layers with increasing heights of the ionosphere. Again, if δ be the time delay of arrival for the signal of this frequency, the (apparent) height of this density layer is

$$h = \frac{\delta}{2} \times c \tag{5.4}$$

c being the velocity of the electromagnetic wave. Since the wave will pass through the ionized region, its group velocity will be reduced, slowing down the wave. This will add to the time-of-flight. Therefore, the apparent or virtual height indicated by this time delay will be greater than the true-height. The difference between true-height and virtual height is governed by the amount of ionization that the wave has passed through. The trace displayed in Fig. 5.1 in the form of a graph showing

---

Box 5.1 MATLAB Exercise to Generate Ionogram for Different Density Profile

In this box section, we shall show using a MATLAB programme how the ionosonde generates an ionogram. Run the programme ionosonde.m to obtain the plot of the height versus frequency graph, i.e. an ionogram. The resultant plot is also shown in the figure below.

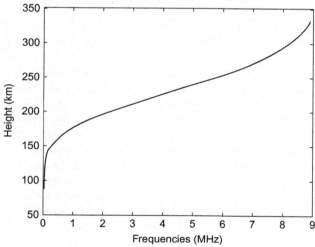

Go through the code to understand the process of imaging the ionospheric vertical electron density profile. Also, run the same programme with different profiles of the ionosphere given in the variable N_h and also for different peak height h_0 and for different peak density N_0. Observe the variation in the ionogram figure. However, in this programme, the effect of reduction of the wave velocity as it passes through the layers of the ionospheric layers is not considered.

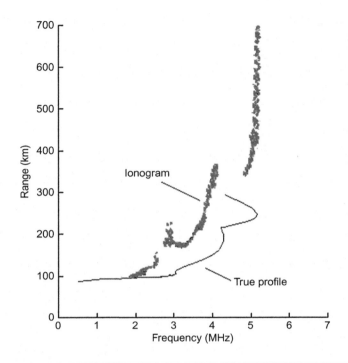

FIG. 5.1

Ionogram.

virtual reflection height (derived from time between transmission and reception of pulse) versus carrier frequency for a given ionospheric condition (Ionograms, n.d.). This is called an ionogram.

As mentioned above, the radio pulses travel more slowly within the ionosphere than in free space. It gets slower with the increasing electron density in the layers. Therefore, the apparent or 'virtual' height is recorded instead of a true-height. Recreating the true-height profile of electron concentration from ionogram data by converting the virtual heights to true-heights is an important aspect of ionosonde measurements. Such a procedure is known as 'True-Height Analysis'. Characteristic values of virtual heights of different layers of the ionosphere, designated as $h'E$, $h'F$, $h'F2$, etc., and the corresponding critical frequencies, designated as f_0E, f_0F1, f_0F2, etc., of each layer are scaled, manually or by computer, from these ionograms. Modern ionosondes with computer-driven automatic scaling procedures routinely scale all the ionograms recorded.

Topside measurement

The ground-based ionosondes has the limitation that they can provide the distribution of the electron density up to the F-layer maximum. This is because, as the density monotonically decreases above this layer, the signals which pass through this do not encounter enough density further for reflection. Therefore, the concept of placing an ionospheric sounder in the satellite orbit to sound the ionosphere from top side is a

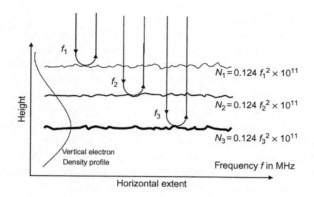

FIG. 5.2

Topside sounding.

logical extension of the conventional technique of sounding the ionosphere from the ground. Like the ground-based sounding, in the topside sounder, the transmission is done vertically down while the reflection from the ionospheric layer goes up. This is shown in Fig. 5.2. It shows a horizontal extent of definite ionospheric layers at different heights with different densities. The thickness of the curve represents its density. The vertical profile is also shown alongside. Emphasis in the figure is on the reflection of the signals of different frequencies from different layers. These reflected signals are used to derive the density down up to the same F2 peak height, the density up to which is provided by the ionosonde. Thus, the combination of the two can provide the full density profile variation. Topside electron density and scale heights are generally lower for latitudes $\lambda > 60$ compared to lower latitudes. Nighttime variation is stable but large irregular variations with latitude observed above this height (Heise et al., 2002).

The satellites carrying topside sounders are typically LEO with heights around 1000 km. These satellites are sometimes augmented with other instruments like mass spectroscope, incoherent electron backscatter radars, or VLF receivers to collect ancillary information. An important difference between an ionosonde on the ground and that in a satellite is that, in the latter case, the instrument is immersed in the plasma of the ionosphere. An attempt to launch radio waves of certain specific frequencies within the plasma leads to a number of resonance effects (Chapman and Warren, 1968).

Small scale structures of the F region above the peak are identified by the diffuseness and the range spreading of the echo pulses. This is similar to the Spread F condition observed from the ground. However, topside spread during daytime is observed only at latitudes more than 55° (Chapman and Warren, 1968), while the night-time spread is observed both at high and low latitudes, mostly for certain parts of the year.

The first topside sounder satellite was Alouette 1, launched in 1962 by Canada. Since then a numbers of ionospheric topside sounder satellites have been launched by different countries. Ionospheric sounder was also placed on the erstwhile Mir Space Station. Nowadays, topside ionospheric estimations are also done using GNSS signals from a LEO satellite.

Incoherent scatter radar

In an ionosonde, the transmitted wave is reflected from a level where the electron density is such that the refractive index reduces to zero for the travelling signal of given frequency. In a regularly structured ionosphere, this condition occurs at a definite level which is extended over a wide area. Therefore, the returned signal is coherent in nature. However, if the medium is irregular and contains spatial variations comparable to the wavelength of the radio wave, there will be scattering from each electron and ion in the irregularity. Now since each scattering particle will scatter independently of the others, the returned signal will be incoherent. The individual particles scattering from the ionosphere are the electrons. The scattered wave returning back can be measured by an incoherent scatter radar for its strength, frequency and bandwidth. The total power received by the receiver is the sum of scattered power received from all the individual electrons. As the amount of scattering from one single electron is theoretically known, the total number of electrons that has contributed to the scattering can be derived from the measured strength of the scattered wave. Thus, using the back scattered component of the signal penetrating into the ionosphere, the vertical profile of the electron density can be derived. Further, the electrons are in random thermal motion and hence a random phase shift and Doppler will be observed in the scattered component. So, due to this random motion in different directions, the received echo will exhibit a spread in the frequency. As the random motion is a measure of its kinetic energy or temperature, in general, the bandwidth spread in the received signal will represent the temperature of the particles in the ionosphere. Finally, the electron plasma in the ionosphere, as a whole, displays drift motion. This will lead to an overall deviation in the frequency in a definite direction, depending upon the physical direction of motion of the plasma. Thus, the drift velocity of plasma can be identified from the mean frequency deviation of the received signal (Incoherent Scatter Radar, n.d.).

5.1.1.2 Derived estimates

GNSS-based measurements

Total electron content (TEC) of the ionosphere is basically the path integral of the ionospheric electron density along the signal path through the ionosphere. So, it represents the total numbers of electrons contained in a hypothetical cylinder of unit cross section along the path and represented by electrons/m^3 (Fig. 5.3). Since the magnitude for ionospheric TEC is quite large, it is more typically represented by TEC Unit (TECU), where 1 TECU = 10^{16} electons/m^3. The GNSS signal can be alternatively used to derive the TEC of the ionosphere, though all alternatively derived estimates of the ionospheric TEC values are, however not essentially, GNSS-based (Acharya, 2014). The navigation signals transmitted by the satellites are in the L band. The signals in this band of frequency experience large ionospheric delay as their group velocity reduces. Therefore, a particular phase of the ranging code embedded in the signal is received after a finite time delay compared to the time it was expected to come, had the medium been vacuum. This delay is inversely proportional to the square of the frequency and directly to the TEC values. The modified

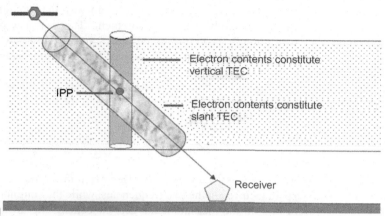

FIG. 5.3

Conversion of TEC from STEC at IPP.

group velocity in the ionosphere is $v_g = \sqrt{(1 - 40.3/f^2 N_e)}$. Therefore, the time of travel of the signal of frequency f_1, as we have seen in the previous chapter, is

$$T = \int \frac{dl}{v_g} = \int \frac{dl}{c}\left(1 + \frac{40.3}{f^2}N_e\right)$$

$$= \tau + \frac{40.3}{cf_1^2}\text{TEC} \tag{5.5a}$$

$$= \tau + \delta\tau$$

In the navigation services, generally two frequencies are used for the transmission of the data from the satellite. However, depending upon the requirements, the receivers can be of single frequency type or dual or multiple frequency type. The former can receive signal only in one of the available frequencies, while the latter types can receive and handle signals in both the frequencies. The ionospheric effect being dispersive, the dual or multiple frequency receivers get a great leverage due to this fact. It enables them to estimate the ionospheric delay by itself. Here is how it is done.

From Eq. (5.5a), it is easy to derive that the ionospheric delay adds an equivalent extra range δr_{ion} to the true geometric range ρ during the range measurements. This delay is hence a function of frequency f and TEC as given in Eq. (5.5a). So, the added extra range can be represented as

$$\delta r_{\text{ion}} = \delta\tau \times c$$

$$= \frac{40.3}{f^2}\text{TEC} \tag{5.5b}$$

So if R_1 be the range measured by a receiver in one of the frequencies f_1 added with ionospheric delay $\delta r_{1,\text{ion}}$ then

$$R_1 = \rho + \delta r_{1,\text{ion}}$$

$$= \rho + \frac{40.3}{f_1^2}\text{TEC} \tag{5.6a}$$

where, ρ is the true geometric range. However, there is no way that this excess path $\delta r_{1,\text{ion}}$ can be derived from this single measurement. This is because, the TEC values along the path through which the signal passes are not yet known and therefore the two components in the right hand side of the equation cannot be separated.

For dual frequency receivers, where there is available a separate and independent measurement in the other frequency f_2, let the range measured there be R_2, then

$$R_2 = \rho + \delta r_{2,\text{ion}}$$

$$= \rho + \frac{40.3}{f_2^2}\text{TEC} \tag{5.6b}$$

Now, if these two measured ranges R_1 and R_2 are differenced, the resultant is the difference in the excess path incurred in the two frequencies, while the common true geometric range present in both the measurements gets cancelled. The difference thus yields,

$$R_1 - R_2 = 40.3\ \text{TEC}\left(\frac{1}{f_1^2} - \frac{1}{f_2^2}\right) \tag{5.6c}$$

In this expression, since all the other parameters save the TEC are known, we can derive the TEC values from it. The exact relation turns into

$$\text{TEC} = \frac{R_1 - R_2}{40.3\left(\dfrac{1}{f_1^2} - \dfrac{1}{f_2^2}\right)} \tag{5.6d}$$

Focus 5.1 Exercise to Estimate TEC From Range Measurements

The objective of this focus is to understand the method used for estimation of the total electron content (TEC) from GNSS-based range measurements. Suppose the ranges measured at two GPS frequencies, $f_1 = 1575.42$ and $f_2 = 1227.6$ MHz, along the slant direction of a specific navigation satellite are 20,217,126.42 and 20,217,135.77 m, respectively. To find the TEC, assuming that the difference is due to the ionospheric delay only, we first write

$$D1 = D + \delta 1 = D + 40.3/f_1^2\,\text{TEC} \quad \text{and}$$

$$D2 = D + \delta 2 = D + 40.3/f_1^2\,\text{TEC}$$

So, $D1 - D2 = \delta 1 - \delta 2 = 40.3\left(1/f_1^2 - 1/f_2^2\right)\text{TEC}$

or, $\text{TEC} = (D1 - D2)/40.3 \times \left(1/f_1^2 - 1/f_2^2\right)^{-1}$

Putting the values we get

$$\text{TEC} = 7.26/40.3 \times \left\{1/(1227.60)^2 - 1/(1575.42)^2\right\}^{-1} \times 10^{12}$$

$$= 890086.64 \times 10^{12}$$

$$= 8.9009 \times 10^{17}$$

$$= 89\,\text{TECU}$$

Here, we have used the conversion between electrons/m^2 to TEC using the relation 1 TEC $= 10^{16}$ electrons/m^2.

Hence, the corrected value of the range is ion for $D1$ is $\delta 1 = 40.3 \times 8.9009 \times 10^{17}/(1575.42 \times 10^6)^2 = 14.45$

So, the corrected range is 20,217,126.42–14.45 m $= 20,217,111.97$ m

Convince yourself with the result by correcting the range for the other frequency.

Therefore, using the navigation links, the total electron content can be derived along the direction of the line of sight of the signal. The TEC thus derived in this slant direction is called the slant TEC (STEC). This can be converted to vertical TEC (VTEC) using a secant conversion formula. The conversion is given by

$$\text{VTEC} = F \times \text{STEC} \tag{5.6e}$$

where $F = [1 - \cos^2 E \times \{R_e/(R_e + h_i)\}^2]^{\frac{1}{2}}$ (RTCA, 1999). Here R_e is the earth's radius and h_i is the effective ionospheric height and generally taken as 350 km. and E is the elevation angle at the receiver location. This VTEC is defined at the ionospheric pierce point (IPP), which is the point at which the line of sight of the signal intersects the equivalent layer of the ionosphere at 350 km. The schematic for this conversion is shown in Fig. 5.3.

Knowing the location of the receiver, and the look angle to the GNSS satellite, the position of the IPP can be derived as (RTCA, 1999)

$$\lambda_{\text{IPP}} = \sin^{-1}\left(\sin \lambda_u \cos \psi_{\text{pp}} + \cos \lambda_u \sin \psi_{\text{pp}} \cos A \right) \tag{5.7a}$$

$$\varphi_{\text{IPP}} = \varphi_u + \sin^{-1}\left(\frac{\sin \psi_{\text{pp}} \sin A}{\cos \lambda_{\text{pp}}} \right) \tag{5.7b}$$

Here λ_{IPP} and φ_{IPP} are the latitude and the longitude of the IPP, respectively, and λ_u and ϕ_u are those of the user receiver. A is the azimuth angle of the receiver. ψ_{pp} is the geocentric angle between the user receiver and the pierce point location and is given by

$$\psi_{\text{pp}} = \frac{\pi}{2} - E - \sin^{-1}\left(\frac{R_e}{R_e + h} \cos E \right)$$

Fig. 5.4A shows the variations of the vertical TEC values with local time for an equatorial region. It is important to note here that, there exist certain techniques by which these discrete and definite path integrals of the electron densities can be utilized to get the total 3D ionospheric density profile.

5.1.2 MEASUREMENT OF SCINTILLATION

It has been illustrated in Chapter 4 that the ionospheric scintillation occurs as a result of the irregularities in the ionospheric electron density. These irregularities cause fluctuations in the phase and amplitude of the received signal from the satellite. However, there we emphasized only on the phase fluctuations for simplicity. We have also derived a theoretical expression for the phase fluctuations causing from

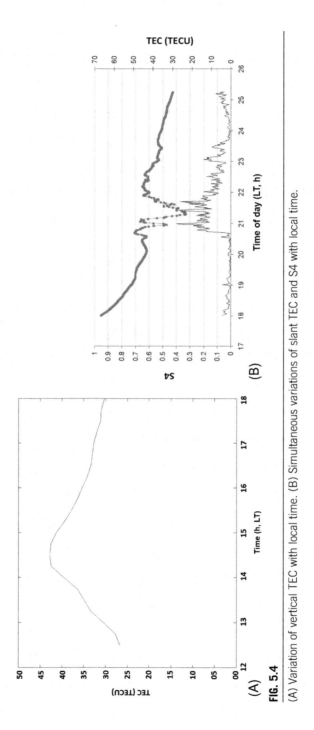

FIG. 5.4

(A) Variation of vertical TEC with local time. (B) Simultaneous variations of slant TEC and S4 with local time.

these irregularities in terms of the spatially varying permittivity of the layer of the medium (Basu, 2003) through which the signal traverses.

There are two main metrics of measuring the strength of scintillation. Amplitude scintillation is measured by calculating the index S_4 and is derived from the signal intensity received from satellites. S_4 index is defined as the standard deviation of the received signal intensity, normalized by the average of the signal intensity over a given period of time. Therefore, if I be the instantaneous received intensity (amplitude squared) of the signal, then,

$$S_4{}^2 = \frac{<I^2> - <I>^2}{<I>^2}$$

(5.8a)

where $<.>$ represents the expected value. In terms of signal amplitude A, S_4 can be represented as (Wernik et al., 2004)

$$S_4{}^2 = \frac{<A^4> - <A^2>^2}{<A^2>^2}$$

(5.8b)

For empirical data, this is computed as a time average over the given period. Being a ratio of identical quantities, S_4 index is dimensionless. S_4 values <0.3 indicate weak scintillation, and as S_4 reduces further and approaches zero, scintillation becomes insignificant and the signal is considered unaffected. Although there is no upper limit for the S_4 coefficient, values between 0.3 and 0.6 indicate moderate scintillation, and values >0.6 are indicative of strong scintillation (Humphreys et al., 2010). The following Fig. 5.4B shows the variation of TEC and S_4 index.

Phase scintillation is the rapid fluctuations in the signal phase. It occurs due to the traversing of the signal through the ionospheric irregularities and follows zero-mean normal distribution. Therefore, it can be completely described by σ_φ.. σ_φ is defined as the standard deviation of a signal's phase from its nominal excursion over a given time interval. In mathematical form, the phase scintillation index is given by (Basu, 2003; Wernik et al., 2004)

$$\sigma_\varphi = \sqrt{\{<\varphi^2> - <\varphi>^2\}}$$

(5.9)

where σ_φ is a measure of phase scintillation with φ representing the detrended phase measurement and $<.>$ denotes expected value, usually replaced by temporal averaging in practise. Phase scintillation is computed empirically and values $>5°$ indicate strong scintillation (Humphreys et al., 2010).

5.1.2.1 Measurement methods

The received amplitude and phase of the satellite signals traversing through the ionosphere are required to be measured for estimating the ionospheric scintillation parameters. Satellite navigation signals provide an excellent means for measuring ionospheric scintillation effects. This is because the signals are continuously available anytime and almost anywhere across the globe and can be measured through many points of the ionosphere simultaneously. But, the scintillation hinders the process of tracking and synchronizing a receiver, leading to a loss of lock in many occasions. This poses the difficulty of measuring the scintillation itself (Curran et al., 2014).

These scintillation-affected signals are still possible to be tracked with certain receiver modifications. Modifications, primarily in the process of decorrelation, like PLL tuning using a reasonably wide bandwidth tracking loops (Van Dierendonck and Hua, 2001) may be done. This enables the receivers to measure the scintillation parameters and compute the indices.

Further, the measured phase is also varied due to the relative range variation between the satellite and the receiver, i.e. due to the Doppler. This gets included in the scintillation measurement process and results in erroneous estimates, unless are precisely removed. The Doppler variation, however, is slow comparative to the scintillation. For this reason, the carrier phase measurement is filtered using a high-pass filter with appropriate cut-off frequency. This removes the low-frequency effects of the Doppler shift and hence the variation is detrended. The standard deviation is then calculated over the user-specified integration period (Ganguly et al., 2004).

Some receivers, which are placed at a fixed known location, use the knowledge of its position and time information to estimate the local—replica signal parameters. After the removal of the deterministic phase variation induced by satellite-user dynamics by relatively cancelling the effect, the incoming signals are compared with the estimated local signals during the process of demodulation. The resultant thus obtained remains only affected by the ionospheric processes in its amplitude and phase which can now be measured. This process is continued in a feed forward fashion (Curran et al., 2014).

It is important to recognize that the scintillation is path-dependent. The elevation and azimuth dependence of scintillation, however, is more complex than that for TEC (total electron content). The irregularities which cause the scintillation are aligned along the direction of the geomagnetic field so that there is a maximum in this direction as it also increases for the lower elevations.

5.1.2.2 Issues

The general concern in estimating the received phase variation are (a) the correctness of detrending, (b) the effects of multipath and (c) the insertion of the receiver phase noise into the measurements. However, questions of usage and interpretation also exist. The phase scintillation index (σ_φ) is the standard deviation of measured phase deviation as the signal passes through the ionosphere. It is often used to characterize the variations in the ionospheric environments that cause the scintillation. But, the measurements are actually influenced by other background factors. So, it is difficult to distinguish between induced artefacts and true ionospheric irregularity conditions.

To appreciate this, it is necessary to recall from what is discussed in the previous section that a fixed filter is used to remove the low-frequency components of the scintillation spectrum. It separates the components originating due to the Doppler. The cut-off frequency is kept at a comparatively lower value than the lowest frequency of the true scintillation spectrum originating due to the outer scale dynamics.

The notion of a fixed (temporal) frequency filter cut-off for Doppler removal etc. during the phase detrending process for variable relative drift speeds can be, however, questioned (Forte and Radicella, 2002). Increase in the relative drift speed with respect to the line of sight between the satellite and the receiver causes the phase fluctuation to occur more rapidly. Consequently, the phase scintillation spectrum moves right more towards the higher frequencies. Similarly, slower movements shift the spectrum towards the lower frequency side. Thus, a fixed detrending cut-off can remove a portion of the low-frequency power from the actual scintillation spectrum. This may lead to erroneous results in estimating scintillation parameters or characterizing ionospheric irregularities.

Second, also remember that the phase of the received signal is an assimilation of the diffracted waves received from the different parts of the wave front emerged out of the irregularity phase screen. These irregularities move in different spatial scales with time. Therefore, the advection length of the irregularity that contributes to the measurement of the scintillation parameter, while averaging over the predefined constant time interval, depends upon the scale of its variation and the rate of their movement. That is, as the relative drift velocity increases, the length of the irregularity, whose features are contained in the measurement, increases. Conversely, slower the relative drift of these ionospheric irregularities, the temporally averaged estimate of the parameter accommodates smaller scale of length of it. Therefore, for the same spatial variation of the ionospheric irregularity, considered frozen in time, there can be different measured indices for different drift velocities, provided the time interval of integration is kept same. In an even more realistic case, the irregularity distribution also evolves in time. As they do so, the unknown but varying outer scale and the motion of the complex and variable multidimensional structures make the quantitative meaning of temporally averaged σ_ϕ quite obscured (Beach, 2006). Therefore, the measured scintillation parameters can mask crucial information of the ionospheric irregularity dynamics of any particular scale sizes. Hence, characterizing ionospheric irregularity by these measured parameters is not issueless.

For relating phase fluctuations to amplitude scintillation, a parameter like the rate-of-TEC index (ROTI) (Pi et al., 1997) has improved performance under typical equatorial conditions (Beach and Kintner, 1999). Another suggested alternative for more reliable representation is to store subsets of phase samples and later use them to estimate S4 from a 'subsampled phase screen' (Beach et al., 2004).

5.2 MODELS OF IONOSPHERE

In this section, we shall discuss the different types of models for ionospheric impairments. However, we need to mention here that our main focus in this book being the propagation of the satellite signals, we shall concentrate on those models only which have direct bearing on the same.

In general, the ionospheric models can be divided into two main categories, viz. the theoretical models and the empirical or semiempirical models. Theoretical models utilize the physics behind the origin and distribution of the electron density to obtain the model output. They are most commonly developed by considering the physical processes including the photochemical reactions, drifts, diffusion, collision with neutral particles, wind, etc. Equations are formed for continuity, momentum energy, etc. to obtain the solution which finally gives spatial electron density as a function of location and time. The popular theoretical models like the Time-Dependent Ionospheric Model (TDIM), Coupled Thermosphere-ionosphere-plasmasphere model (CTIP), Global Theoretical ionosphere model (GTIM), Sheffild University Plasmasphere—ionosphere Model (SUPIM), etc. (AIAA, 1998).

The empirical models, on the other hand, avoid the uncertainties of the physical understanding. They depend upon the pragmatic data, partially or totally, to obtain the variational form of the electron density of the ionosphere. Due to their empirical nature, they are as good and representative as the data used to develop the model (Acharya, 2014).

Depending upon the extent of validity of the model, it can either be global or regional or local in nature. Local ionospheric models are mainly developed to accommodate localized phenomena and hence are mostly restricted to a definite zone.

5.2.1 DELAY MODELS

5.2.1.1 Ne quick

i. *Introduction and characteristics*. The Ne-Quick model is a global empirical model for ionosphere and gives ionospheric electron density as a function of time and location (Di Giovanni and Radicella, 1990). The model defines the analytical function for variation of the electron density distribution. For that, it uses a combination of Epstein layers, referenced to different anchor points, and reproduces the vertical profile of the electron density distribution. The anchor points are the peaks of the different ionospheric layers, viz. D layer, E layer, F1 and F2 layer. The features of these anchor points are predefined from the characteristics routinely scaled from ionograms (f_oF_2, $M(3000)F_2$, f_oF_1, f_oE). The model is particularly suited for the study of trans-ionospheric radio propagation effects of interest to satellite navigation and positioning (Radicella and Leitinger, 2001).

ii. *Basic i/o structure*. The basic inputs of the NeQuick model are position, time and solar flux (or sunspot number), while the output is the electron concentration at the given location in space and time. The position inputs provided as geographic latitude and longitude are converted to dip latitudes before estimation. Solar activity is represented by either the sunspot number or the 10.7 cm solar radio flux or even by the ionization factor. These values may be either directly measured or alternatively obtained by using representative functions. Time includes both season (month) and time (Universal Time UT or local time LT). The model, at its output, thus provides 4D variation of the electron density, i.e. three-dimensional density for a particular given location and its variation with given time. The densities at points along a definite signal path may be integrated to obtain the TEC along the path.

iii. *Variants*. The original Ne-Quick model has been modified and adopted by various users. Particularly, the successful development of the NeQuick model has been adopted by the International Telecommunication Union, Radio-communication Sector in its recommendation (ITU-R, 2016) as a suitable method for TEC modelling. The ITU recommended model, NeQuick 2, predicts electron density from analytical profiles, depending on the solar activity-related input values: R12 (12—month smoothed sunspot number) or F10.7 (solar flux at 10.7 cm), month, geographic latitude and longitude, height and UT. Another very important variant of the Ne Quick model is the NeQuick-G which is adopted for ionospheric corrections in the single frequency operation of the European GALILEO satellite navigation system (EU, 2015). Here, instead of using F10.7 or SSN, a surrogate number called effective ionization level, A_z, is used. A_z is location-dependent and can be obtained from the relation

$$A_z(\lambda) = a_0 + a_1\lambda_M + a_2\lambda_M^2 \tag{5.10}$$

where λ_M is the Modified Dip latitude (MODIP), a geomagnetic coordinate (Rawer,1963) given by $\lambda_M = \mu_g/\sqrt{(\cos\lambda)}$, where μ_g is the geomagnetic dip latitude and λ is the geodetic latitude. The coefficients a_0, a_1, a_2 vary from day to day and are assumed to be known by the user. In a navigation system, these coefficients are broadcast to the user by the service provider to allow A_z calculation at any wanted location.

iv. *Model characteristics.* Ne-Quick makes use of three anchor points: E-layer peak (at a fixed height of 120 km), F_1 and F_2 peak, where E, F_1 and F_2 are different layers of the ionosphere, as previously introduced. These anchor points are defined by their heights, h_mE, h_mF_1 and h_mF_2 and their respective critical frequencies, f_0E, f_0F_1 and f_0F_2.

Among these anchor points, f_0E is obtained from A_z or R, while the peak layer of E, viz. h_mE, is assumed to be fixed at 120 km. Once the E-layer anchor point is defined, f_0F_1 is derived as a function of f_0E. Finally, the f_0F_2 and $M(3000)F_2$ are separately obtained for a given place and time using spherical harmonics with provided coefficients. Then the peak of the F_2-layer, termed as the h_mF_2, is derived from $M(3000)F_2$. The F_1-layer peak h_mF_1 is finally obtained as the arithmetic mean of h_mF_2 and h_mE. The peak densities are derived from the respective critical frequencies using Eq. (4.24b).

Once the points are defined, the profiling of the ionospheric electron density with height is carried out by dividing it into two major components as given below.

Bottom side. The bottom side of the profile in the model is defined for the heights below the peak of the F_2-layer. The bottom side density function, as described by the model, is constituted by the Epstein functions, one for each of the layers, E, F_1 and F_2. At the given height, each layer contributes a density component, which is an Epstein function of its peak density and the difference between this height and the corresponding peak height. It is also a function of the corresponding layer thickness. However, the amount of contribution of each layer at any particular height is not constant, but varies with time and the individual strength.

Top side. The topside of the profile in the model is defined for the heights above the F_2-layer peak. The topside of Ne-Quick is a semi-Epstein layer with a height-dependent thickness parameter B. A correction factor adjusts vertical TEC values to take into account exosphere electron density in a simple manner. The original model has been proposed for improvement by many researchers (Radicella and Zhang, 1995). The improvements mainly focus on more accurate topside building of the ionosphere. Fig 5.5 shows the TEC derived over the Indian region using the NeQuick Model. Note the prominent manifestation of the anomaly crest around 20° N latitude represented by dark shades.

5.2.1.2 Klobuchar model
The Klobuchar model is the global parametric model for the TEC and being used in the navigational applications for correcting ionospheric delay in system like GPS. Here the total TEC is represented as a summation of half cosine terms with some prespecified quasistatic parameters. This model gives the vertical ionospheric delay δ at any local time t given in seconds of day as

$$\delta(t) = a + b \cos\left\{2\pi\left(\frac{t-c}{d}\right)\right\} \qquad \text{when} \quad 2\pi\left|\frac{t-c}{d}\right| < \pi/2$$
$$= a \qquad \text{when} \quad 2\pi\left|\frac{t-c}{d}\right| \geq \pi/2 \qquad (5.11a)$$

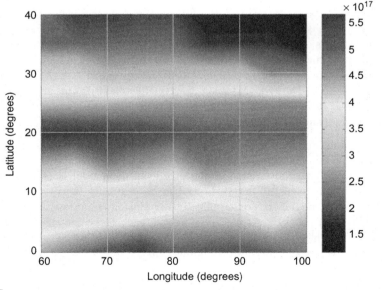

FIG. 5.5

Output of Ne quick model.

The coefficient 'a' is a constant representing the constant bias. The value of 'a' is 5×10^{-9} s and it is the minimum delay offered by the ionosphere over a day. 'c' is the time of peak and its value is taken as 50,400 s. The parameters 'b' and 'd' are location-dependent and are derived using the latitude and longitude of the location and some specific coefficients which varies from day to day. 'b' and 'd' are given respectively as,

$$b = \sum_{0}^{3} \alpha_n \lambda_m{}^n$$
$$d = \sum_{0}^{3} \beta_n \lambda_m{}^n \qquad (5.11b)$$

λ_m is the geomagnetic latitude. α_n and β_n are coefficients which vary from day to day. The delay is for vertical direction which can be converted to slant delay by multiplying it with the obliquity factor given by;

$$F = 1.0 + 16.0(0.53 - E)^3 \qquad (5.11c)$$

Where E is the elevation angle of the slant path and measured in semicircles. This conversion from the vertical to the slant delay occurs at the IPP, which is already defined in Section 5.1.1.2 (Klobuchar, 1987, 1996; Spilker, 1996).

The variation of the vertical TEC with time derived from the Klobuchar model for a given location is shown with all the parameters marked in the Fig. 5.6.

5.2.2 SCINTILLATION MODELS

Scintillation occurs in the satellite signal when it passes through the ionospheric irregularities. These irregularities are the results of complex electrodynamic phenomena with the drivers originating in the ionospheric region or due to the space weather effects. Many theoretical and statistical models for scintillation have been

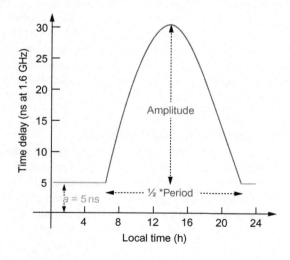

FIG. 5.6

Variation of ionospheric delay with local time by Klobuchar model.

Box 5.2 Klobuchar Estimation

In this box section, we shall run the MATLAB programme klobuchar.m to generate the figure representing the Klobuchar estimation of ionospheric delay for a specific location. Notice the variation of the vertical delay amplitude with time and also the width of the half period of the variation. The amplitude and the width vary from place to place.

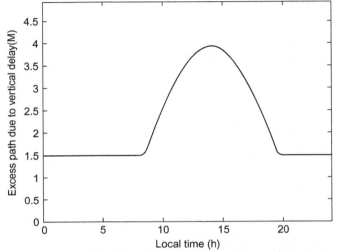

Run the same programme with different locations in terms of latitude and longitude and observe the changes.

---

developed based on recorded GNSS data, scintillation monitors and observations of space weather and atmospheric activity. There are few models currently used in practise. However, the most commonly used theoretical models out of them are the ones using the Phase Screen approach of the ionosphere. We have already discussed in Chapter 4 the fundamentals of the phase screen model. Therefore, we shall describe here only one of the prominent scintillation models, called the wide band model (WBMOD), for ionospheric scintillation

5.2.2.1 Wide band scintillation model (WBMOD)

WBMOD is a comprehensive computer model for ionospheric scintillation developed at the North-West Research Associates (NWRA). It predicts the levels of scintillation activity at a given time and location. Using the satellite and the receiver location parameters, time and date and geophysical conditions and signal frequency, WBMOD estimates the scintillation parameters. These output parameters include, the spectral index of phase scintillation, p, the spectral strength of phase, T, occurrence statistics and the amplitude and phase scintillation indices, S_4 and σ_ϕ, respectively, among others.

WBMOD essentially consists of two parts;

(i) a collection of empirically derived models of the global distribution and characteristics of ionospheric irregularities. These models are used to describe the geometry and orientation of the irregularities including their height, extent and axial ratio. The strength and motion of the irregularities are found as a function of location (latitude, longitude), date, time of day, solar activity level (sunspot number, SSN) and geomagnetic activity level (planetary K-index, K_p).

(ii) a power law phase screen propagation model which allows the strength of scintillation activity to be calculated in a user-defined system.

This power law phase screen propagation model provides the effect of the ionospheric irregularities on the propagating signal. The propagation model takes the description of the electron density irregularities from the ionospheric irregularity model and uses it to calculate the expected scintillation effects on a user-defined system, based on a phase screen theory, discussed previously. The model specifies the expected effects on the signal's phase and amplitude to characterize the scintillation power density spectrum and the total variance due to scintillation in terms of σ_ϕ and S_4.

These parameters, estimated by the model, include the in situ spectral slope, p, the height-integrated irregularity strength, $C_k L$, the in situ drift velocity of the irregularities, v_d, the phase screen height, etc. One of the very important and basic parameters generated by this model is the height-integrated electron density irregularity strength, denoted $C_k L$. The $C_k L$ parameter is actually the product of C_k and L, the power spectral density of the 1 km cross-section of the ionospheric irregularity and the thickness of the irregularity layer, respectively. Combined together, it represents the measure of the total 'power' in the electron density irregularities along a vertical path passing through the entire ionosphere. Of the eight parameters used within the model to characterize the electron density irregularities, $C_k L$ is the most dynamic and affected by the time of year, local time, location, solar cycle and geomagnetic activity. (Secan, 1996; Secan et al., 1995, 1997)

Finally, WBMOD provides either of two types of output

(a) the percentage of time that a specified level of scintillation activity is exceeded,

(b) the level of scintillation activity associated with a given percentile of occurrence.

5.2.2.2 Empirical/statistical models

An empirical model of scintillation captures the characteristics of real scintillation, without necessarily addressing the physical processes that gave rise to those characteristics. Such a model is developed based on analysis of a large volume of empirical scintillation data. We have already discussed in Chapter 4 about the spectral model of scintillation (Eq. 4.39) which is essentially empirical in nature.

Various statistical models have also been designed to represent the ionospheric scintillation. The signal scintillation can be effectively modelled as a complex process of multipath where associated with each path is a random propagation delay and attenuation factor. Therefore, the total scintillating signal may be expressed as,

$$S_{sc}(t) = \sum_n \left(a_n e^{j\delta\varphi_n} \right) \tag{5.12}$$

where α is the amplitude factor for the signal received through n-th path and $\delta\varphi_n$ is the random phase deviation due to propagation delay in that path. The above incoherent summation of the components results in a scintillating signal. It can be typically assumed to have the form $S_{sc}(t) = z + \xi$, as mentioned in Eq. (5.12). Here, z is the direct unperturbed complex component while $\xi(t)$ is the time varying multipath component. ξ is also referred to as the fading component (Humphreys et al., 2009). It is a complex variable and so results in the variation of both amplitude and phase of the total signal. Here, we shall take a simple approach considering the fluctuating component ξ to be small compared to z, i.e., *weak scintillation*. At any instant, ξ may be decomposed into components ξ_i and ξ_q, respectively the in-phase and quadrature component of ξ with respect to z. For small ξ, the in-phase component changes the amplitude while the quadrature component deviates the phase. ξ is a Gaussian, zero mean stationary random process with standard deviation σ_ξ, and each of its components independently behaves so. Therefore, for $z \gg \xi$, the phase variation of the total signal, $\Delta\varphi = \tan^{-1}\{\xi_q/(z + \xi_i)\}$ is also zero mean Gaussian. The amplitude variation, is obtained by vector addition of a Gaussian random variable ξ with constant z, and is Rician. For small ξ, this turns out to be $\Delta a = z + \xi_i$. This evidently produces a Gaussian distribution about the mean z, and is exactly what happens to a Rician distribution when the fluctuating component reduces largely compared to the constant. The autocorrelation function R_ξ of this fading component is represented in terms of its characteristic correlation time, τ_0, as

$$R_\xi(\tau) = R_\xi(0) \exp\left(-\frac{\tau^2}{\tau_0^2}\right) \tag{5.13}$$

where τ_0 is the signal decorrelation time. τ_0 is defined as the time lag in which the auto correlation function of a scintillated signal is reduced by a factor of $1/e$. It also defines the spectral width of the scintillation. Due to zero mean nature of ξ, $R_\xi(0) = \frac{1}{2} E\{\xi^*(t)\cdot\xi(t)\} = \sigma_\xi^2$. Again, for this wide sense stationary process, the auto-correlation and power spectral distribution forms Fourier transform pair, $\sigma_\xi^2 = \int S_\xi(f) \, df$, where $S_\xi(f)$ is the power spectral density of ξ.

Using the empirical data, the statistical model parameters, can be established. However, when the parameters are derived empirically, measurements are to be done using practical receivers. The phase errors, measured at the PLL becomes dependent upon the noise of the signal, i.e.

$$\sigma_\xi^2 = f(C/N_0) \tag{5.14}$$

where C/N_0 is the carrier to noise density ratio. During periods of strong scintillation and canonical fading, due to the nonlinear properties of the PLL the model relies on deriving a statistical model parameter based on empirical observations during periods of varying scintillation intensity. For moderately intense to intense scintillation, the power spectrum looks like a complex valued white noise passing through a low pass filter. Accordingly the original autocorrelation function, $R_\xi(\tau)$, is replaced by a second-order Butterworth filter function represented by:

$$R_\xi(\tau) = \sigma_\xi^2 e^{-\beta|\tau|/\tau_0}\{\cos(\beta\tau/\tau_0) + \sin(\beta\tau/\tau_0)\} \tag{5.15a}$$

where $\beta = 1.2396464$ was selected to satisfy $R_\xi(\tau_0) = \exp(-1)$, satisfying the definition of the decorrelation time τ_0 (Humphreys et al., 2009).

The amplitude coefficient distribution, characterized by the PDF of a Rician fading channel also has a good fit of Nakagami-m distributions to empirical data (Humphreys et al., 2009). Further, σ_ξ^2 is related to K, the Rician K-parameter by:

$$K = \frac{z}{2\sigma_{sc}^2} \tag{5.15b}$$

where K is related to S_4 by:

$$K = \frac{\sqrt{1 - S_4^2}}{1 - \sqrt{1 - S_4^2}} \quad \text{when } S_4 \leq 1 \tag{5.15c}$$

5.3 IMPAIRMENT MITIGATIONS

5.3.1 MITIGATION OF GROUP DELAY/PHASE ADVANCEMENT

The ionospheric effects are typically experienced by comparatively lower frequencies of the satellite signals, for example, L and lower S bands. Since the L band is allotted for the navigation applications, where the propagation time is of importance, the effect severely influences the applications. This makes it necessary to device some mitigation techniques for these ionospheric effects. Therefore, we shall primarily discuss here the mitigation of the ionospheric effects for the navigation signals.

The mitigation techniques, in satellite navigation signals, are mainly accomplished through few standard methods. In these methods, the group delay (or the phase advancement, as applicable), is either estimated or directly measured and then the signal propagation time is corrected accordingly. Apart from the direct derivation of this delay from dual frequency range measurements, the estimation of the delay, for single frequency receivers, can also be done by using models or being aided by correction data provided by a reference receiver or by an augmentation system (Acharya, 2014).

5.3.1.1 Techniques of error mitigation

The range between the receiver and the satellite is measured by using the time delay of arrival of the signals from the satellite at the receiver. The range values thus obtained through measurement cannot be directly used at the receiver for position fixing. This is because the delay in propagation through the ionosphere adds an additional equivalent path to the measured range. So, a wrong solution of user position will be obtained unless all the error components in the measured ranges used for position estimation are removed. Therefore, in navigation receivers, it is required to remove the excess error terms from this pseudorange to get the true solution. Correction of ionospheric delay terms is generally made in two ways as discussed in the following subsections.

Dual frequency measurement-based correction

We have learnt in previous section how the TEC is directly measured using the GNSS measurements. However, such measurements require dual frequency operation of the receivers, which are essentially expensive. In this method, the TEC values at any instant along any definite path are derived from the pseudorange measurements in two different frequencies by using the Eqs (5.6a) through (5.6b). Once the TEC value is known, the true range ρ can be derived by correcting the delay. Therefore, starting from the measured ranges in two frequencies and using Eqs (5.6a), (5.6b), we get,

$$\rho = R_1 - \frac{40.3}{f_1^2} \text{TEC}$$

$$= R_1 - \frac{40.3}{f_1^2} \times \frac{R_2 - R_1}{40.3\left(\frac{1}{f_2^2} - \frac{1}{f_1^2}\right)} \tag{5.16a}$$

$$= R_1 - \left(\frac{R_2 - R_1}{f_1^2/f_2^2 - 1}\right)$$

Calling (f_1^2/f_2^2) as μ_f, we get

$$\rho = R_1 - \frac{R_2 - R_1}{\mu_f - 1} \tag{5.16b}$$

Correction in single frequency receivers

Reference-based Correction. In reference-based correction, there exists a reference station or a network of such station, which does the measurement of the ionospheric delay. These stations are equipped with dual frequency receivers which can readily measure the corresponding delay added ranges in two frequencies along the signal path observed by it. Using them, the reference station can derive the delay values and updates this information as frequently as required.

These updated ionospheric delay values are then disseminated to the users around him. The dissemination is done broadly in two ways:

Disseminating the exact delay: Disseminating the exact delay measured by the reference station along the particular line of sight to the satellite is done typically for standalone reference stations when the reference station and the user receivers are very closely located. In this case, it is considered that the ionospheric spatiotemporal variation is small and hence the same delay values, as observed by the reference station can be used as the ionospheric delay error for the user receivers too, for any particular satellite. Therefore, knowing the delay along a particular signal path, the corresponding delay error can be eliminated. However, the important thing to remember here is that, by virtue of their vicinity, we consider that both the user and the reference station 'sees' the same set of satellites. We know from previous descriptions that the ionospheric delay determined by the reference station is linearly related to TEC and also varies with space and time in a correlated manner. So, a more generalized approach is to express the delay at the user, δ_u, at any particular time instant in terms of the reference delay, δ_r, using Taylor's Series expansion, as

$$\delta_u = \delta_r + \frac{d}{dx}\delta_r|_{X_r}dl + \frac{1}{2}\frac{d^2}{dX^2}\delta_r|_{X_r}dl^2 + \ldots \tag{5.17}$$

where dl is the distance between the reference and the user. For correlated values, the derivatives are deterministic in nature. Besides space decorrelation, the time latency between the estimation instant and the instant of applying it plays an important role in determining the accuracy of the ionospheric corrections. Thus, knowing the derivatives in both time and space, in addition to the absolute delay, we can derive the range errors due to the ionosphere at the user receiver. This enables the user to correct his measured range utilizing those at the reference receiver. The accuracy of the correction, however, depends upon the accuracy of the measured delays and the derivatives estimated. Nevertheless, this is still only valid for the paths for the common satellites viewed by the reference and the user.

Disseminating the total ionospheric model: Reference-based relative ionospheric error corrections as described above are best when the total area required to be served is small. The limit of the spatial extent that a single reference station can serve is based upon the distance for which the ionospheric correlation exists. Similarly, the temporal correlation of these errors determines the latency of using the corrections of these errors. For larger extent of service area, the above methods start showing up large errors (Kee, 1996) and thus the position estimates get worse with increasing distance between the reference station and the user. To alleviate the problems faced, instead of providing the delays along the individual discrete paths, the total ionospheric delay model is first recreated from the measured delays. Then the information regarding this model is disseminated to the users. This is done, either by sending out the coefficients representing the models or by providing the exact delays at the regularly placed predefined grid points. The dissemination of these information is done through a separate source accessible to all users instead of the reference station directly doing so. So, wherever the user receiver is, it can estimate the vertical delay values along the path of its own line of sight using this information. This approach totally liberates the user from adhering to the same satellite as that of the reference station and also from the constraint of remaining in vicinity of the latter. However, this is at the cost of additional computational and modelling activities at the reference side.

Model-based correction. Single frequency receivers can also correct ionospheric errors using the models of ionosphere. The relevant models using which the estimations of the ionospheric group delay or phase advancement can also be made are discussed in Section 5.2.1

In such receivers, the corrections are obtained by using parametric models, which provides either the ionospheric delay or the TEC for any location and time, using predetermined coefficients. The two important models are the Klobuchar model and the NeQuick model. In Klobuchar model (Klobuchar, 1987), the vertical ionospheric delay is obtained as a function of the local time and geographic location of the receiver. While in NeQuick model, the electron density is obtained from similar inputs which may be integrated along the required path to get the TEC. Since the TEC and the Group delays are linearly related through some constant terms, these models are popularly used for correcting signals in existing satellite navigation receivers.

However, it does not need any special mention that the dual frequency receivers estimate the ionospheric corrections better than the single frequency receivers, irrespective of whether the latter is using the reference-based corrections or model-based estimations for the ionospheric delay.

5.3.2 MITIGATION OF SCINTILLATION

Scintillations are caused by ionospheric irregularities and affect the amplitude, phase and hence the received power and related parameters of the affected signals. Since the effect shows inverse variation with the frequency, relatively lower portions of the band used in satellite links like the L band is mostly affected. In addition to communication, navigation applications are also affected by scintillation, particularly at near equatorial and high latitudes. It results in either loss of lock or positioning errors leading to unavailability of the service. Thus, it is important to mitigate this effect.

Scintillation has adverse effect on VHF/UHF satellite systems and also on satellite navigation systems. The fundamental receiver activity of signal acquisition and tracking is affected due to the introduction of the jitter in receiver PLL's due to the scintillation. It can result in considerable variance in the phase discriminator and, for sufficiently strong scintillation conditions, may drive the PLL beyond its bandwidth leading to loss of carrier phase lock. This eventually results in the degradation of the performance in terms of availability and continuity. The accuracy of the navigation systems are also affected. In such receivers, mitigation can be implemented by considering certain factors during the designing of the system or via receiver hardware modifications. The latter can be implemented by making the phase tracking more robust against scintillation, or by means of intelligent software/firmware.

5.3.2.1 System-based mitigation approach

There are several actions that can be taken at the time of system designing to lessen the impact of scintillation. These are more avoidance type than corrective. Since the occurrence of scintillation is specific to the region through which the signal is passing, increasing the satellite numbers will create an orbit diversity and consequently increase the probability of more signals remaining unaffected by it. Since scintillation is location-specific and regionalized in nature, it affects only few satellites whose signals come through this path. Thus, more satellites a user receiver locks to, before the onset of scintillations, more options it has at its disposal during scintillation and hence it is more likely the receiver will retain performance during a scintillation event. Therefore, incorporating as many satellites as possible to increase the numbers of unaffected satellites visible to the receiver, is an effective means of mitigation. However, in many cases, orbit diversity is not preferred due to its prohibitive cost.

A more practical solution is to exploit the options of space diversity on ground. Therefore, putting ground receiving system spaced beyond the correlation distance of the scintillating fades can be combined to effectively mitigate the scintillating effects. Space diversity is not an option for a navigation system between the space and user segment, although it can be effective for communicating between the space and control segment. Some systems use largely separated second frequency of operation, say in L and S band for the same services. In such cases, frequency diversity acts as an effective technique to combat ionospheric scintillation fading as S band suffers much lesser scintillation than L band.

For satellite mobile communication services utilizing L band or lower S band, time diversity techniques can be used which can take several different forms. But they are all based on the fundamental notion that a fading channel will have positive link margin at least some of the time over the scintillation period. For those times of positive margins, it is possible to transmit information with a low probability of error. Coding, with or without interleaving, can also be considered as a form of mitigating the scintillation effects.

5.3.2.2 Receiver-based mitigation approach

There are two general approaches for mitigation of the scintillation effects at the receiver. It is either accomplished by making the receiver PLL more robust to these errors or by using software algorithms to mitigate the effect.

Modification of the tracking Loop

To obtain the best carrier phase measurement and processing, the receiver should employ the pure PLL loop for carrier tracking. The bandwidth of this loop should be small in order to minimize noise. However, small loop bandwidth cannot accommodate the variation in the phase caused due to the scintillation and consequently lose lock. So, this drawback of this architecture, i.e. lack of robustness in the presence of ionospheric scintillation, can be removed by widening the PLL loop bandwidth. However, strong scintillations are associated with amplitude fades. So, increasing the bandwidth will allow these amplitude fluctuations to contribute further reducing of the lower SNR inherent to the wideband PLL. Improvement in the receiver performance can be also achieved by using properly designed FLL-assisted PLL. Although this will improve receiver robustness, it will be at the cost of introducing additional noise into the phase measurement.

To obtain a better solution to this problem, an adaptive carrier tracking loop may be used. The loop normally operates using only the PLL portion with a tight bandwidth. When phase and amplitude scintillation are detected, loop bandwidth is adaptively increased to accommodate new signal conditions. When a deep amplitude fade occurs, there is no signal to track, which causes PLL alone to lose lock. At this point, the FLL portion of the loop is activated in order to facilitate recovering the signal after it re-emerges from the fade (Strangeways and Tiwari, 2013). Another option to combat the scintillation effects is by using Kalman filter based phase tracking.

In order to understand, control, and mitigate scintillation effects, it is necessary to have access to tracking loops and associated components such as local oscillator, discriminator and phase counter. However, excepting a few commercial receivers, access to the internal workings is not provided to users, in general (Ganguly et al., 2004).

Using software-based algorithms at processing
The aforementioned mitigation method has been found to be very successful in reducing errors during ionospheric scintillation conditions. However, the receivers with a software-based approach are also suitable for ionospheric monitoring as well as for reliable operation during scintillating conditions.

Software-based techniques can be implemented in navigation receivers by leaving out the satellites in the positioning calculation whose paths to the receiver have been severely affected by scintillation. However, it reduces the useable satellite range measurements for positioning. The same can be achieved if a communication system has satellites in orbit diversity.

In navigation receivers, software-based mitigation technique is also implemented by weighing all the satellite to receiver paths inversely according to the scintillation present on each. Scintillation is alleviated by attaching a preference to the signal paths with more weights in all the subsequent calculations. This is fairly effective as it allows the receiver to have more options while selecting satellites for positioning. This method has been found to be successful when scintillation indices and spectral parameters for the received signals for all the satellite paths are available at each epoch. But this requires extra computations, if not additional hardware, to measure the scintillation at each instant of time.

5.4 APPLICATIONS

Theory of electromagnetic wave propagation through ionosphere is now mature enough with thorough understanding and insight. By virtue of this understanding, ionospheric impairments like the group delay (or equivalently the phase advancement) and the scintillation can be used intelligently as a powerful tool for probing the characteristics of ionosphere. These parameters are also useful for studying ionospheric response to various stimulus. Below, we briefly mention a few applications of using the ionospheric impairments beneficially for examining the attributes and analysing other important facets.

One of the important applications of the measured ionospheric TEC is using them to find the electron density through tomography. Tomography is the technique by which the spatial distribution of a given physical quantity is estimated from its path integral by dividing the total space into smaller sections. In short, it is the method of imaging a physical quantity by sectioning.

The ionospheric delay is a direct surrogate for the measure of the TEC along the signal path. These TEC values along the discrete paths may be used to estimate the 3D electron density distribution over the total space using the techniques of Ionospheric Tomography.

There are two broad genres of tomography being implemented for the electron density estimation. These two types are mainly distinguished by the approach of sectioning the total space, where the electron density is required to be estimated. The types are as described below.

Ray tomography. In this kind, the total ionospheric space is divided into evenly distributed physical volume elements in a 3-dimensional space, called voxels. These voxels are assumed to have a constant electron density and hence acts as the units of the ionospheric electron density content, which are required to be determined.

It is considered that the measured TEC values are along such discrete paths that almost all the voxels are ray-traced. That is, each voxel has at least one signal path passing through it. Then the density of each voxel is solved to get the electron density distribution such that the density of the respective voxels along the signal path adds up to give the measured TEC. However, in ray tomography, because of the limited view angle, resolution in the vertical direction is poor (Sutton and Na, 1995).

Modal decomposition tomography. In this algorithm, the ionospheric density distribution is reconstructed as the weighted sum of a set of orthonormal basis. Each basis is represented by a suitable known function which spans over the whole space of interest. Therefore, here the functions are known but only their respective weights are required to be determined. This is done by equating the sum of the path-integrated values of the component functions to the measured TEC and finding out the respective weight coefficients for each. Once the weights are determined and the basis functions being known, the distribution of the electron density can be estimated (Hansen, 2002; Acharya, 2004). The resolution performance of the orthogonal decomposition algorithm depends upon the geometry used for the reconstruction (Sutton and Na, 1995).

The basic schematic of tomographic estimation using ray tomography is given in Fig. 5.7. The tomographic reconstructions, in general, are dependent, not only on geometries, but also on algorithms, measurement accuracies, noise, bias, and multipath, among other things. Relative roles of different factors are not clearly

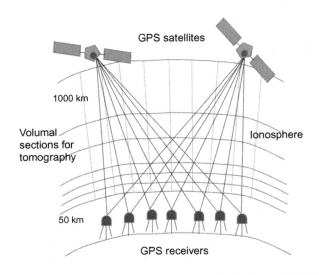

FIG. 5.7

Schematic for tomographic estimation.

understood (Brown and Ganguly, 2001). The ionospheric tomographic estimations have its applications in observing the ionospheric electron density variations, especially to the scientific community. The dynamics of the ionospheric plasma in the equatorial and polar region can be studied from the derived electron density variations. The spatiotemporal density variation can be related to the electrodynamic forces to study the ionospheric response to various such forcing factors and forces acting upon it and prompted by geophysical and other agencies. For example, in the equatorial locations, the electron density shows a sudden enhancement in its values at certain heights just near the sunset and is known as the Prereversal enhancement (PRE). This can be observed through the tomographically derived values of densities and correlated to the pertinent factors. The derived distributions are also used for observing the equatorial fountain and plasma transport which are characteristics of equatorial ionosphere.

The tomographic estimates, generally obtained using GNSS measurements, however, also have uses for GNSS systems. The knowledge of complete spatial electron density distributions, derived from few line of sight TEC values, can be used to obtain the TEC along any arbitrary direction over the entire volume encompassed by the process. It can also be used for estimating the Grid Ionospheric Vertical Delays (GIVD) for the augmentation systems (Acharya, 2014) and also for correcting the ionospheric delays in applications like altimetry, radar ranging, etc. The accurate tomographic ionospheric model can be used to solve in real-time the ambiguities in a phase-based range estimations, L1 and L2, for the reference stations and for the rover, as well (Hernandez et al., 2000) in real time kinematics (RTK).

Other than the tomographic estimations and their utilizations, the TEC values in its integrated form can also be used for deriving the ionospheric electrodynamics and for related parameter estimations. The TEC measured in the equatorial region have been efficiently used for estimation of the equatorial electrojet (Acharya et al., 2011). Ionosphere is also sensitive to the variations in the interplanetary magnetic fields (IMF) incident upon the earth's magnetosphere. The disturbances and IMF effects can be traced out using the measured ionospheric variations. Scintillations measured in GNSS signals and obtained over a large extent, using scintillation measuring Network, can be successfully used for the short-term forecast of scintillation and related ionospheric conditions. Systems like the communication navigation outage forecasting system (C/NOFS) will measure the characteristics of ionospheric irregularities from similar measurements. Apart from this, it will also attempt to forecast scintillation by using in situ measurements of plasma instability drivers, models and ionospheric irregularities.

CONCEPTUAL QUESTIONS

1. While estimating ionospheric TEC with dual frequency range measuring GNSS receiver, would you prefer to choose the two frequencies close to each other or wide apart?
2. Is the sampling rate during a scintillation measurement important? If so, why?
3. Between the polar and the mid-latitude regions, where will the ionospheric models are expected to perform better? List out the ionospheric conditions when the reference-based differential corrections of delay will fail to perform satisfactorily.
4. For the same latitude value and at the same instant, does the Klobuchar model give the same delay value for the Northern and the Southern hemispheres?
5. On which factors do the choice of the basis functions in a modal decomposition tomography depends?

REFERENCES

Acharya, 2004. Tomographic estimation of ionosphere over Indian region. In: Proceedings of the 12th International Conference on Advanced Computing and Communications—ADCOM, Ahmedabad, India, pp. 564–567.

Acharya, R., 2014. Understanding Satellite Navigation. Academic Press, Elsevier, Waltham, USA.

Acharya, R., Roy, B., Sivaraman, M.R., Dasgupta, A., 2011. Estimation of equatorial electrojet from total electron content at geomagnetic equator using Kalman filter. Adv. Space Res. 47 (6), 938–944.

AIAA, 1998. Guide to Reference and Standard Ionospheric Models. ANSI/AIAA, Washington, DC. G-034.

Basu, S., 2003. Ionospheric scintillation: a tutorial, space vehicle division. In: Air force research laboratory, CEDAR Workshop, June.

Beach, T.L., 2006. Perils of the GPS phase scintillation index. Radio Sci. 41, http://dx.doi.org/10.1029/2005RS003356. RS5S31.

Beach, T.L., Kintner, P.M., 1999. Simultaneous global positioning system observations of equatorial scintillations and total electron content fluctuations. J. Geophys. Res. 104 (22), 553.

Beach, T.L., Pedersen, T.R., Starks, M.J., Su, S.Y., 2004. Estimating the amplitude scintillation index from sparsely sampled phase screen data. Radio Sci. 39, http://dx.doi.org/10.1029/2002RS002792. RS5001.

Brown, A., Ganguly, S., 2001. Ionospheric tomography: issues, sensitivity and uniqueness. Radio Sci. 36 (4), 745–755.

Chapman, J.H., Warren, E.S., 1968. Topside sounding of the earth's ionosphere. Space Sci. Rev. 8, 846–865.

Curran, J.T., Bavaro, M., Fortuny, J., Morrison, A., 2014. Developing an Ionospheric Scintillation Monitoring Receiver, Inside GNSS, September–October 2014.

Di Giovanni, G., Radicella, S.M., 1990. An analytical model of the electron density profile in the ionosphere. Adv. Space Res. 10 (11), 27–30.

European Union, 2015, Ionospheric Correction Algorithm for Galileo Single Frequency Users, Issue 1.1, June 2015.

Forte, B., Radicella, S.M., 2002. Problems in data treatment for ionospheric scintillation measurements. Radio Sci. 37 (6), 1096. http://dx.doi.org/10.1029/2001RS002508.

Ganguly, S., Aleksandar, J., Brown, A., Kirchner, M., Zigic, S., Beach, T., Groves, K.M., 2004. Ionospheric scintillation monitoring and mitigation using a software GPS receiver. Radio Sci. 39, http://dx.doi.org/10.1029/2002RS002812. (RS1S21).

Hansen, A., 2002. Tomographic Estimation of Ionosphere Using Terrestrial GPS Sensors, Ph. D. dissertation at Stanford University, Available from was.stanford.edu/papers/Thesis/AndrewHansenThesis02.pdf.

Hargreaves, 1979. The Upper Atmosphere and Solar Terrestrial Relations. Van Nostrand Reinhold Co. Ltd, New York, USA.

Heise, S., Jakowski, N., Wehrenpfennig, A., Reigber, Ch., Luhr, H., 2002. Sounding of the topside ionosphere/plasmasphere based on GPS measurements from CHAMP: initial results. Geophys. Res. Let 29 (14), 44-1–44-4. http://dx.doi.org/10.1029/2002GL014738.

Hernandez, M.P., Juan, J.M., Sanz, J., 2000. Application of ionospheric tomography to real-time GPS carrier-phase ambiguities resolution, at scales of 400-1000 km and with high geomagnetic activity. Geophys. Res. Lett. 27 (13), 2009–2012.

Humphreys, T.E., Psiaki, M.L., Hinks, J.C., O'Hanlon, B., Kintner Jr., P.M., 2009. Simulating ionosphere-induced scintillation for testing GPS receiver phase tracking loops. IEEE J. Sel. Top. Signal Process. 3 (4), 707–715.

Humphreys, T.E., Psiaki, M.L., Kintner Jr., P.M., 2010. Modelling the effects of ionospheric scintillation on GPS carrier phase tracking. IEEE Trans. Aerosp. Electron. Syst. 46 (4), 1624–1637.

Incoherent Scatter Radar, n.d. www.haystack.mit.edu/atm/mho/instruments/isr/isTutorial.html Accessed 21 October 2016.

Ionograms, n.d., NOAA, http://www.ngdc.noaa.gov/stp/iono/ionogram.html Accessed 12 October 2016.

ITU-R, 2016. Ionospheric Propagation Data and Prediction Methods Required for the Design of Satellite Services and Systems, Rec. ITU-R P. 531–13.

Kee, C., 1996. Wide area differential GPS. In: Parkinson, B.W., Spilker Jr., J.J. (Eds.), Global Positioning Systems: Theory and Applications, vol. 1. AIAA, Washington, DC, USA.

Klobuchar, J.A., 1987. Ionospheric time delay algorithm for single frequency GPS users. IEEE Trans. Aerosp. Electron. Syst. 23 (3), 325–331.

Klobuchar, J.A., 1996. Ionospheric effects on GPS. In: Parkinson, B.W., Spilker Jr., J.J. (Eds.), Global Positioning Systems: Theory and Applications, vol. 1. AIAA, Washington, DC, USA.

Pi, X., Mannucci, A.J., Lindqwister, U.J., Ho, C.M., 1997. Monitoring of global ionospheric irregularities using the worldwide GPS network. Geophys. Res. Lett. 24, 2283.

Radicella, S.M., Leitinger, R., 2001. The evolution of the DGR approach to model electron density profiles. Adv. Space Res. 27 (1), 35–40.

Radicella, S.M., Zhang, M.L., 1995. The improved DGR analytical model of electron density height profile and total electron content in the ionosphere. Ann. Geophys. 38 (1), 35–41.

Rawer, K., 1963. Meteorological and astronomical influences on radio wave propagation. In: Landmark, B. (Ed.), Propagation of Decameter Waves. Academic Press, New York. p. 221 (Chapter 11).

RTCA, 1999. Minimum Operational Performance Standards for GPS/WAAS Airborne Equipment, RTCA/DO 229 B. .

Secan, J.A., 1996. The WBMOD Ionospheric Scintillation Model, http://www.nwra.com/ionoscint/wbmod.html, NorthWest Research Associates.

Secan, J.A., Bussey, R.M., Fremouw, E.J., Basu, Sa, 1995. An improved model of equatorial scintillation. Radio Sci. 30, 607–617.

Secan, J.A., Bussey, R.M., Fremouw, E.J., 1997. High-latitude upgrade to the wideband ionospheric scintillation model. Radio Sci. 32 (4), 1567–1574.

Spilker Jr., J.J., 1996. GPS navigation data. In: Parkinson, B.W., Spilker Jr., J.J. (Eds.), Global Positioning Systems: Theory and Applications, Vol. 1. AIAA, Washington, DC, USA.

Strangeways, H.J., Tiwari, R., 2013. Prediction and mitigation of ionospheric scintillation and tracking jitter for GNSS positioning. In: 55th International Symposium ELMAR-2013, Croatia.

Sutton, E., Na, H., 1995. Comparison of geometries for ionospheric tomography. Radio Sci. 30 (1), 115–125.

Van Dierendonck, A.J., Hua, Q., 2001. Measuring ionospheric scintillation effects from GPS signals. In: Proceedings of the 57th Annual Meeting of the Institute of Navigation (2001), Albuquerque, NM, pp. 391–396.

Wernik, A.W., Alfonsi, L., Materassi, M., 2004. Scintillation and its effects on systems. Acta Geophys. Pol. 52 (2), 237–249.

FURTHER READING

ITU-R, 1999. Choices of Indices for Long-Term Ionospheric Predictions, Rec. ITU-R P.371-8, 1999.

Tropospheric propagation and impairments

6

We already had the preliminary idea about the atmospheric structure from the overview in the opening chapter of this book. It described the vertical temperature profile of the atmosphere and classified it into different regions on its basis. The troposphere was the lowest of these regions. In this chapter, our focus is on the propagation of the satellite signals through the troposphere and on the impairments tendered by this region. However, in order to understand the origin and characteristics of these impairments, it is required to explain the basic composition of this region. Therefore, first we shall briefly refer to the relevant components and then deal with the associated impairments.

6.1 TROPOSPHERE—CONSTITUTION AND CHARACTERISTICS

Troposphere is the lowest layer of the atmosphere surrounding the earth surface. It extends up to 10–12 km on average above the earth surface (NASA, 2016). The vertical height of the troposphere varies spatially and depends upon the latitude. The height is maximum over the equator and minimum at the poles and shows variation with season. Troposphere is in contact with the earth surface with negative temperature gradient with height and the temperature reduces at about 6.5 K/km. This layer is flanked at the top by the region where the inversion of the temperature gradient takes place. This inversion from negative to positive gradient with height occurs as a result of the presence of the ozone in the layers above. Ozone absorbs the UV radiation from the sun and thus results in a rise in temperature at these heights.

Troposphere has the maximum contribution to the atmospheric mass and has the highest mass density amongst the other atmospheric layers. However, inside the troposphere due to the negative gradient of the temperature profile, the layers nearer to the earth surface get quickly heated up causing rarefaction in density. Mass from this region hence rises up causing stir in the tropospheric layers. This layer is named as Troposphere, where 'Tropos' means change and hence the name of this layer justifies the conditions of this layer, which are constantly changing. The permanent constituents of the troposphere are nitrogen, oxygen, argon and carbon dioxide. Their relative concentration remains practically unchanged in this layer (Mitra, 1972). The different percentages by volume of these constituents are as given in Table 6.1.

Satellite Signal Propagation, Impairments and Mitigation. http://dx.doi.org/10.1016/B978-0-12-809732-8.00006-5

Table 6.1 Constituents of troposphere (NASA, 2016)

Main constituent	% by volume
Nitrogen	78
Oxygen	21
Argon, water vapour and carbon dioxide and trace gases	1

6.1.1 STANDARD REFERENCE ATMOSPHERE

In order to cite the atmosphere for the propagation purpose, it requires to define a standard atmosphere which may be used as a reference. With respect to this reference, all the variations of the atmosphere may be expressed. We shall refer to the standard atmosphere defined by the International Telecommunication Union (ITU) in terms of temperature, pressure and water-vapour pressure as a function of altitude. The following tropospheric variables, thus, define the reference standard atmosphere, which are the representations of the annual mean profiles when averaged across the globe (ITU-R, 2012).

6.1.1.1 Temperature

For the standard atmosphere, the variation of temperature along the vertical height is given by a piecewise linear relationship. That is, in this model, the vertical temperature variation can be divided into seven distinct layers. The temperature within each layer is a function, varying linearly from the base of the layer up to the base of the next layer. Therefore, the temperature $T(h)$ at height h is given by

$$T(h) = T_i + L_i(h - H_i) \text{ K} \tag{6.1}$$

where H_i is the height of the base of the ith layer and $T_i = T(H_i)$ is the temperature at the base. L_i is the temperature gradient for that layer. The values for H_i and L_i are given in Table 6.2 (ITU-R, 2012). Here, $i = 0$ and $i = 1$ represent the temperature variations in the troposphere and the tropopause, respectively (Fig. 6.1).

Table 6.2 Standard atmospheric temperature profile.

Layer (i)	Base height, H_i (km)	Temperature gradient, L_i (K/km)
0	0	−6.5
1	11	0.0
2	20	1.0
3	32	2.8
4	47	0.0
5	51	−2.8
6	71	−2.0
7	85	

Adapted from ITU-R, 2012. Recommendation ITU-R P.835-5, Reference Standard Atmospheres. ITU.

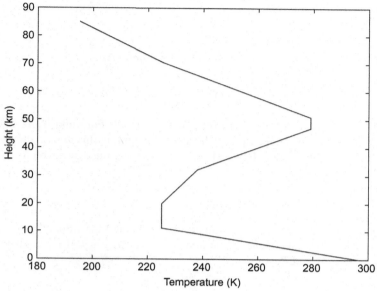

FIG. 6.1

Standard atmospheric temperature profile.

6.1.1.2 Pressure

The atmospheric pressure can be estimated using the barometric equation which can be easily derived from the ideal gas equations. The ideal gas equation can be written as

$$p = \frac{\rho}{M} RT \qquad (6.2)$$

where p is the pressure of the gas, ρ is the density, M is the molar weight, R is the gas constant and T is the absolute temperature of the gas. Again, considering the equilibrium condition, we get

$$dp = -\rho g \, dh \qquad (6.3)$$

It is important to note that we have assumed a decreasing pressure with increasing height in Eq. (6.3). Now, dividing Eq. (6.3) by Eq. (6.2), we get

$$dp/p = -\frac{Mg}{RT} dh \qquad (6.4)$$

But, temperature T is a function of height h. So, writing $T(h) = T_i + L(h - H_i)$, where the parameters are as defined in Eq. (6.1), we get

$$dT = L \, dh$$
$$\text{or,} \quad dT/L = dh \qquad (6.5)$$

Thus, we can write Eq. (6.4) as

$$dp/p = -(Mg/RL)dT/T \qquad (6.6a)$$

$$\text{or,} \quad \ln(p/p_i) = -(Mg/RL) \ln(T/T_i) \qquad (6.6b)$$

Eq. (6.6b) can be written as

$$p/p_i = (T_i/T)^{(Mg/RL)} \qquad (6.7)$$

where p_i is the pressure at the reference base height $h = H_i$. For troposphere, L is negative. Therefore, as the temperature reduces with height, in this region, the pressure reduction occurs monotonically with it. Again, replacing the expression for T_i using Eq. (6.1), we get

$$p = p_i \left[\frac{T_i}{T_i + L_i(h - H_i)} \right]^{\frac{Mg}{RL_i}} \qquad (6.8a)$$

The molar weight M of air is approximately 28.954 g and the value of gas constant $R = 8.314 \text{ J K}^{-1} \text{ M}^{-1}$, while the acceleration due to gravity is 9.81 m/s^2. Then, using the values of g, M and R, we get

$$p(h) = p_i \left[\frac{T_i}{T_i + L_i(h - H_i)} \right]^{\frac{34.163}{L_i}} \text{hPa} \qquad (6.8b)$$

However, this expression is valid only when the temperature gradient L is nontrivial. When $L \to 0$, using L'Hospital rule on the logarithm of the above equation, the expression for the pressure p at height h may be expressed as (ITU-R, 2012)

$$p(h) = p_i \exp\left[-34.163 \left(\frac{h - H_i}{T_i} \right) \right] \text{hPa} \qquad (6.8c)$$

The ground-level standard temperature and pressure are taken as, respectively, $T_0 = 288.15$ K and $p_0 = 1013.25$ hPa. Using these values, the variation of pressure at other tropospheric heights may be derived. However, it should be remembered that atmosphere containing water does not behave as an ideal gas, and hence, these equations only give approximate values and not the exact results. At About 85 km altitude, local thermodynamic equilibrium of the atmosphere starts to break down and hence the above equations are no longer valid (ITU-R, 2012). The vertical profile of the atmospheric pressure for the standard atmosphere is shown in Fig. 6.2.

6.1.1.3 Water-vapour pressure

Water vapour is a condensable gas found in the troposphere. It can also convert from the gaseous state to ice through sublimation (Visconti, 2001). Although the distribution of water vapour in the atmosphere is generally highly variable, the vertical

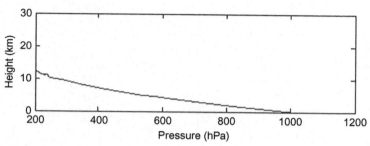

FIG. 6.2

Vertical profile of atmospheric pressure for standard atmosphere.

profile approximately follows the exponential variation. An approximate expression for the same is given by the equation:

$$\rho(h) = \rho_0 \exp(-h/h_0) \text{ g/m}^3 \tag{6.9}$$

where the scale height $h_0 = 2$ km, and the standard ground-level water-vapour density is taken as $\rho_0 = 7.5$ g/m^3. To obtain the water-vapour pressure profile in terms of its density, we use the gas equation $p = (\rho/M) RT$ and replace the relation in the above equation to get

$$\begin{aligned} e(h) &= \rho(h)\frac{RT}{M} \\ &= \rho_0 \exp(-h/h_0)\frac{RT}{M} \end{aligned} \tag{6.10}$$

where $e(h)$ is the height profile of the water-vapour pressure. Now, R/M is the gas constant per unit molar weight and is called the specific gas constant, R_{specific}. Thus, the gas equation gets converted to

$$e(h) = \rho R_{\text{specific}} T \tag{6.11a}$$

The value of R is 8.314 J/mol/K in MKS, making the R_{specific} for water vapour as 1/2.167. This gives the expression for pressure in Newton/m^2, i.e. in Pascal. Therefore, the pressure in hPa can be obtained by dividing it by 100 and we get (Fig. 6.3),

$$e(h) = \frac{\rho(h)T(h)}{216.7} \text{ hPa} \tag{6.11b}$$

6.2 PROPAGATION IN TROPOSPHERE

What makes the troposphere different from free space for an EM wave? We have seen in Chapter 3 that for a signal of any given frequency, the refractive index (*RI*) depends upon the permittivity of the medium. Depending upon the nature of the permittivity, the *RI* is changed accordingly. The permittivity can be real for

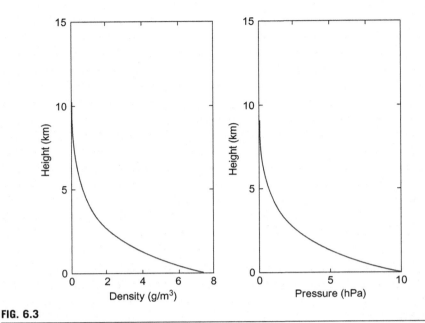

FIG. 6.3

Vertical profile of water-vapour density and pressure for standard atmosphere.

the troposphere for a given frequency, while it may also be complex for other frequencies. It will also vary from location to location and with height. It may also vary under different atmospheric conditions. When the medium has permittivity which is real and is different from that of the vacuum (which it always is), the refractive index is also real. It results in the change in the phase velocity of the wave in the medium compared to that of the vacuum. This situation is observed when the signal passes through dry air and gases, in general. If the permittivity and hence the *RI* is a function of frequency, it leads to a group velocity, different from the phase velocity. Consequently, there is a group delay experienced by the signal. When the *RI* is complex, it has a real part and an imaginary part. The real part leads to the change in phase velocity as discussed above. The imaginary part leads to the decay in the signal amplitude on propagation. The signal amplitude decays exponentially leading to attenuation of the wave. When the imaginary part is large, the wave amplitude decays down heavily within a short distance from its entry to it.

The permittivity and hence the *RI* is typically isotropic. That is, its value remains the same irrespective of its direction. However, it can also be anisotropic, as in certain structures, the electrical features are different across different directions. In such cases, the attenuation or the phase change of a traversing wave remains different for different polarization of incidence of the wave.

In this chapter, from the following section, we shall have a treatise on the factors affecting the propagation of the radio waves through the troposphere. We shall discuss the individual elements which affect the propagation and their characteristics. The fundamental physics behind the effects, which we have mostly covered in previous chapters, will be either referred or reiterated, wherever necessary. In the next chapter, we shall learn how to measure the related impairments arising out of these factors and the methods of alleviating them for the satellite signal.

6.3 FACTORS AFFECTING TROPOSPHERIC PROPAGATION

In this section, we shall first identify the different conditions of the troposphere, and for each of them, associate the factors affecting the propagation and will discuss their characteristics. We shall start with the clear sky condition and then shall proceed for conditions dominated by clouds and rains.

6.3.1 CLEAR SKY

The formal definition of clear sky is when the sky is free of clouds and other obstacles from the point of observation. It is to be understood that the propagation through clear sky is characteristically different from the free space propagation. The reason is now obvious to us as we have learnt in Chapter 3 that the material medium which the troposphere is comprised of, due to their permittivity, will have different characteristics in offering to the wave, which is distinctly different from vacuum. Clear sky also includes water vapour. It is one of the prime constituents of the troposphere which alters the propagation characteristics under the clear sky condition. It also largely varies in content both temporally and spatially, thus proffering large variability to the propagation of the wave. The medium changes the effective RI and hence the effective path length experienced by the traversing wave. The excess path Δl, thus, extended by the medium is

$$\Delta l = (n-1)\delta l \tag{6.12}$$

where n is the effective refractive index of the medium and δl is path of the signal through it. This function being a function of the effective RI 'n' and also the geometric path through it, it is dependent upon the extent of gases, primarily upon the water vapour content in air. The total water vapour content along a path can be thus used for calculating the excess path length. The water vapour is also responsible for absorptive attenuation. This happens when the attenuation is conspicuous for frequencies near the absorptive resonance frequencies. The amount of this attenuation is related to the total water vapour content through its specific mass absorptive coefficient. We shall learn about this in details in our next chapter.

The water vapour content is defined as the mass of the water vapour column over an unit area and is expressed in kg/m^2. It is also equivalently represented in mm of precipitable water. The spatial variation of the water-vapour density ρ, or its temporal profile, can be obtained using different techniques, the most popular being the use of the instrument Radiosonde, which is also discussed in the next chapter. However, other techniques like sounding rockets, sounding satellites, radio occultation and GNSS-based measurements are also being used. These values vary with height and over seasons and locations. The vertical variation of water vapour can be obtained by using Eq. (6.9). The ITU recommendation P-836 (ITU-R, 2012) also provides a method for finding the approximate average global water vapour content when the local measurements are not available.

6.3.1.1 Key parameters

The propagation of radio waves through the troposphere during the clear sky condition is mostly influenced by the change in refractive index as the latter determines the change in phase and amplitude of the wave during the propagation. Refractive index is the most important parameters in the study of tropospheric propagation.

The dry air has relatively constant *RI*. Only a considerable change in temperature and pressure can induce some amount of variations in it. It is the water vapour present in the air which has more contribution to the variations of the tropospheric *RI*. Yet, the variation term of the tropospheric refractive index is small to the extent that it is difficult to handle the variations of such small quantity in all mathematical calculations. So, we define a term called the refractivity. We have seen before that the RI, '*n*' of a medium compared to that of vacuum, i.e. ($n - 1$) is an important metric in all calculations. Therefore, we define *Refractivity* as the scaled and adjusted value of the refractive index with respect to unity. Thus, refractivity can be expressed as

$$N = (n - 1) \times 10^6 \tag{6.13}$$

where '*N*' is the refractivity and '*n*' is the true refractive index of the medium. Since the absolute refractive index '*n*' is a function of pressure and temperature, so does the term *N*. *N* can be considered to be composed of two individual components, the dry component N_{dry} due to the dry air and the wet component N_{wet} due to the water vapour. So, $N = N_{dry} + N_{wet}$. The variation can be modelled in terms of *P* and *T* as (Allnutt, 2011):

$$N = 77.6 \frac{p}{T} + 375,000 \frac{e}{T^2} - 5.6 \frac{e}{T} \tag{6.14}$$

Here *p* is the pressure in hPa, *T* is the absolute temperature in Kelvin and e is the water-vapour pressure, also in hPa. The first term in the equation is due to the dry component of the air, while the rest of the terms, containing the term '*e*', are due to the wet component. This equation is expected to be correct for the values of *p* between 200 and 1100 hPa, for the values of *T* between 240 and 310 K and for $e < 30$ hPa. The value of water-vapour pressure '*e*' may be substituted by a more convenient form in terms of temperature *T* and density ρ as in Eq. (6.11b). The standard variation of *N* for tropospheric heights is shown in Fig. 6.4.

N varies with geographic locations, height and time as do the terms ρ and *T*. For any particular time, the height variation is more rapid than the spatial variation.

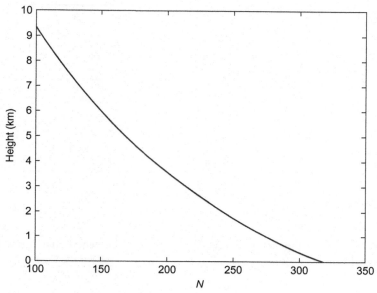

FIG. 6.4

Vertical profile of *N* for standard atmosphere.

6.3.1.2 Effects

The effects of clear air on the propagating EM wave due to the refractive index of a medium and its variation are now discussed in the following subsections.

Ray bending

As the ray of an EM wave passes through the medium, it experiences changing refractive index. Consequently, the ray bends due to refraction following the Snell's law. The troposphere may be assumed to be horizontally stratified with changing *RI*. Therefore, as the ray passes from one layer of the atmosphere to the adjacent with refractive indices n and $n - dn$, respectively, the infinitesimal change dn in the refractive index leads to the resultant bending of the ray by an angle $d\theta$. The scenario is illustrated in Fig. 6.5.

Therefore, we can write from the Snell's law,

$$n \cdot \sin \theta = (n - dn) \sin (\theta + d\theta)$$
$$\text{Or,} \quad = (n - dn)(\sin \theta \cos d\theta + \cos \theta \sin d\theta) \tag{6.15a}$$

Here, θ is the incident angle of the ray on the interface of two layers and 'n' is the refractive index of the first layer which changes by dn abruptly between the adjacent layers. Notice that the *RI* decreases with height and consequently the angle of refraction increases. For small $d\theta$, $\cos d\theta$ can be approximated to 1 and $\sin d\theta$ as $d\theta$, and neglecting the terms containing both dn and $d\theta$, we get

$$n \sin \theta = n \sin \theta + n \cos \theta d\theta - \sin \theta \, dn$$
$$\sin \theta dn = n \cos \theta \, d\theta \tag{6.15b}$$

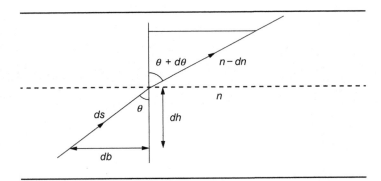

FIG. 6.5

Bending of rays.

Consider 'h' to be the height above the earth surface where the refraction is occurring. Then

$$dn/dh = \frac{n}{\tan\theta}(d\theta/dh) \tag{6.15c}$$

As $db/dh = \tan\theta$, we can write $dh\,\tan\theta = db$, so

$$d\theta/db = \frac{1}{n}\,dn/dh \tag{6.15d}$$

Now, for the given geometry, $db = ds\,\sin\theta$ and $ds/d\theta = r$, the radius of curvature, and since n is very close to 1, Eq. (6.15d) is simplified as

$$\frac{1}{r} = dn/dh\,\sin\theta \tag{6.15e}$$

It is therefore clear that, for vertical incidence ($\theta = 0$), no curvature is observed, while for any other given value of θ, if the vertical gradient is constant, 'r', the radius of curvature for the bending is also constant. Then, the trajectories are arcs of a circle. Evidently, the bending is large when r is small which occurs when the gradient of 'n' is large and vice versa.

We have previously seen that the spatial variation of tropospheric temperature, pressure and water vapour is not constant over different locations. Hence, the RI also varies spatially. Therefore, the propagating wave 'sees' changing RI as it moves through it. Consequently, it experiences infinitesimal bents as it propagates which cumulatively gives the total bending along the path. Therefore, the total amount of bent is obtained from the path integral of the above expression given in Eq. (6.15d). Since, for lower elevation angles rays traverse through longer paths, they are largely affected. Alternatively, from Eq. (6.15e), we get that for a given vertical gradient of RI dn/dh, curvature $1/r$ varies as $\sin\theta$. So, for lower elevation angles, as θ is large, the curvature is also large. As a result, significant ray bending occurs for elevation below 10°. The precise calculation of total amount of ray bending for a total distance traversed may be obtained by integrating the infinitesimal bents occurring along the ray path. The latter, in turn, would require the knowledge of the profile of the RI along the whole path. However, this is not always practical to have the total path profile of RI measured, especially when the spatial extent of the ray path is large. It is here when the models of these parameters get useful. Conversely, sometimes the total bending angle is known and it is required to get the profile of the refractive index, in order to retrieve the atmospheric parameters. For such cases, it requires the inverse problem to be solved. This is the basic principle of the Radio occultation technique.

Since the atmosphere is never truly stationary, the ray of the signal which is coming to the receiver antenna through the atmosphere goes on fluctuating in their apparent 'Angle of Arrival'. For antennas with very narrow beamwidth, this will lead to variations in the effective antenna gain experienced by the signal.

The spatial variation in RI may happen in a manner such that a ray reaches the receiver antenna passing through different conjugate paths. Thus, the signal received at the receiver becomes a collection of rays having different phase and amplitude leading to refractive multipath. Then, several possible signal components coexist, resulting in mutual interference and thereby causing multipath associated fading. For satellite signals, it is more likely to happen for low elevation angles. An elaborate treatise on multipath fading is given in Chapter 8. Unlike reflective multipath, no abrupt change of the phase of the signal occurs here and the phase difference between the components is solely due to the difference in their equivalent paths.

Modified phase and group velocity

The phase and group velocity of the wave is dependent upon the refractive index, n. The latter again is a function of the permittivity of the medium, ε. We have already derived in Chapter 3 the refractive index as a function of medium permittivity, given by $n = \sqrt{\varepsilon}$. As the permittivity of the medium is different from that of the vacuum, the waves move with different phase velocity in the medium than in the vacuum. The phase velocity of the wave in a medium of permittivity ε becomes

$$v_p = \frac{c}{\sqrt{\varepsilon}} = \frac{\omega}{k} \tag{6.16}$$

where $k = nk_0$ and k_0 is the propagation constant for vacuum. Reiterating Eq. (6.12), the effective excess path experienced by the signal during the propagation is

$$\Delta l = (n-1)\delta l$$

where, δl is the geometric path through the medium. Therefore, the excess change in phase over this length is

$$\begin{aligned} \delta\varphi &= \frac{2\pi}{\lambda} \Delta l \\ &= (n-1)\frac{2\pi}{\lambda}\delta l \\ &= Nk\delta l \times 10^{-6} \end{aligned} \tag{6.17}$$

Box 6.1 Matlab Exercise

Matlab program 'Addl_phase.m' was run to obtain the additional phase changes when the signal of a given frequency passes through a medium of refractivity N. The given plot is obtained for frequency 30 GHz and over a range of N values from 100 to 350 and for a given length of the medium. Notice the linearity of additional phase experienced with increasing N in the upper panel. The lower panel illustrates the phase changes with angular representations.

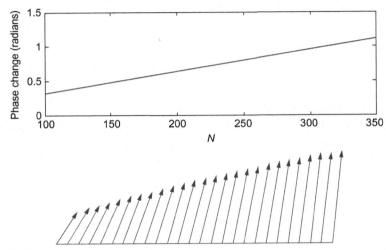

FIG. M 6.1

Excess phase change on traversing through a medium.

Change the values of frequency f and observe the results. Also see what happens if the length of the path traversed changes

---

From Chapter 3, we have learnt that the permittivity, in general and particularly near the resonant frequency of the component charges, becomes function of frequency, i.e. $\varepsilon = \varepsilon(\omega)$. So, when n is frequency-dependent, the effective propagation constant $k = (2\pi/\lambda)\sqrt{\varepsilon(\omega)} = (\omega/c)\sqrt{\varepsilon(\omega)}$ is also a function of frequency. Therefore, for a signal consisting of a spectrum of frequencies, the phase difference will be different for different components. The medium, hence, becomes Dispersive. The group velocity and the phase velocity are respectively given by the expressions

$$v_p = \omega/k$$
$$v_g = d\omega/dk$$

(6.18)

Due to the presence of the frequency-dependent term $\sqrt{\varepsilon(\omega)}$, neither the phase velocity, nor the group velocity is equal to c any more and is now actually a function of frequency now. So, there is a finite group delay experienced by the signal depending upon the frequency of the signal. For practical purposes, the dispersive effects may impose serious limitations, in the windows of the spectrum near the centres of major

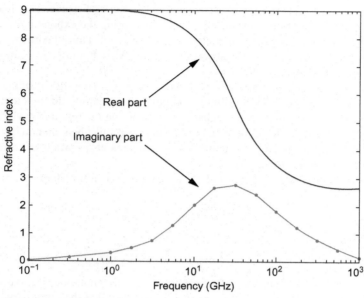

FIG. 6.6

Refractive index of water.

resonance lines. The *RI* for water in the Ka band frequencies is given in Fig. 6.6. Notice the variation in both its real and imaginary part near the resonance frequency at 22.23 GHz.

Gaseous absorption

The atmospheric gases and water vapours present in the troposphere when exhibit complex permittivity and can be written as $\varepsilon = \varepsilon' + j\varepsilon''$. The complex values are conspicuous when the frequency of the traversing wave gets near the resonant frequency of the oscillating bound charges in the medium, resulting in considerable movement of the bound charges. This complex permittivity results in complex value of the refractive index consisting of a real and an imaginary part, with $\sqrt{\varepsilon} = \sqrt{(\varepsilon' + j\varepsilon'')} = (n_r + jn_i)$. Putting this expression for the complex *RI* in the wave equation, we get

$$
\begin{aligned}
E(t) &= E_0 \exp\{j(\omega t - kx)\} \\
&= E_0 \exp\{j(\omega t - k_0 nx)\} \\
&= E_0 \exp\{j(\omega t - k_0 n_r x)\} \exp(-k_0 n_i x)
\end{aligned}
\tag{6.19}
$$

Therefore, due to the imaginary part of '*n*', the field strength of the wave reduces exponentially on propagating through this medium and thus causing the attenuation due to absorption. You may revisit Section 3.3 to recall the theory pertinent to this phenomenon.

We have seen in Chapter 3, the real part defines the phase constant $k = (\omega/c) \times n_r = k_0 \times n_r$, representing the rate of change in signal phase with distance. Therefore, larger the value of n_r, larger the phase change over a given distance. The imaginary part defines the extinction constant $\alpha = jk_0n_i$, representing the exponential decay of signal field strength, i.e. amplitude with distance traversed. Thus, the imaginary part of the refractive index defines the absorption factor. Also, both n_r and n_i are dependent upon frequency.

Also recall, in Chapter 3, we have seen that this decay in the signal strength is the manifestation of the absorption of the signal power propagating through a medium. The power is utilized in moving the charges, overcoming the resistance while oscillating in response to the incident signal field. However, this phenomenon is not only dispersive, but also selective in nature and maximizes at a particular critical frequency, causing resonance.

Utilizing the expression for ε' and ε'', which we derived in Chapter 3, we get

$$\varepsilon' = \varepsilon_0 K_r = \varepsilon_0 + \left(\frac{N_e e^2}{m}\right) \frac{\omega_0^2 - \omega^2}{\left(\omega_0^2 - \omega^2\right)^2 + (\omega k_m)^2} \tag{6.20a}$$

$$j\varepsilon'' = \varepsilon_0 K_i = \left(\frac{N_e e^2}{m}\right) \frac{-j\omega k_m}{\left(\omega_0^2 - \omega^2\right)^2 + (\omega k_m)^2} \tag{6.20b}$$

where ω_0 is the natural frequency of the electrons and other notations are as described in the mentioned chapter. So, it can be readily observed that the imaginary component of the permittivity ε'' maximizes at $\omega = \omega_0$ while the real part $\varepsilon' = \varepsilon_0$ at that frequency.

Practically, the gases exhibit more than one resonance frequency corresponding to each of the degrees of freedom it possesses. The atmospheric water vapour has an absorption line at 22.23 and 183.3 GHz, while the Oxygen resonates at 60 and 118.75 GHz. The effective ε of the gas has contributions from all of these. However, at frequencies, far enough from the resonant frequency, the imaginary part of ε becomes negligible, which is evident from its expression given above. Then, the RI can be considered to be real with only phase change occurring during the propagation, but no absorption. However, for all practical gases, when the contribution from one component becomes completely imaginary, the other component with far off resonant frequency remains real. Therefore, the incident signal does never get completely evanescent.

Scintillation

Scintillation is the rapid fluctuation in the amplitude and phase of the signal as observed from a receiver. Tropospheric scintillation is basically caused by small-scale inhomogeneities of the refractive index appearing in the path of the propagating wave. On satellite links, the significant scintillation effects under clear sky condition may be attributed to this inhomogeneity.

The atmospheric refractive index in the troposphere, as defined in the standard reference atmosphere, exhibits horizontal strata with RI decreasing gradually with altitude. However, this is true for calm and clear sky conditions. But the troposphere is very prone to dynamics. Presence of thermal gradients and other reasons causes motions in the atmosphere which leads to mixing of the different layers. However, this mixing is slow and uneven. Typically, strong turbulence in vertical winds results in inhomogeneous mixtures of constituents. The wet components of the air, like the water vapour, are not uniformly distributed over the atmospheric regions. This uneven mixing and irregular distribution of wet components of the air cause random smaller and larger regions of variations in the *RI*, both spatially and temporally. These parcels of different random densities and hence of different *RI* move with the wind and intersect the line of sight of the signal from transmitter to the receiver. Therefore, as the wave passes through these regions of small-scale fluctuations, it results in rapid and randomly changing *RI*. It does not only cause anomalous refraction, but also renders the wave passing through it vary in phase across its wavefront. These components of the signal, after passing through the inhomogeneous region gets diffracted to reach the receiver from multiple conjugate paths with arbitrarily different amplitudes and phases. They combine either constructively or destructively leading to the rapid and random variation in signal amplitude and phase, resulting in the phenomenon of Scintillations. Therefore, the basic theory for scintillation remains the same as it was mentioned in Chapter 3 and described for the ionosphere in Chapter 4.

Tropospheric scintillation intensity increases with increasing carrier frequency. So, the fluctuations are more conspicuous in higher frequencies like in Ka band than in lower bands like C and X. Scintillation effects are always much more pronounced for lower elevation angles, i.e. near the horizon than near the zenith, as in the former case the signal wavefront has to pass through a larger extent of inhomogeneity. Therefore, the satellites, directly overhead, usually scintillate less than those which are near the horizon. The effect also diminishes with increasing antenna size.

The propagating signal when undergoes scintillation may show both reduction and enhancement in its strength resulting in what is called the 'Scintillation Fades'. However, it is worth noting that variance in scintillation value for depreciation and enhancement is not equally likely. Signal depreciation had a long tail compared to enhancement and thus the shape is not symmetrical (Van de Kamp, 1998).

Scintillation fades can have a major impact on the performances of low margin communication systems, for which the long-term availability is sometimes predominantly governed by scintillation effects rather than by rain (Bertram et al., 2002). In addition, the dynamics of scintillation may interfere with tracking systems or Tropospheric fade mitigation techniques.

6.3.2 CLOUDS

Clouds are the visible mass of condensed watery vapour floating in the atmosphere due to the buoyant forces acting upon it, typically high above the general level of the ground. As the water vapour goes higher in the sky, it experiences very low temperature. As a result, the water vapour condenses into water on suspended particles like dust, ice, salt, etc., called seeds. These particles hold the condensed water and keep them afloat. Each individual droplet constituting the cloud is so small that 1 m^3 of air can accommodate more than 10^8 of such droplets.

Clouds are generally classified on the basis of their shape and their height in the sky. Out of these, the Nimbostratus and the Cumulonimbus clouds are of importance to us. Nimbostratus is a multilevel stratiform cloud that forms in the low or middle level of the troposphere. It has large horizontal extent, and therefore, produces precipitation over a wide area. Nimbostratus usually has a thickness of about 2000 m and is found more commonly in the middle latitudes. Nimbostratus generally produces light to medium rain and by itself is not associated with thunderstorms. However, thunderstorm-producing cumulonimbus clouds may be embedded within the usual nimbostratus.

Cumulonimbus clouds are cumuliform clouds, characterized by their large vertical extent, and are large in size with dense content. Its vertical growth is aided by strong updraft of air current. Such clouds are formed at altitude from below 5 km and can grow up to a height of 10 kms or even to the level of the Tropopause. Its base can extend over several kilometres. These types of cloud can form alone or in clusters and are typically accompanied by smaller cumulus clouds. Generally, cumulonimbus clouds precipitate larger droplets and more intense downpours of convective type including thunderstorms (Houze, 1993) (Fig. 6.7).

6.3.2.1 Effects

Cloud attenuation: Clouds are made up of condensed water vapour produced on selected seeds and kept afloat by the buoyant force. These individual particles are <0.1 mm in diameter. The constituent particles thus induce significant scattering to the electromagnetic waves of the satellite signals passing through it. This results in cloud attenuation. Although the amount of attenuation, even at higher frequencies, being ~1 dB, is much lower compared to that due to the rain, etc. in the same band, it can turn out to be significant for low margin systems.

The water droplets of the clouds which essentially carry out scattering of the propagating signals have dimensions much smaller than the wavelength of the signal of the popular satellite communication bands, like C, Ku, Ka, etc. So, Rayleigh

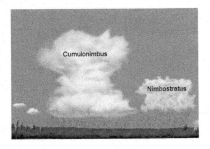

FIG. 6.7

Cumulonimbus and nimbostratus clouds.

scattering is predominant here. Recall that the Rayleigh scattering is directly proportional to the sixth power of diameter of the scattering particles and inversely to the fourth power of the wavelength. So, the higher frequencies suffer more scattering than the comparatively lower frequencies. Further, the effect becomes more pronounced the larger the water droplets get. The amount of the signal power getting scattered also increases with the number of water droplets present in the path, and hence, depends on the droplet density and height of the clouds. When clouds are thin, only a small portion of the signal power is scattered while most of the power passes through it. But as their thickness increases, the cloud gathers more water droplets. Therefore, more power it scatters, resulting in lesser penetration of the signal as it passes all the way through it. It is due to this reason that the specific attenuation of signal through the clouds is proportional to the columnar mass density in addition to being a function of frequency. The same phenomenon happens in optical frequencies also. It makes the bottom of tall watery clouds look dark grey than white. The taller the clouds become, the darker they look from the bottom.

6.3.3 RAIN

Rain is the precipitation of the liquid water originating from the cloud. It is the major cause for the signal impairment, and hence it is also the most important tropospheric parameter to be considered for signal propagation. Normally, when the radio waves are incident on the precipitating rain drops, they encounter an abrupt change in *RI*. So, as the waves traverse through it, there is a considerable absorption and scattering of the wave in addition to the rain induced scintillation. The amount of absorption and scattering depends upon the features of the rain, viz. the rain rate, rain drop size, distribution, etc. We shall discuss here the two most important aspects of rain, viz. the key rain parameters and the associated rain effects. The first is to understand the inherent features of rain along with its spatial and temporal behaviour. The latter is to understand how it influences the propagation of the signals.

6.3.3.1 Key parameters

The rainfall is characterized by various parameters like the rain rates, relative occurrences of different rain rates, total cumulative rain, rain drop distribution, the structure of the spatial extent of the rain, etc. The rainfall characteristics, determined by these parameters, are not uniform, but vary with geographic region and location on the earth, local geophysical features and many other factors. These characteristics of the rainfall, their relative occurrences, their vertical and horizontal variations, etc. all influence the rain-induced propagation impairments along a given satellite to ground link (Allnutt, 2011). Nonetheless, given a definite characteristics of rain, the fundamental physical process responsible for the generation of the rain impairments and the definite role of different rain parameters in determining it remain the same. The phenomena of absorption, diffraction, relative phase delay, etc. play the key role in producing the impairments.

Rain types

Rain can be broadly classified into two main types, viz. Stratiform and Convective. These two types of rain have clearly distinguishable features and origin. Stratiform rain falls from nimbostratus clouds, while convective rains fall from cumulonimbus clouds (Houze, 1993). Stratiform clouds have extensive horizontal development, but show relatively little development in the vertical direction. They cover large areas as opposed to the convective clouds and, as the name suggests, are stratified in structure.

The vertical air velocity within the clouds plays a crucial role in the formation of two types of precipitation and is the major feature by which they are distinguished. The vertical air velocity or updraft associated with stratiform clouds like the Nimbostratus is about 10 cm/s. This updraft velocity is low compared to the fall velocity of the condensed water particles. This slow updraft results in smaller precipitating drops and unelevated cloud structures, as well. Further, stratiform precipitation, in general, is relatively continuous and uniform in intensity.

Convective rains, on the other hand, originate from the cumulonimbus clouds. These clouds have large vertical growth resulting from buoyant ascent, due to the vertical air velocity resulting in strong updraft. Convection occurs when a fluid under gravity is heated from below or cooled from the top at such a rate that molecular diffusion cannot redistribute the modified density fast enough to maintain equilibrium (Houze, 1981) and thus creating a vertically elevated structure. The upward transported vapour thus becomes buoyantly unstable and thus overturns macroscopically. Now, these clouds have updraft velocity magnitude of more than 1 m/s and can be as high as 10 m/s which is few orders of magnitude larger than that of the stratiform clouds. So, due to this strong updraft, the condensed overturned particles stay more time as they pass through the clouds. In the process, some of them collect more water mass to get heavy. This growth by accretion continues until they become heavy enough and finally begin to fall relative to the ground and precipitate. Thus, in such convective clouds, growth of the particles by accretion of liquid water is the dominant mechanism followed by collision, coalescence and breakup of drops during precipitation. Raindrop growth in a convective cloud is faster, rainfall is intense and the drops are larger. Further, convective precipitation is of shorter duration than stratiform precipitation (Houze, 1981, 1993).

Spatial rain fall distribution

The variation of the rain profiles along time or space coordinates are important parameters contributing to the signal impairments. We shall find in the next chapter that, when rain-induced attenuation is to be determined, it requires the spatial rain intensity variation to be considered. Therefore, for such a need, the rain rate profile is required to be known.

The rain profile must be described for both the horizontal and vertical variations. Ideally, spatial variations are determined with direct measurements of rain with rain gauge networks and radar. The current satellite-based measurements are also used for the purpose. It is important to understand that a rain profile for any location cannot be derived from a single event and neither any characteristic profile is applicable to any single event of rain. Any variation in rain profile across time or space is a statistical representation of its larger ensembles. This makes it impossible to determine profile, specific for a particular event, by any means, other than in situ measurements. Therefore, the rain rate profile obtained from a large collection of a-priori measurements is only useful in statistical predictions of the parameters.

Vertical variation of rain: The vertical structure of the precipitation can be divided into two distinct regions. The lower region, extending from the ground to a definite constant height, consists of liquid precipitation, i.e. rain, while the upper region consists of a mixture of ice and snow. The transition height between the two regions corresponds approximately to the height where nonprecipitating hydrometeors get converted to form the precipitating water. This demarcates one region from the other. This is the height of the 0°C isotherm and is also considered as the basis to fix the rain height. It is also known as the melting layer or freezing layer. This region appears as a bright band in the observations of radar reflectivity below which a uniform structure of the rain exists. The height of the melting layer can be statistically derived from the position of the bright band in the radar reflectivity. However, it was later found that the 0°C isotherm is actually a few hundred metres above the bright band. The 0°C isotherm height varies with latitude and with the season of the year. ITU-R P.839 (ITU-R, 2013) gives an estimate of the mean annual rain height for any given location.

It can be assumed that the rain structure is uniform from the ground up to an effective rain height H_r (Stutzman and Dishman, 1982). For all practical purposes, the rain height may be considered to be approximately constant and equal to the height of the 0°C isotherm for low rain rates of stratiform rain. As rain rate increases, however, the rain height also increases. This increase is mainly because the liquid water is carried well above the 0°C isotherm level by updrafts (Goldhirsh and Katz, 1979). The effective rain height, therefore, can be obtained by using the 0°C isotherm height for low rain rates and adding a rain rate-dependent term to this height for higher rain rates. A simple relationship has been proposed by Stutzman and Dishman (1982) which relates the effective rain height H_r (in km) to the rain rate by a logarithmic equation and is given by,

$$H_r = H_i \qquad \text{for } R_0 < 10 \text{ mm/h}$$
$$= H_i + \log\left(\frac{R_0}{10}\right) \qquad \text{for } R_0 \geq 10 \text{ mm/h} \qquad (6.21a)$$

where R_0 is the ground rain rate in millimetres per hour and H_i is the 0°C isotherm height in kilometres at the given location. H_i is a latitude-dependent term and can be approximated as (Crane, 1979)

$$H_i = 4.8 \text{ km} \qquad \text{for } \lambda_g < 30$$
$$= 7.8 - 0.1\lambda_g \text{ km} \qquad \text{for } \lambda_g \geq 30 \qquad (6.21b)$$

where λ_g is the geographic latitude in degrees. The following contour plot in Fig. 6.8 shows the variation of the effective rain height with latitude for the Northern hemisphere and with rain rate. Notice the invariant nature of the plot up to 30° in latitude and up to 10 mm/h rain rate in the plot.

The effective rain height can also be obtained from the grid values given in the current ITU recommendation ITU-R P.839 (ITU-R, 2013).

Horizontal variation of rain: Rain precipitation systems are combinations of both stratiform and convective rain structures with proportions varying from place to

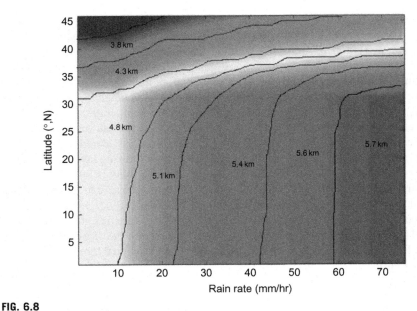

FIG. 6.8

Contour plot of rain height variation with latitude and rain rate.

place and from event to event. Most precipitation is characterized by large areas of low rates with a number of smaller regions of high rain rates. Convective cells imbedded in stratiform rain render the distribution of rain highly nonuniform in the horizontal direction and result in such structure. It is the presence of these imbedded convective rain columns which makes the spatial distribution of rain difficult to describe and model.

Structurally, the rain consists of the central core region termed as 'Cell' (Crane, 1977) with high intensity of rain and a limited radius of around 3 km in average. This central core or the 'Rain Cell' has exponential profile of the rain inside with rotational symmetry (Capsoni et al., 1987). The adjacent larger region of comparatively lighter rain intensity surrounding the core is referred to as the 'Debris' (Crane, 1977). Convective cells tend to cluster with an intercellular separation of the order of 5–6 km. The most intense cells in the rain structure tend to elongate in the direction of the motion of the rain cells, thereby creating asymmetry in the cell axis. Terrain structure plays important role in shaping up the structure of the cells. The average lifetime of rain ranges in order of minutes. However, intense rains for higher time periods are also observed. The probability of continuation of rain above any predefined rate decreases exponentially with higher duration of rain.

Rain drops

Drop shapes: The shape of a raindrop is determined by the size of the drop. It is a complex function of its surface tension and the aerodynamic forces acting upon it. Very small drops are essentially spherical in shape as the surface tension is excessively large compared to that of the total gravitational force and of any other external force acting upon it. With increase in size, the drops experience considerable vertical air drag during its fall and consequently become progressively flattened at the base. The droplets, therefore, get the shape of oblate ellipsoid (Ishimaru, 1991).

The drops, however, do not always fall vertically, but with a certain canting angle. Canting angle is the angle between the minor axis of the ellipsoid that the drop shape has assumed and the vertical direction. It is the shape and relative orientation of the drop with respect to the direction of propagation of the wave and the polarization of the signal, which has important contributions during the scattering and absorption of the signal. Canting, therefore, plays a major role in this process.

Drop sizes: The rain drop size also plays an important role in determining the propagation effects. Absorption and scattering takes place in an individual raindrop, leading to the attenuation of the wave. The amount of attenuation is determined by the drop size and the frequency of the wave. The raindrop diameter typically varies between 0.5 and 5.0 mm. The size distribution density $n(D)$ of the rain droplets is the number of raindrops present per unit volume of space within the diameter range from D to $D+dD$. Here, the diameter means that of the equivolumal spherical drop. The distribution depends upon the rain rate, R, and can be well-represented by definite empirical expressions as well as by standard statistical models. Broadly, three types of statistical distributions are used for its representation, viz. Exponential distribution, Gamma Distribution and Lognormal Distribution. The distribution also shows significant variation between different climatic regions.

Laws and Parsons (1943) conducted the first systematic measurements of drop size distribution who observed that the number of raindrops 'n' diminishes with increasing drop diameter, D. They have also showed that the median rain drop size shifts with increasing rain rate. Quantitatively, $n(D)$ can be represented from their measurements as an exponentially decaying function of D as (Marshall and Palmer, 1948)

$$n(D) = n_0 \exp(-\Lambda D) \tag{6.22}$$

where $n_0 = 8 \times 10^3$ m^{-3}/mm, $\Lambda = 4.1R^{-0.21}$ and the rain rate R is in mm/h. However, the model gives considerable deviations. Due to its negative exponential nature, the distribution predicts more number of smaller particles than the larger drops. These numbers monotonically increase with decreasing drop sizes, even when the drop diameter approaches zero. This does not truly represent the pragmatic situation and the model overestimates the smaller size droplets. Very obviously, when mean drop size is estimated, due to the abundance of the lower droplets, it remains inappropriately skewed towards the lower side in the distribution.

A more general expression for size distribution is the Gamma distribution (Diermendjian, 1969; Ishimaru, 1991) given by:

$$n(D) = n_0 D^\alpha \exp\left(-\Lambda D^\beta\right) \tag{6.23}$$

where n_0, Λ, α and β are constants and location-dependent. This expression may be generalized to the Marshall-Palmer distribution of Eq. (6.22) by putting $\alpha = 0$ and $\beta = 1$ in Eq. (6.23). This expression apparently corrects the problem of the negative exponential distribution of Marshall and Palmer; however, the difficulty lies in the use of this model (Karmakar, 2012).

It has been observed that at tropical regions due to the abundance of convective rainfall over stratiform, the larger drop sizes dominate the drop distribution (Green, 2004). It has been found that for these regions, the lognormal model is more appropriate than the other empirical models. The lognormal model was proposed by Zainal et al. (1993) and is given in generic form as

$$n(D) = \frac{n_0}{D} \exp\left\{ -\tfrac{1}{2}\frac{(\ln D - \mu)^2}{\sigma^2} \right\} \tag{6.24}$$

where n_0 is a constant, D is the drop diameter and μ and σ are, respectively, the mean and the standard deviation of the diameter D (Karmakar, 2012). The values of the parameters in the model, n_0, μ, and σ are to be obtained empirically. Different researchers have fitted their measurements of raindrops carried out in the tropical region to this lognormal model and derived the values of the parameters n_0, μ and σ (Verma and Jha, 1996; Ajayi and Olsen, 1985; Pontes et al., 1990).

Box 6.2 Matlab Exercise

Matlab program 'Drop_distbn.m' was run to obtain the plots for the rain drop distribution with three different models as discussed above. Notice that, except Marshall and Palmer distribution, both the other two models show a bent of the distribution function towards the triviality as the drop diameter approaches zero.

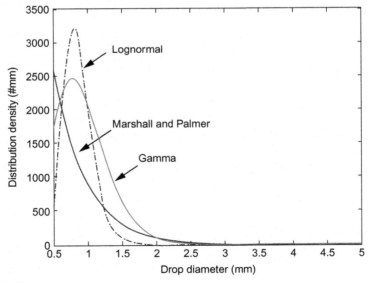

FIG. M 6.2

Distribution of drops for different sizes.

Change the different parameters in the model and run the program again to see how they influence the distribution.

Terminal velocity: The size of the raindrop determines the terminal velocity with which it falls on the ground. When the droplet starts precipitating, its gravitational force is greater than the upward drag. Thus, there is effective downward acceleration acting upon the drop. This causes the drop velocity to increase as it traverses vertically down. With the increment in velocity, the upward drag also increases and reduces the acceleration. Finally, at a point, the two forces get balanced and the drops attain a limiting velocity. If the drag coefficient is c_d, the upward dragging force experienced by the drop is

$$F = \tfrac{1}{2}\rho_0 v^2 S c_d$$
$$= \tfrac{1}{2}\rho_0 v^2 \pi \frac{D^2}{4} c_d \tag{6.25}$$

where, ρ_0 is the density of the medium through which the drop moves, v is the velocity, S is the cross-sectional area in the direction of the velocity projected onto the dragging force. Again, the net downward force acting upon the drop, considering the gravitational pull and the buoyant force, is

$$G = \frac{4}{3}\pi \frac{D^3}{8}(\rho - \rho_0)g \tag{6.26}$$

Equating (6.25) and (6.26) for the balanced condition, we can get the expression for v as

$$v^2 = \frac{4}{3}gD\frac{\rho - \rho_0}{\rho_0 c_d} \tag{6.27}$$

Thus, it becomes a function of the drop diameter 'D' and density of water ρ, in addition to the drag coefficient 'c_d' and the air density, ρ_0. However, the drop terminal fall velocity is also measured in laboratory and empirical relationship has been found. The most popular among them is the measurements done by Gunn and Kinzer (1949), the plot for which is given in Fig. 6.9.

Different quantitative models have been proposed. It was found that the raindrop terminal fall velocity $v(a)$ can be well-represented by (Kerr, 1951)

$$v = 200.8 a^{\frac{1}{2}} \text{ m/s} \tag{6.28}$$

with 'a' as the equivolumetric radius of the drops expressed in metres.

It has also been observed that not all the drops of certain size fall with the expected velocity. Speed versus size measurements of drops during natural rainfall events show that many raindrops fall up to an order of magnitude faster than expected. These 'superterminal drops' are actually differently sized fragments of a recent break-up from larger size drops. Although they have reduced in size due to the break up, they continue to move with the speed of the parent drops. This break up is further supported by the fact that there is a preponderance of superterminal drops in the presence of large raindrops, i.e. during periods of high rainfall rates (Houze, 1993; Martinez et al., 2009).

Rainfall rate

For a given interval of time, the height of the rain water collected over a unit area is termed as the total rainfall in the given interval. The total rainfall occurring in unit time is the rainfall rate. It is thus obtained as

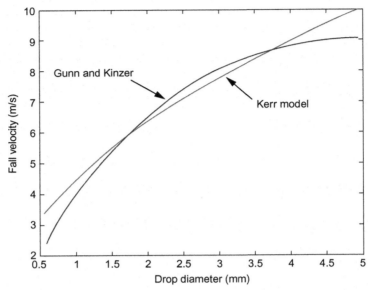

FIG. 6.9

Variation of fall velocity with drop diameter.

$$R = \frac{V}{S\,\delta T} \tag{6.29}$$

where V is the volume of water collected over a given surface area S over a given time interval δT. The volume of water collected over the time depends on the raindrop size distribution and the drop terminal fall velocity. If $n(D)$ is the drop size distribution and 'D' is the equivolumetric diameter of the drops, $n(R,D)dD$ is the number of droplets per unit volume with diameter between D to $D + dD$. If $v(D)$ is the drop terminal fall velocity, then the rain rate is given by

$$R = \int \frac{4}{3}\pi \frac{D^3}{8} n(D)v(D)dD \times 36 \times 10^{-4} \text{ mm/h} \tag{6.30}$$

where, D and dD is in mm, $n(D)$ is in numbers/m^3/mm and $v(D)$ is in m/s.

Box 6.3 Matlab Exercise

Matlab program 'Reltv_cntbn.m' was run to obtain the relative contribution of different rain drop sizes to the total rain rate.

This program estimates the relative contribution of different rain drop diameters to the total rain rate. The x axis in the plot represents the drop diameters, while the y axis represents the relative contribution of drops of the discrete diameters considered. This can be considered as the rain rate contribution density of drop diameters and hence its unit is (mm/h)/mm.

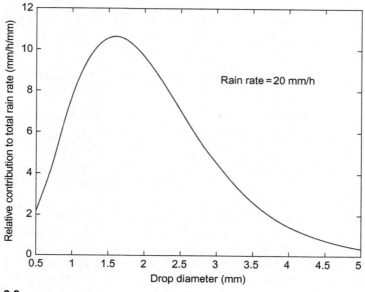

FIG. M 6.3

Contribution of differently sized raindrops to total rain rate.

To get the total rain rate, this density distribution is to be integrated over the total range of diameters. Here, since the interval is taken very small, the total rainfall is obtained by summing up the contributions and multiplying each with the interval.

Change the values of rain rate R and observe the difference. The diameter of the maximum contributing drop also changes. For 20 mm/h, the maximum diameter of the drop is between 1.5 and 2.0 mm, whereas for 150 mm/h this value approaches 5 mm.

There are different temporal models which provide the statistics of the occurrences of different rain rates. We shall briefly mention a few of them in this section. But, before that, it would be useful to define the term 'Exceedance', which we shall be using frequently hereafter. By definition, '*Exceedance*' value of a time-distributed occurrence, for $p\%$ of time, assuming values represented by variable 'x', is that value of x which is exceeded during the occurrence for only $p\%$ of time cumulatively over a year.

Rice and Holmberg model. Rice and Holmberg model estimates the annual statistics of the rainfall rate from the total rainfall accumulation measured over an average year. This model divides rainfall into two components, viz. the Mode 1 type of rain and Mode 2 type of rain. Mode 1 ($M1$) represents the total convective rainfall, consisting a core and including thunderstorm. Mode 2 ($M2$) represents the rest of the rains including the uniform stratiform type. Therefore, the total rain M can be assumed to be the sum of $M1$ and $M2$ type of rains, i.e.

$$M = M1 + M2 \tag{6.31}$$

The key parameter required in R–H model is Thunderstorm ratio (β), which can be computed as the ratio between $M1$ rain fall and total rain, M, i.e.

$$\beta = M1/M \tag{6.32}$$

Then the time, T_1, in hours that a predefined rainfall rate R is exceeded over a year is given by

$$T_1(r > R) = M\left\{ (0.03\beta e^{-0.03R}) + 0.2(1 - \beta)(e^{-0.258R} + 1.86e^{-1.68R}) \right\} \text{ h} \tag{6.33}$$

Here, R is the rain rate (mm/h), while the output T_1 is obtained in hours. However, for this model to work, meteorological measurements are required. Although it may not work so well at high rainfall rate and hence low percentage time portion of the cumulative statistics, it is excellent for predicting low to medium rainfall rates (Allnutt, 2011).

ITU-R model. ITU model initially used to classify the global distribution of rain rate into different rain zones depending upon the similarities of their rain characteristics. Very obviously, this classification was coarse. Now, using the measured data, ITU has proposed a grid-based distribution of the characteristic rain rate over the entire globe. In this model, the necessary meteorological parameters concerning the rainfall are given in the definite regularly spaced grid points on the surface of the earth. Using these values and the algorithm provided by ITU, the rain exceedance values can be derived for any point on the earth. The required meteorological parameters are available in respective files provided by ITU, while the algorithm is described in ITU Recommendation ITU-R, P.837.

Moupfouma model. Recent analysis suggests that the rain rate distribution is better described by a model which approximates a lognormal distribution at the low rates and a gamma distribution at high rain rate. This kind of model was developed by Moupfouma et al. (1990). The model is good for both tropical and temperate climates. The Moupfouma model has a lognormal behaviour for mean and low rain rates and a Gamma behaviour for high rain rates has the following form:

$$p(r \geq R) = 10^{-4} \left(\frac{R_{0.01}}{R} + 1 \right)^b e^{u(R_{0.01} - r)} \tag{6.34}$$

where p represents the fraction of time over a year that the observed rain rate r exceeds a rate R (mm/h). $R_{0.01}$ is the rain intensity in mm/h exceedance value for 0.01% of time in an average year. b is an empirical constant approximated by the value $8.22R_{0.01}^{-0.584}$. The parameter u in Eq. (6.34) governs the slope of rain rate cumulative distribution and depends on the local climatic conditions and geographical features.

Crane model. Weather radar observations show that rain is always spatially inhomogeneous with regions of relatively heavy rain occurring surrounded by lower rain rates. The 'Two-Component model' (Crane, 1977) for the prediction of attenuation due to rain separately addresses the contributions of both the concentrated central heavy rain showers and the larger regions of lighter rain fall surrounding the showers. The region of central heavy shower is called 'Cell', while the term 'debris' is used to refer to the larger region of lighter rain rate surrounding a volume cell. A simple approximation to the rain rate probability within the cell has an exponential distribution. The portion of the distribution function not accounted for by the exponential volume cell component distribution is attributed to rain debris. The debris rain rate probability is nearly lognormal over its entire range of time. Putting it in mathematical form,

$$p_C(r > R) = P_C e^{-R/R_c}$$
$$p_D(r > R) = P_D N \left\{ \frac{ln(R) - ln(R_D)}{\sigma_D} \right\}$$

(6.35)

where $p(r > R)$ is the probability that the observed rain rate, r, exceeds the specified rain rate, R. $p_C(r > R)$ and $p_D(r > R)$ are the probability functions for cells and debris, respectively, where R and r are rain rates in mm/h, P_C and P_D are the probability of occurrence of a cell and debris, respectively, while R_C and R_D are, respectively, the average cell and debris rain rate. N is the normal distribution function and σ is the standard deviation of the logarithm of the debris rain rate. Since the occurrence of these two components is assumed to be independent, the total probability is given by

$$p(r > R) = p_C(r \geq R) + p_D(r \geq R)$$

(6.36)

The described model with a combination of an exponential plus a lognormal distribution also fits in a fairly precise manner for rain rate probability distribution functions for the other rain climate regions (Crane, 1979, 1982).

6.3.3.2 Effects

In general, the rain effects on frequencies below 1 GHz are negligible. With increasing frequency, the effects also increase. The important effects of rain on the propagating EM wave are rain attenuation, depolarization, scintillation and noise enhancement. The rain parameters which affect these impairing effects are the rain rate, its spatial and temporal extent and other rain features like the rain composition, size, shape and orientation of the rain drops, etc.

In order to study the effects, it is important to understand and correlate the pertinent rain and signal parameters. We have already seen the relations of the rain intensity with the rain parameters like drop size velocity, etc. Here, we shall have the overview of how these parameters influence the rain-induced impairments. In the next chapter, we shall learn about the models of these impairments.

Rain attenuation

Rain attenuation occurs due to absorption and scattering of the radio waves by the constituent rain drops, most significantly by scattering. The rain attenuation is not of significance for frequencies below 10 GHz. However, the attenuation being dependent upon the rain intensity, significant rain attenuation has been observed at certain intervals of heavy rains for links operating around 10 GHz and obviously at frequencies above it.

As the wave passes through the rain, the wave energy is incident upon the rain drops. This incident wave initiates movement of the charged particles inside the drop. In each drop, a part of the incident energy is utilized in moving the charges. A part of this energy is absorbed and gets converted to heat. Therefore, this irreversible loss from the waves results in attenuation. Further, a part of the incident ray is scattered. This energy is distributed back to space, however, in an incoherent manner, such that it does not contribute back to the original incident signal. Common to both these phenomena is the fact that in both the cases, the droplets receive energy from the incident wave which is never replenished back to the wave. Even the forward scatter also does not get coherently added to the wave. Therefore, this causes loss of the energy in the original wave. This phenomenon happens in each drop. The absorption and scattering are determined by the dielectric constant of the medium, i.e. water and also dependent upon the size of the drop and frequency of the wave. We have already seen how these factors influence the scattering and absorption cross-section in Chapter 3.

Recall the discussion on the scattering cross-section in Chapter 3. Further, consider that the rain drops are in the range 0.5–5.0 mm. The size of the drops determines its contribution to the attenuation process for a given frequency of the signal. At frequencies below 6 GHz, ($\lambda = 5$ cm) most rain droplet sizes satisfy the condition for the occurrence of Rayleigh scattering. Even at 10 GHz, ($\lambda = 3$ cm), Rayleigh solution is the appropriate scattering process except for heavy rain, when the larger drop population dominates. At frequencies higher than 10 GHz, it is the Mie scattering that needs to be considered in order to obtain the total scattering loss. However, since the scattering is the dominant phenomenon, the smaller drops, having negligibly small scattering cross-section, do not have significant role to play in it when frequency is in the lower side. With increase in frequency, the wavelength of the signal decreases and also with increase in diameter, the extinction cross-section increases and hence attenuation due to the Mie Scattering turns significant, even for smaller drops. The total cross-section of the drops, for a given wavelength and drop distribution, may be estimated from the scattering theory. We have mentioned the it in Eq. (3.34) and shall rework with it in our next chapter while describing the specific attenuation due to the rain.

For determining the total rain attenuation due to scattering and absorption by the rain drops, two parameters are necessary. First, is the rain drop density distribution, which determines the number of drops of different diameters that the wave is incident to, in an unit volume of space and for a given rain rate. From this, the total extinction cross-section offered by a definite length of rainy path to a signal of certain frequency can be obtained. This allows one to estimate the specific attenuation γ, i.e. the amount of attenuation experienced by the wave of a given frequency and for the given rain rate per unit distance. Second is the effective path of the signal through rain. For a given location, this will be determined from the rain height and the receiver—satellite geometry. Since rain is an inhomogeneous process in both time and space, the effective rainy path will vary with location, time and rain rate. Hence, this parameter is found statistically more often than not.

To summarize the whole thing up, with the increase in the rain rate, as the interaction with the number of drops increases, the total extinction cross-section of the rain drops for a given frequency also increases, in turn. This results in increase in attenuation. Similarly, the attenuation is also increased as the total path increases through the rain due to higher rain heights or lower elevation angles. Higher frequencies also lead to increased cross-section and thus results in higher attenuation. In addition to spatial variation, the rain attenuation also exhibits temporal variation. For temporal variation, the statistical attenuation exceedance value $A_{0.01}$ is an important parameter. It represents the attenuation value, which is exceeded only for 0.01% of time over a complete year. It readily gives the amount of fade margin required in a satellite system for the standard availability figure of 99.99%.

Rain depolarizations

The propagating electromagnetic wave consists of an electric field and a magnetic field, oscillating in mutually perpendicular direction, both of which are also perpendicular to the direction of propagation. The polarization of a radio wave refers to the direction of the oscillation of the electric field in space as it propagates. An unpolarized wave has no preferred orientation sense of the electric vector, while a linearly polarized wave has a fixed direction in which its electric field oscillates. Sometimes the effective field strength of a signal remains constant, but its direction rotates on the plane containing the electric and the magnetic field describing a circle. Accordingly, the magnetic field also rotates with it. Then the wave is said to be 'Circularly polarized'. When the sense of the rotation of the electric vector with respect to the direction of propagation is such that looking in the direction of propagation, the rotation appears clockwise, it is called 'Right Hand Circular Polarization' or RHCP. Else, if the rotation appears to be anticlock wise, it is called 'Left Hand Circular Polarization', or LHCP. Circular polarization may be considered as the combination of two orthogonal linearly polarized waves with phase difference of $\pi/2$. Thus, in RHCP, one linear component leads the other by $\pi/2$, while in an LHCP the condition is just the reverse. Sometimes, the electric fields of these components are not of equal strength and this makes the polarization of the wave 'Elliptical'.

If two signals are perfectly orthogonal in polarization, then they can propagate independently without interfering each other. Therefore, the vertically polarized wave does not interfere with the horizontally polarized wave. Similarly, the RHCP wave can coexist with the LHCP wave without interference. But, in all practical cases, due to the propagation effects, one of the components may experience a deviation in its polarization and thus create a finite projection onto its orthogonal polarization direction. As a result, there will be some power contributed by one polarization to the other with orthogonal polarization, however small. This is called 'Depolarization'. The limiting value of such cross-polarization interference is ensured through proper designing of the system. The amount of such transfer of power across the polarization is quantitatively expressed with the help of two terms, the cross-polarization discrimination (XPD) and the cross-polarization isolation (XPI) (Fig. 6.10).

Cross-polarization discrimination (XPD). For a polarized wave, the ratio of the power in the direction of original polarization, i.e. copolar power, to the power that is projected onto the orthogonal direction, i.e. cross-polar power, is called the XPD. So, if 'a' be the field strength of a linearly polarized signal and if due to depolarization its field strength in the copolar direction and cross-polar directions becomes

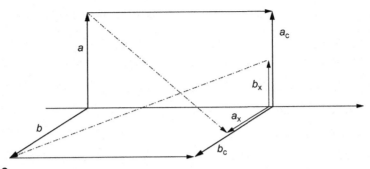

FIG. 6.10

XPD and XPI.

'a_c' and 'a_x', respectively, with the corresponding powers given by 'P_c' and 'P_x', respectively, then XPD is defined as (Maral and Bousquet, 2002)

$$XPD = 10\log\left(\frac{P_c}{P_x}\right)$$
$$= 20\log\left(\frac{a_c}{a_x}\right) \qquad (6.37)$$

Cross-polarization isolation (XPI). If there be two orthogonally polarized waves, then the ratio of the copolar power of any one of them to the cross polar power of the orthogonal signal, available in the direction of the first, is called the cross-polarization isolation (XPI). So, if 'a' and 'b' are the field strengths of these two linearly polarized signals in two orthogonal directions and if 'a_c' and 'b_x' be the copolar component of 'a' and the cross-polar component of 'b', respectively, and 'P_{ac}' and 'P_{bx}' be their corresponding powers, then XPI is defined as (Maral and Bousquet, 2002),

$$XPI = 10\log\left(\frac{P_{ac}}{P_{bx}}\right)$$
$$= 20\log\left(\frac{a_c}{b_x}\right) \qquad (6.38)$$

Thus, as the depolarization increases, the effect of the cross-polar component increases and hence it eventually results in reduction of both of these two terms.

Under nominal conditions, the XPD generally remains more than 30 dB. That is, less than a thousand part of energy of a signal is projected upon the orthogonal polarization direction. But, as a wave passes through a medium, depolarization occurs due to the anisotropy of the medium. The direction of the polarization gets deviated from the one it is supposed to be in. For example, when a linearly polarized wave passes through a raindrop, its electric vector decomposes into components along the two principal axes of the drop. If the drop is symmetrical, then both the component waves would propagate through exactly equal equivalent path. Therefore, when they emerge out of the drop, each of them would have experienced equal attenuation and phase shift. So, upon combination at the other end of the drop, the polarization direction would be exactly the same as it were at incidence. For all practical cases, however, most of the drops get distorted in shape due to the action of hydrodynamic forces acting upon them, making the shape ellipsoid. Further, it may have some finite canting as they fall. So, if the drop that the wave is incident upon has asymmetry along its two orthogonal axes, and the wave is incident upon it in such a way that the cross-section of the drop, as 'seen' by the wave, is asymmetric, then the attenuation and the phase shift experienced by the two components, decomposed along the two principal axes of propagation, will be different. Therefore, when they recombine after they exit from the drop, the effect will be a deviation in the polarization angle of the wave. The effect of differential attenuation and phase shift causing the depolarization is illustrated below.

Focus

We wish to show here the fact that whenever a signal passes through a medium, in the form of two orthogonal components with differential phase change or attenuation, it results in a depolarization of the wave.

Let us consider that a wave with electric filed $E = E_0 \cos(\omega t)$ is incident on an asymmetric water drop. Let this field be making an angle of τ with one of the principal axes of the drop. Therefore, when the field propagates through it, the field may be considered to be decomposed into two orthogonal components. These components are:

$$E_1 = E_0 \cos \tau \cos(\omega t) \text{ and}$$
$$E_2 = E_0 \sin \tau \cos(\omega t)$$

Now, when these components move through the medium, let E_1 experience an attenuation of $\alpha 1$ while E_2 experience an attenuation of $\alpha 2$. Similarly, due to difference in the effective path, the phase changes due to the propagation are, respectively, kx and $kx + \delta$.

Using the above considerations, the two components of the field, after moving out of the medium, can be expressed as

$$E_1 o = E_0 \alpha 1 \cos \tau \cos(\omega t - kx) \text{ and}$$
$$E_2 o = E_0 \alpha 2 \sin \tau \cos(\omega t - kx - \delta)$$

Now, when they recombine, the resultant component in the direction of its original polarization is

$$E oc = E_0 \ \alpha 1 \cos^2 \tau \cos(\omega t - kx) + E_0 \alpha 2 \sin^2 \tau \cos(\omega t - kx - \delta)$$

Similarly, the resultant component in the direction perpendicular to its original polarization direction is

$$E ox = E_0 \alpha 1 \cos \tau \sin \tau \cos(\omega t - kx) - E_0 \alpha 2 \sin \tau \cos \tau \cos(\omega t - kx - \delta)$$

Now, we shall consider two different cases,

Case 1: $\delta = 0$

For such a case,

$$E oc = (\alpha 1 + \alpha 2) E_0 \cos(\omega t - kx) \text{ and}$$
$$E ox = (\alpha 1 - \alpha 2) E_0 \cos(\omega t - kx)$$

Due to the nontriviality of the cross-polar component $E ox$ at the output, we can conclude that there is certain amount of rotation of the polarization of the field on propagation through the medium. Notice that the cross-polar component only vanishes when the attenuations along both components $\alpha 1$ and $\alpha 2$ are equal.

Case 2: $\alpha 1 = \alpha 2 = \alpha$ and $\delta > 0$

For such a case,

$$Eoc = \alpha E_0 \left[\cos^2\tau \cos(\omega t - kx) + \sin^2\tau \cos(\omega t - kx - \delta))\right]$$
$$= \alpha E_0 \left[\cos^2\tau \cos(\omega t - kx) + \sin^2\tau\{\cos(\omega t - kx)\cos\delta + \sin(\omega t - kx)\sin\delta\}\right]$$

Now, as δ is very small, we can approximate $\cos\delta = 1$ and $\sin\delta = \delta$, then

$$Eoc = \alpha E_0 \left[\cos(\omega t - kx) + \sin^2\tau \sin(\omega t - kx)\delta\right]$$

Similarly, the cross-polar component becomes

$$Eox = \alpha E_0 \cos\tau \sin\tau \left[\cos(\omega t - kx) - \cos(\omega t - kx - \delta))\right]$$
$$= \alpha E_0 \cos\tau \sin\tau \left[\cos(\omega t - kx) - \cos(\omega t - kx)\cos\delta - \sin(\omega t - kx)\sin\delta\right]$$

Using the approximation of small δ as before, we get

$$Eox = \alpha E_0 \cos\tau \sin\tau \left[-\sin(\omega t - kx)\delta\right]$$

So, using the same argument as before, since there remains a finite nonzero component in the cross-polar direction at the output, it can be inferred that a finite depolarization has happened to the wave. Again notice that, here, the orthogonal component vanishes when $\delta = 0$.

However, for the same drop, for certain incidence angle, the cross-section may appear symmetric to the wave, e.g. for vertical incidence of the wave on an elliptical drop compressed vertically and without canting. Then there would be no depolarization effect.

The amount of the depolarization is related both to the amount of differential attenuation and to the differential phase between two orthogonal components that the traversing wave brakes up on entering the drop. Therefore, we can, at this point, appreciate the fact that the amount of depolarization is dependent upon the relative orientation of the drop and the wave, which is determined by a few factors described below.

Canting angle: As we have learnt that a precipitating raindrop will experience vertical drag and will get distorted in shape, the resultant force will compress the drop along the direction of the flow and will effectively make it ellipsoidal. When it falls in stagnant air, the major axis of the ellipsoidal drop will remain horizontal while the minor axis will remain vertical. Even when there is a horizontal component of wind and that the wind speed is constant with height, then the hydrodynamic forces acting upon it will still be irrotational, i.e. have zero curl. This keeps the orientation of the drop unaltered, although there can be a horizontal movement of the drop as a whole. However, when the wind velocity changes with height, there will be a differential force acting of the top and the bottom side of the drop and it creates a torque, however infinitesimally small. This torque is still capable to turn the orientation of the drop, rotating the axes. Consequently, the minor axis will not remain vertical any more, but will make a finite lean angle with the vertical direction. This angle is called the canting angle. Not only the canting angle offers different cross-section areas to the incident wave, but its rapid variation makes the depolarization highly unpredictable.

Tilt angle and elevation: For a given satellite, the signal polarization at the transmission antenna has a definite orientation with respect to any preselected frame of reference. However, this orientation will be different for a receiving antenna on the ground. For example, for geostationary satellite, a vertically polarized wave is transmitted with the electric field oriented in the direction of the meridian on which the satellite is located. Therefore, at the nadir point of the satellite, on the equator and over the earth, this signal will appear to have its electric fields along the local horizontal plane in the N–S direction. However, for locations away from the nadir, the situation is different. For a location on the same meridian but due North, due to the relative curvature of the spherical wave front and of the earth's spherical shape, there generates a certain included angle between the plane of the wave and the local horizontal plane over the earth. Therefore, the signal polarization will have a finite vertical component, which will increase with increasing latitudes. For other locations, this polarization will have different tilt. This tilt determines the angle with which the wave is incident upon the precipitating drop.

As the change in perspective exposes different cross-section areas of the drop to the wavefront of the traversing signal, the elevation angle also plays a role here. For all practical canting angles, high elevation angle gets the 'top view' of the drops and sees approximate axial symmetry about the direction of its propagation. But for low elevation, the wave sees the 'side view' of the descending drop and hence observes asymmetry in the drop dimension. This results in a large depolarization for low elevation angles with consequent reduction in XPD.

Frequency: Depolarization is also dependent upon the frequency of the wave. Since the total amount of attenuation is frequency-dependent and also the differential phase change is dependent upon the phase constant '*k*' and hence on the wave frequency, the latter plays a prominent role in determination of the depolarization.

The variation of depolarization is illustrated in Fig. 6.11 along with the corresponding signal fade due to rain.

Rain scintillation

When an electromagnetic wave, propagating through the atmosphere, traverses a medium which varies randomly in its characteristics, or whenever a wave encounters abrupt small-scale perturbations in the refractive index of that medium in space and

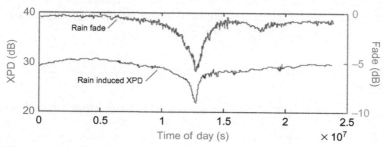

FIG. 6.11

Rain-induced fade and depolarization.

time, scintillation in the received signal is experienced both in amplitude and in instantaneous phase. The resultant fluctuations are short-term and random. Rain, consisting of discrete water drops falling through the atmosphere, causes abrupt and random refractive index perturbations. The consequent fast signal fluctuations along an Earth–satellite path is termed as rain scintillation. The scintillations occur both preceding and during a rain event. This scintillation of signals, although much less severe than the rain attenuation, can significantly degrade the quality of the received signal in terms of received signal to noise power ratio and hence affecting the quality of service, especially of the low-power-margin communication systems. The effect is relatively small for vertical polarization, but for high frequencies and horizontal polarization, it can lead to significant scintillations when both precipitation rate and raindrop canting angle variations contribute significantly. The effect is less intense at lower frequencies (Haddon and Vilar, 1986).

The atmospheric noise also increases during the rain. As the plane wave enters the region with rain, it is both absorbed and scattered. The sum of the powers thus lost through these two processes from the incident powers constitutes the total rain attenuation. The lost power raises the effective radiation temperature of the system which thus radiates more incoherent noise as per the laws of radiation. We shall read about this in details in Chapter 8. Of all these radiated noise power, only that which is directed forward is picked up by the receiving antenna within its beam width. This will lead to an increment in the overall noise in the receiver system. Consequently, due to the increase in noise, the signal quality deteriorates.

CONCEPTUAL QUESTIONS

1. Is it correct to say that near resonant frequency, the dielectrics behave like a conductor for electromagnetic fields?
2. Derive Eq. (6.30), i.e. the rain rate as a function of terminal velocity of the drops and the drop size density distribution.
3. Rain causes both attenuation and depolarization. How can you distinguish these two effects when you see the copolar power of a signal is depreciated? Draw a polar plot of the electric field vector variation with increasing rain.
4. How can you say that the received signal fluctuation is due to scintillation and not due to satellite transmission power fluctuation?

REFERENCES

Ajayi, G.O., Olsen, R.L., 1985. Modelling of a tropical raindrop size distribution for microwave and millimetre wave applications. Radio Sci. 20, 193–202.
Allnutt, J.E., 2011. Satellite-to-Ground Radiowave Propagation, second ed. The Institution of Engineering & Technology, London, ISBN: 978-1-84919-150-0.

Bertram, A.R., Martellucci, A., Fiser, O., Habetha, J., 2002. COST Action 255 Final Report: Radiowave Propagation for SatCom Services at Ku-Band and Above. In: Harris, R.A. (Ed.), ESA Publications Divison, Noordwijk, The Netherlands, ISBN: 92-9092-608-2.

Capsoni, C., Fedi, F., Magistroni, C., Pawlina, A., 1987. Data and theory for a model of the horizontal structure of the rain cells for propagation applications. Radio Sci. 22 (3), 395–404.

Crane, R.K., 1977. Prediction of the effects of rain on satellite communication systems. Proc. IEEE 65, 456–474.

Crane R.K., 1979. A global model for rain attenuation prediction. In: Paper Presented at Electronics and Aerospace Systems Convention (EASCON), Arlington, USA.

Crane, R.K., 1982. A Two Component Rain Model for the prediction of Attenuation statistics. Radio Sci. 17 (6), 1771–1787.

Diermendjian, D., 1969. Electromagnetic Scattering on Spherical Polydispersion. Elsevier, New York, NY.

Goldhirsh, J., Katz, I., 1979. Useful experimental results for earth-satellite rain attenuation modelling. IEEE Trans. Antennas Propag. 27 (3), 413–415.

Green, H.E., 2004. Propagation impairments on Ka band Satcom links in tropical and equatorial regions. IEEE Antennas Propag. Mag. 46 (2), 31–45.

Gunn, R., Kinzer, G.D., 1949. The terminal velocity of fall for water droplets in stagnant air. J. Meteorol. 6, 243–248.

Haddon, J., Vilar, E., 1986. Scattering induced microwave scintillations from clear air and rain on earth space paths and the influence of antenna-aperture. IEEE Trans. Antennas Propag. 34 (5), 646–657.

Houze Jr., R.A., 1981. Structures of atmospheric precipitation systems: a global survey. Radio Sci. 16 (5), 671–689. http://dx.doi.org/10.1029/RS016i005p00671.

Houze, R.A., 1993. Cloud Dynamics. Academic Press Inc., San Diego, CA.

Ishimaru, A., 1991. Electromagnetic Wave Propagation, Radiation, and Scattering. Prentice Hall, Upper Saddle River, NJ. ISBN: 0132490536, 9780132490535.

ITU-R, 2012. Recommendation ITU-R P.835-5, Reference Standard Atmospheres. ITU.

ITU-R, 2013. Recommendation ITU-R P. 839-4, Rain Height Model for Prediction Methods. ITU.

Karmakar, P.K., 2012. Microwave Propagation and Remote Sensing: Atmospheric Influences with Models and Applications. CRC Press, Boca Raton, FL.

Kerr, D.E., 1951. Propagation of short radio waves. MIT Radiation Laboratories Series, McGraw-Hill Book Company, New York, NY (Chapter 8).

Laws, J.O., Parsons, D.A., 1943. The relation of raindrop size to intensity. Eos Trans. AGU 24, 452–460.

Maral, G., Bousquet, M., 2002. Satellite Communications Systems, Systems, Techniques and Technology, fourth ed. John Wiley and Sons, Ltd, Chichester.

Marshall, J.S., Palmer, W.M., 1948. The distribution of rain drops with size. J. Meteorol. 5, 165–166.

Martinez, G.M., Kostinski, A.B., Shaw, R.A., Garcia, F., 2009. Do all raindrops fall at terminal speed? Geophys. Res. Lett. 36. L11818, http://dx.doi.org/10.1029/2008GL037111.

Mitra, A.P., 1972. Tropospheric Environment. In: Mitra, A.P. (Ed.), Report No. 3, Advanced Courses on Tropospheric Propagation and Antenna Measurements, vol. 1. Centre for Research on Troposphere, Radio Science Division, NPL, New Delhi.

Moupfouma, F., Martin, L., Spanjaard, N., Hughes, K., 1990. Rainfall rate characteristics for microwave systems in tropical and equatorial areas. Int. J. Satell. Commun. 8. 151-lh.

Troposphere, 2016. NASA Spaceplace, 2016. http://spaceplace.nasa.gov/troposphere/en Accessed 2 October 2016.

Pontes, M.S., Mello Silva, L.A.R., Migliora, C.S.S., 1990. Ku Band slant path radiometric measurements at three locations in Brazil. Int. J. Satell. Commun. 8, 239–249.

Stutzman, W.L., Dishman, W.K., 1982. A simple model for the estimation of rain-induced attenuation along earth-space paths at millimetre wavelengths. Radio Sci. 17 (6), 1465–1476.

Van de Kamp, MMJL, 1998. Climatic Radiowave Propagation Models for the Design of Satellite Communication Systems.

Verma, A., Jha, K.K., 1996. Raindrop size distribution model for Indian climate. Indian J. Radio Space Phys. 25 (1), 15–21.

Visconti, G., 2001. Fundamentals of Physics and Chemistry of the Atmosphere. Springer-Verlag, Berlin, Heidelberg. http://dx.doi.org/10.1007/978-3-662-04540-4.

Zainal A.R., Gover I.A., Watson P.A., 1993. Rain rate and drop size distribution measurements in Malaysia. In: *Proceedings of the International Geo-science and Remote Sensing Symposium*, pp. 309–311.

FURTHER READING

ESA, 2002. Radiowave Propagation Modelling for Satcom Services at Ku Band and Above. COST Action 255, ESA Publications, Noordwijk.

Tropospheric impairments: Measurements and mitigation

7

The impairments offered by the tropospheric elements to a propagating electromagnetic wave are considerably high for frequencies above 10 GHz. These impairments are not only functions of the signal frequency, but also their magnitude is determined by the characteristics of the medium through which they pass. Among them, the attenuation or fade, depolarization and scintillation are most prominent ones. We have already seen in the last chapter that these impairments are mainly affected by rain, cloud, water vapour as well as by the dry gases in the air. Most of these impairments increase with the frequencies and cause deterioration in the performance of the application the affected signals are used for. This calls for mitigating the effects, which needs a-priori provisions for effectively accomplishing the same. For example, for frequencies in Ka band and above, the total attenuation is so intense that it cannot be compensated by static power margin only. Static margins are fixed additional powers over the minimum requirement in a satellite link, used to compensate the power loss in the signal during the propagation through impeding atmosphere. The part of the attenuation above this amount that cannot be compensated by static margin is to be balanced by other adaptive techniques in real time. Therefore, additional resources are necessary to be arranged for mitigating such impairments, both fixed and adaptive type. So, estimating them quantitatively becomes a mandatory requirement in the process.

Estimations are done in the form of long-term statistics of the impairments and as near real-time predictions, as well. For both of the kinds, measurement of the impairment variable is necessary. It may be direct measurement of the impairment or of a more conveniently measurable meteorological parameter related to it, whose relation with the required impairment is already defined through a model. Such readily available meteorological models are easy to use. However, one cannot undermine the importance of direct measurement of the impairment in the mitigation process. Measurements are also necessary to develop the model itself.

In this chapter, we shall first learn about measuring these parameters. Starting with the brief idea about the few measuring instruments, mainly meteorological, we shall explain how the corresponding measurements are used in models to derive the impairment parameters of our interest. In the process, we shall also study the models of impairments, expressed as a function of these meteorological parameters. Finally, we shall learn in brief about a few techniques for mitigating the fade in real time. This will complete our chapter.

Satellite Signal Propagation, Impairments and Mitigation. http://dx.doi.org/10.1016/B978-0-12-809732-8.00007-7

7.1 MEASUREMENTS OF PARAMETERS

7.1.1 ATMOSPHERIC AND METEOROLOGICAL MEASUREMENTS

7.1.1.1 Radiosonde

A radiosonde is an instrument for measuring atmospheric parameters and consists of different sensors combined with a telemetry instrument package. This instrument is carried to different predetermined altitudes of the atmosphere, usually by a weather balloon, to measure the various target parameters which are then transmitted to a ground receiver.

Modern radiosondes measure directly or derive the atmospheric variables from associated measurements and record their variations with altitude. These variables are pressure, temperature, relative humidity, wind (both wind speed and wind direction), cosmic ray readings at high altitude. These measurements are tagged with geographical position of latitude, longitude, altitude and time.

Radiosondes are essential source of atmospheric and meteorological data. Hundreds of such instruments are launched all over the world daily. However, modern remote sensing by satellites, aircraft and ground sensors are currently also increasing in numbers. Although they can be the alternative sources of atmospheric data, Radiosondes measurements are in situ while satellite or airborne measurements are of remote sensing type. Therefore, very naturally, the accuracy and resolution offered by the radiosonde cannot be matched in general, by these alternative sources. Hence, radiosonde observations still remain essential to modern meteorological activities and atmospheric research.

7.1.1.2 Ground-based radiometers

All material particles at finite temperature absorb a finite amount of energy. Under equilibrium condition, it also radiates equal amount of energy back in all frequencies and directions. Therefore, the atmospheric particles, in equilibrium, radiate uncorrelated noise like radiations of finite power to the atmosphere. We shall learn these in much more details in the next chapter. The radiometer records this radiation that it can 'see' within its antenna beam width. For any given finite bandwidth, the received noise spectrum may be assumed to be constant and the total power collected within this band is then proportional to the corresponding noise temperature. A ground-based radiometer is an instrument that looks towards the sky and measures this total noise temperature including the atmospheric noise temperature. Once the atmospheric noise temperature is known, the amount of absorption expected at this temperature by the constituting particles may be estimated and hence the absorptive loss of any propagating signal may be obtained. Notice that, here we are only considering the conditions when the scattering is negligible so that the absorption dominates as a cause for attenuation. The relation between the absorptive attenuation and the measured temperature can be related by using the radiative transfer equation under the condition of no scattering. The relation is given by

$$A = 10 \log \left(\frac{T_\mathrm{m} - T_\mathrm{c}}{T_\mathrm{m} - T_\mathrm{A}} \right) \tag{7.1}$$

where, T_m is the atmospheric mean radiating temperature (K) and T_c is the cosmic background temperature, usually chosen as 2.7 K. In general, T_m depends upon frequency and attenuation through the physical processes that produce the attenuation. For attenuation values less than 6 dB and frequencies below 50 GHz, T_m can be approximated, with a weak dependence on frequency. Its value can be chosen as 265 K in the 10–15 GHz range and as 270 K in the 20–30 GHz range. When surface temperature is known, a first estimate of T_m can be also obtained by multiplying the surface temperature by 0.95 in the 20 GHz band and by 0.94 for the 30 GHz window. T_A is the total noise temperature at the antenna.

7.1.1.3 Rain gauge

A rain gauge is a meteorological instrument to measure the precipitating rain in a given amount of time per unit area. The instrument consists of a collection container which is placed in an open area. The precipitation is measured in terms of the height of the precipitated water accumulated in the container per given time and is expressed in millimetres. Since the same amount of rain precipitation is assumed to be occurring around the container, the area of collection is not a factor. However, it should not be too small, neither should it be too large. Due to spatial uniformity of rainfall, 1 mm of measured precipitation is the equivalent of 1 L of precipitated rain water volume per metre squared.

A tipping bucket rain gauge consists of a pair of rainwater collecting buckets. It is covered by a funnel, with an open collector area at the top where A is the area of collection. The buckets are so placed on a pivot that only one bucket remains under the funnel at a time. During rain, rain water is collected in the collecting bucket, through the funnel. When the water fills up to a known point of the bucket, say having a volume v, the bucket tips, emptying the water. When one bucket tips, the other bucket quickly moves into place to collect rainwater. Each time a bucket tips, an electronic signal is sent to a recorder which is registered by the instrument with time stamp.

To calculate the total rainfall in a given interval, the total number of tips occurring in the interval is observed. If this number be 'N', then in that time the total volume of water collected is $N \times v$, where v is the volume of the bucket necessary to be filled for tipping. Now, if A be the area of collection of the rain at the funnel top, then the total rain amount RA, occurring in the given time in terms of height is

$$RA = \frac{Nv}{A} \tag{7.2}$$

Therefore, the least count of the instrument is $k = v/A$ which is the rainfall measured for a single tip. The same arrangement may also be used to calculate the rain rate. If two subsequent tips have occurred in an interval δt, and $N \times k$ mm of total rain has occurred in this time δt, then the rain rate occurring in this instant is

$$R = \frac{Nk}{\delta T} \tag{7.3a}$$

To express the rain rate in the standard form of mm/h, the following expression can hence be used

$$R = \frac{Nk}{\delta t} \times 3600 \, \text{mm/h}$$

$$= \frac{Nv}{A\delta t} \times 3.6 \, \text{mm/h}$$

(7.3b)

where k is in mm and is given by $k = v/A \times 10^{-3}$, where v is in cubic centimeter (cc) and the area is in metre squared. δt is in second.

The tipping bucket rain gauge is especially good at measuring drizzle and light rainfall events. The resolution is better when the least count k is small and the clock measuring δt is precise. However, the least count cannot be made arbitrarily small, as for such case there will be many tips of the bucket during heavy rain that some of the rain will go uncollected during the bucket transition leading to an underestimate of the rainfall and inaccuracy. Fig. 7.1 shows a rain gauge of tipping bucket type.

7.1.1.4 Disdrometer

The models used for predictions of rain attenuation are mainly a function of rainfall rate and rain drop size distribution for each rain rate, in addition to the signal path length through rain. Hence, for modelling purposes, it is very important to measure both rainfall rate and rain drop size distribution (DSD). A 'Disdrometer' is an

FIG. 7.1

Tipping bucket rain gauge.

instrument that is used to measure and record rain drop counts for different sizes in a given interval and derive rainfall rate and raindrop size distributions from it. This instrument typically consists of two main units, the sensor, and the processor. The sensor is exposed to the raindrops, the distribution of which is to be measured. For every drop sensed by the sensor, corresponding electrical information is sent to the processor. Typically, the sensors are either acoustic or optical type, which estimates the diameter of the drops. In the first kind, the drop diameter is estimated from the impact. The terminal fall velocity can be derived from the measured momentum, which in turn is a function of the drop diameter. In the optical type, it is done by finding the time of passage of the drop through a laser beam of definite width. The instrument also counts the number of drops encountered by the sensor in a predefined interval of time for every drop diameter and records the information in a suitable format. To convert it to the Drop Size Distribution (DSD), we only need to follow few simple mathematical steps.

For deriving the drop size distribution, the basic assumption taken is that, the terminal drop velocity of the raindrops are known as the function of their diameters. The terminal drop velocities provided by Gunn and Kinzer (1949) is typically used for the purpose. Let v_D be the terminal velocity of the drop of diameter D. Also, let N_D be the number of drops of that size striking the sensor area A in the predefined interval of time, δt. This means, there were N_D numbers of drops of diameter D in a volume $v_D \times \delta t \times A$ of the rain when the instrument did the measurement. Therefore, the number of drops per unit volume is

$$n_D = \frac{N_D}{v_D \delta t A} \tag{7.4}$$

This is the density for the drops of diameter D. Similarly, the density for any drop can be determined. If the measurement resolution of the instrument for drop diameter be δD, then the drop size density distribution may be given as

$$n(D) = \frac{N(D)}{v(D) \delta t A \delta D} \tag{7.5}$$

where $n(D)$ is the density distribution function, $v(D)$ is the function of the terminal velocity of the drop as a function of diameter D and $N(D)$ is the number of drops measured in the diameter range D to $D + dD$.

7.2 PROPAGATION IMPAIRMENT MODELS
7.2.1 LONG-TERM IMPAIRMENT MODELS

The propagation impairments on an Earth-space path are contributed by different tropospheric agencies. The most important out of them considered for the purpose of satellite-based communication and other similar applications are:

- attenuation by atmospheric gases
- attenuation by clouds
- attenuation and depolarization by rain
- scintillation and multipath effects

Each of these contributions has its own characteristics. Their variations can be represented as a function of frequency, geographic location and elevation angle. However, at elevation angles above 10°, gaseous attenuation, rain and cloud attenuation and scintillation will be significant depending on propagation conditions. In order to mitigate the effects of these impairments, it is necessary to have corresponding impairment models. These impairment models are nothing but the mathematical representations which relate either the instantaneous or the long-term statistical values of the impairments experienced by the satellite signal with those of the meteorological parameters. The following features can be considered to be the desirable features of a prediction method (Fedi, 1981),

i. Simple: The model should be easy to apply to communication system calculations. The unnecessary introduction of new parameters and mathematical complexity is to be avoided.

ii. Physically sound: As much as possible, the model should be checked against directly observed physical data, such as spatial rain behaviour.

iii. Tested. The model should be tested against measured data from many different regions and events. Emphasis should be given to the data with low percentages of occurrence probability in time, which typically represent larger impairment values and are of most interest to system designers.

iv. Flexible: As more data become available and a deeper understanding is obtained, model refinements are surely to follow. The model should be structured to accept modifications.

In this section of the chapter, we shall learn about a few models of the impairments. We know that under the clear sky condition, the constituent gases are the only elements to introduce any impairment to the propagating radio waves. Therefore, we shall start with the models for gaseous absorption.

7.2.1.1 Gaseous attenuation model

Attenuation by atmospheric gases which is entirely caused by absorption depends mainly on frequency, elevation angle and on the altitude above sea level, which determine the pressure and the particle density of the absorbing gas. At frequencies below 10 GHz, the gaseous attenuation may normally be neglected but its importance increases with frequency above 10 GHz, especially for low elevation angles. Oxygen and water vapour are the two major candidates participating in the absorption process where the strength of their absorption depends upon their relative content. At a given frequency, the oxygen contribution to atmospheric absorption is relatively constant (ITU-R, 2013a). However, both water vapour density and its vertical profile are quite variable and are mainly responsible for the variation in the gaseous absorption. Therefore, the maximum gaseous attenuation typically occurs during the season of maximum rainfall (ITU-R, 2013b).

We have seen in Chapter 3 that, theoretically the gaseous absorption is selective in nature and maximizes at the resonant frequencies. We have also seen that, there may be more than one resonant frequency, each corresponding to a degree of freedom of the molecules. Since the resonant frequencies are situated far apart, for radio wave propagation with frequency near to any one of the resonant frequency, the effects of the nearest one are dominant while those of the others are comparatively negligible. The variation of the absorbent terms with frequency is called the shape factor. At any arbitrary frequency, the total effect is determined by the shape factor at that particular frequency for each of the component resonant term.

In a more generalized way, the specific attenuation of the tropospheric gases may be differentiated into specific attenuation due to dry air and that due to the wet component of water vapour. Both dry air and water vapour-induced specific attenuation can be evaluated at any value of pressure, temperature and humidity, primarily by means of a summation of the individual resonance lines at the given frequency. The specific gaseous attenuation is given by:

$$\gamma = \gamma_D + \gamma_w = 0.1820 f N''(f) \text{ dB/km} \tag{7.6}$$

where γ_0 and γ_w are the specific attenuations (dB/km) due to dry air and water vapour, respectively. f is the frequency (GHz) and $N''(f)$ is the imaginary part of the frequency-dependent complex refractivity, combined over all components contributing to the absorption process and calculated at the required frequency. This expression can be derived from Eq. (3.16) and is left as an exercise to the readers. N'' can be represented as,

$$N''(f) = N_D''(f) + \sum_i S_i F_i \tag{7.7}$$

$N_D''(f)$ is the continuum due to pressure-induced dry nitrogen absorption and the Debye spectrum. S_i the strength of the ith line, F_i is the line shape factor at the required frequency and the sum extends over all the lines that affect the signal's frequency. The expressions for the strengths and shape factor are given in ITU-R recommendation ITU-R, P.676 (ITU-R, 2013a).

Estimating the specific attenuation by summing over the line shape factors, as per the above procedure, is accurate but quite cumbersome. We can represent the specific attenuation for gaseous absorption in much more simpler empirical expressions, however with lesser accuracy. For such representations, the specific attenuation for oxygen is given as (Ippolito, 1986)

$$\gamma_o = \left[\frac{6.6}{(f-0)^2 + 0.33} + \frac{9}{(f-57)^2 + 1.96} \right] f^2 10^{-3} \qquad \text{for } f \le 57 \text{ GHz}$$

$$= 14.9 \qquad \text{for } 57 < f < 63 \text{ GHz} \tag{7.8a}$$

$$= \left[\frac{4.13}{(f-63)^2 + 1.1} + \frac{0.19}{(f-118.7)^2 + 2} \right] f^2 10^{-3} \qquad \text{for } f \ge 63 \text{ GHz}$$

Here, we can see different resonant frequencies originating due to the natural frequencies of oscillation of the gas molecules for different degrees of freedom. Similarly, the water vapour also contributes to the absorption through the function (Ippolito, 1986)

$$\gamma_w = \left[0.067 + \frac{2.4}{(f-22.3)^2 + 6.6} + \frac{7.33}{(f-183.5)^2 + 5} + \frac{4.4}{(f-323.8)^2 + 10} \right] f^2 \rho 10^{-4}$$
$$\text{for } f < 350 \text{ GHz} \tag{7.8b}$$

Here, f is the frequency of the signal used in GHz and ρ is the water vapour density in g/m³. The corresponding variations of γ_o and γ_w with frequency are shown in Fig. 7.2.

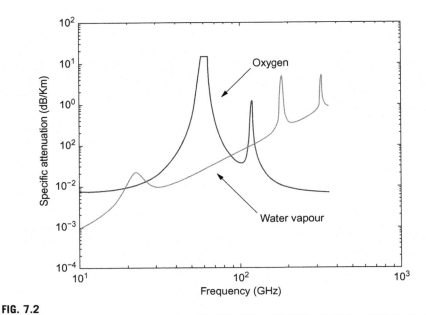

FIG. 7.2

Variation of gaseous absorption with frequency.

To get the total attenuation, the specific attenuation is to be integrated along the effective signal path. Since the specific attenuation varies with pressure, density, etc. it is a function of height h. Therefore, in carrying out the integration of specific attenuations to calculate the total attenuation for a satellite link, it is necessary to know the specific attenuation at each point of the path in the link. The effective path responsible for the attenuation may be accurately determined simply by dividing the atmosphere into horizontal layers. Each individual layer has its definite set of values of the meteorological parameters of pressure, temperature and humidity. For each such layer, the corresponding meteorological parameters are used to get the corresponding N''. Then, as the integration is done along the path through the atmosphere, for each segment of the path, the corresponding specific attenuation, thus calculated by adding up individual absorption lines, are used as the integrand. In the absence of local profiles from radiosonde data or any other sources, or the reference standard atmospheres in Recommendation ITU-R P.835 may be used.

The total slant path gaseous attenuation, $A_G(H, \theta)$, from a station with altitude, H, and elevation angle, θ, can be calculated as follows when $\theta \geq 0$:

$$A_G(H, \theta) = \int_H^\infty \frac{\gamma(h)}{\sin\theta_{\text{eff}}} dh \qquad (7.9)$$

It is to be noted here that, due to the refractive effect, the beam bends, as it moves from one layer to the other with different refractive indices. The effective elevation angle θ_{eff} at height h changes from that of the ground, θ (Fig. 7.3).

Box 7.1 Gaseous Absorption Model

In this box section, we shall run the MATLAB program absorb.m to generate the figure representing the specific attenuation due to the oxygen and water vapour. The figure shows gradual variation of the function with peaks arising at the resonant frequencies. Notice that more than one peaks appear corresponding to different degrees of freedom. The total combined specific attenuation is shown in black dots.

Also, observe the functions have a constant added with the selective term in the denominator. These terms arise due to collisions and other dissipative forces and determine the width of the peaks. Change these values and see the variations in the peak. Similarly, change the density factor ρ in the program to any pragmatic value and note the changes.

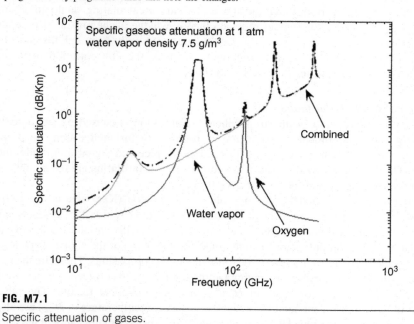

FIG. M7.1

Specific attenuation of gases.

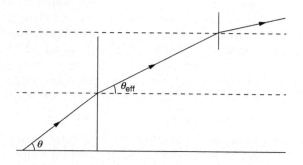

FIG. 7.3

Path locus of refracted wave through troposphere.

θ_{eff} can be determined from the geometric elevation angle θ, considering the ray bending that occurs on a spherical earth. The value of θ_{eff} can be determined from the elevation angle θ, using the relation based on Snell's law in polar coordinates as,

$$\theta_{\text{eff}}(h) = \cos^{-1}\left[\frac{(R_e+H)\,n(H)\cos\theta}{(R_e+h)\,n(h)}\right] \tag{7.10}$$

where R_e is the radius of the earth and $n(h)$ is the atmospheric radio refractive index, calculated from pressure, temperature and water vapour pressure at a height h along the path. Pertinent ITU recommendation is available (ITU-R, 2012a) which is useful in this regard.

It is to be noted carefully that the above method for calculating slant path attenuation by gases and water vapour relies on the knowledge of the profile of gaseous density and water vapour pressure (or density) along the path. Although the concentration of atmospheric oxygen is relatively constant, the concentration of water vapour varies both with location and with time.

7.2.1.2 Scintillation model

We have learnt previously that tropospheric scintillation is caused by small-scale refractive index inhomogeneities induced by tropospheric turbulence along the propagation path. It results in the rapid fluctuations of received signal amplitude and phase. The parameters used for the characterization of scintillation are the amplitude and phase deviations and their respective variances.

There are various models for tropospheric scintillation (Karasawa et al., 1998; ITU-R, 2015a; Van de Kamp et al., 1999; Otung, 1996; COST Action 255, 2002), which predicts the statistical nature of the scintillation. They give the variance values of the fluctuations in the amplitude or log amplitude and the phase of the signal. Typically, every scintillation model has few of the following input components. These are the frequency f, the antenna aperture averaging factor G, total antenna efficiency η, elevation angle θ, the turbulent layer height H_t and the meteorological factors. The other meteorological-dependent conditions like the surface air temperature or N_{wet}, are also sometimes required as input. However, although turbulence is a major cause of the scintillation, no model has yet compensated for wind velocity (COST Action 255, 2002).

Here we shall discuss the ITU model for the tropospheric scintillation for elevation above $5°$ and for low to medium fade depths. It is based on monthly or longer averages of temperature and relative humidity and reflects the specific climatic conditions of the site. As the average surface temperature and average surface relative humidity vary with season, the scintillation fade depth distribution varies accordingly. The model is valid up to at least 20 GHz. Since the standard deviation σ of the scintillation is correlated to the N_{wet} component, the first task in the model is to represent the σ for a reference signal with the N_{wet} term. It is given as (ITU-R, 2015a)

$$\sigma_{\text{ref}} = (3.6 + 0.1N_{\text{wet}}) \times 10^{-3} \text{ dB} \tag{7.11}$$

N_{wet} can be derived from the measured values of the meteorological parameters. For a given value of T, where T is the average surface ambient temperature (°C), the saturation water vapour pressure, e_s (hPa), can be calculated. Then the wet term of the radio refractivity, N_{wet}, can be derived corresponding to e_s, for the given T and average surface relative humidity H_R in percentage. The relations are given in Recommendation ITU-R P.453. Both T and H_R are the average values at the site over a period of one month or more.

Now, depending upon the elevation angle θ, the effective path length through the turbulence may be derived as

$$L_{eff} = \frac{2h_L}{\sqrt{(sin^2\theta + 2.35 \times 10^{-4})} + sin\theta} \quad m \tag{7.12}$$

where h_L, the height of the turbulent layer and taken as 1000 m. Once the effective length is obtained, the antenna averaging factor g is obtained as

$$g(x) = \left[3.86(x^2+1)^{11/12} sin\left(\frac{11}{6}tan^{-1}\frac{1}{x}\right) - 7.08(x^{5/6})\right]^{1/2} \tag{7.13}$$

where $x = 1.22 D_{eff}^2 (f/L_{eff})$ and $D_{eff} = D\sqrt{\eta}$, η being the antenna efficiency factor. 'f' is the carrier frequency in GHz. For values of $x > 7$, the argument of the square root is negative. It means the antenna averaging compensates the fluctuation due to propagation and then the predicted scintillation fade depth for any time percentage is taken as zero and the following steps are not required.

Finally, the standard deviation of the signal for the applicable period and propagation path becomes

$$\sigma = \sigma_{ref} f^{\frac{7}{12}} \frac{g(x)}{(sin\theta)^{1.2}} \tag{7.14}$$

The function $g(x)$ is a decreasing function of x asymptotically reaching the value 0 for large x. Thus, the scintillation decreases with increasing D_{eff} and increases with L_{eff}. Due to the presence of the term $f^{7/12}$, the total term σ increases with f.

The time percentage factor, $a(p)$, for the time percentage, p, in the range between $0.01\% < p < = 50\%$ is obtained as a polynomial of $log(p)$. If we represent $log(p)$ as y, then

$$a(p) = -0.061y^3 + 0.072y^2 - 1.71y + 3 \tag{7.15a}$$

Then the scintillation fade depth, $A_S(p)$, exceeded for $p\%$ of the time is obtained as:

$$A_S(p) = a(p)\sigma \quad dB \tag{7.15b}$$

7.2.1.3 Cloud attenuation

The satellite signals passing through the clouds get scattered and hence attenuated. The attenuation due to clouds is small compared to the other elements of impairments like rain. Yet, it is important to have an idea of the cloud attenuation as it may be a factor of importance for those systems where the availability requirement is high and the margin is minimal due to very tight power economy.

The attenuation due to suspended water droplets contained in atmospheric clouds can be determined with considerable accuracy. In a way similar to other path attenuations, the total attenuation is considered as the product of the specific attenuation and the effective path through the cloud. The specific attenuation may be determined from the Rayleigh model of electromagnetic wave scattering. For clouds consisting entirely of small droplets, generally less than 0.1 mm, the Rayleigh approximation is valid for frequencies below 100 GHz. Each particle in the cloud scatters a part of the signal which thus loses a portion of its power to get attenuated. The specific attenuation, i.e. the attenuation per unit distance, is related to the number of particles present in unit volume of the cloud. This is again proportional to the total water content per unit volume. Therefore, the specific attenuation for cloud can be expressed in terms of liquid water content per unit volume and can be written as

$$\gamma_c = K_1 M \quad \text{dB/km} \tag{7.16}$$

where γ_c is the specific attenuation in dB/km within the cloud, K_1 is the specific attenuation coefficient in $(\text{dB/km})/(\text{g/m}^3)$ and M is the liquid water density in the cloud expressed in g/m^3.

Specific cloud attenuation coefficient is derived from a mathematical model based on Rayleigh scattering. It uses a Double-Debye model for the dielectric permittivity $\varepsilon(f)$ of water which is expressed in complex form as

$$\varepsilon(f) = \varepsilon' + \varepsilon'' \tag{7.17a}$$

where ε' and ε'' are functions of frequency given by

$$\varepsilon'(f) = \frac{\xi_0 - \xi_1}{1 + (f/f_p)^2} + \frac{(\xi_1 - \xi_2)}{1 + (f/f_s)^2} + \xi_2 \tag{7.17b}$$

$$\varepsilon''(f) = (f/f_p)\frac{\xi_0 - \xi_1}{f_p\left[1 + (f/f_p)^2\right]} + (f/f_s)\frac{\xi_1 - \xi_2}{1 + (f/f_s)^2} \tag{7.17c}$$

Here, $\xi_0 = 77.66 + 103.3\,(\Theta - 1)$, $\xi_1 = 5.48$, $\xi_2 = 3.52$ and the parameter Θ is a function of absolute temperature, T given by $\Theta = 300/T$ (ITU-R, 1999).

f_p and f_s are the principal and secondary relaxation frequencies, respectively, and are given as

$$f_p = 20.09 - 142(\Theta - 1) + 294(\Theta - 1)^2 \text{ GHz} \tag{7.18a}$$

$$f_s = 590 - 1500(\Theta - 1) \text{ GHz} \tag{7.18b}$$

Knowing ε' and ε'' in terms of temperature profile and frequency, we can get to calculate the value of the specific attenuation coefficient K_1 for frequencies up to 1000 GHz, as

$$K_1 = \frac{0.819f}{\varepsilon''(1 + \tau^2)} \text{ (dB/km)}/(\text{g/m}^3) \tag{7.19}$$

where f is the frequency in GHz and $\tau = (2 + \varepsilon')/\varepsilon''$.

Finally, as we get the specific attenuation in terms of K_1 and M, it is possible to express the total attenuation in terms of these frequency-dependent terms and the total water content per unit volume and total length through the cloud. The total cloud attenuation A_C, over a slant path of the signal may be obtained as (ITU-R, 1999):

$$A_C = \frac{\gamma_c l}{\sin\theta} \quad \text{dB} \quad \text{for } 90° \geq \theta \geq 5°$$

$$= \frac{MK_1 l}{\sin\theta} \quad\quad\quad\quad\quad\quad\quad (7.20)$$

$$= \frac{m_L K_1}{\sin\theta}$$

where θ is the elevation angle and $m_L = M \times l$ is the total columnar content of liquid water, expressed in kg/m^2. When $\theta = 90$, i.e. for vertical direction, the specific attenuation coefficient, $K_1 = A/m_L$, i.e. attenuation per unit columnar mass of the cloud. This is also known as the mass absorption coefficient and the total vertical attenuation may be expressed as

$$A_{Cv} = K_1 m_L \quad\quad\quad\quad\quad\quad\quad (7.21)$$

Thus, to obtain the attenuation due to clouds for a given probability value, the statistics of the total columnar content of liquid water m_L (kg/m^2) or, equivalently, mm of precipitable water for a given site must be known. This approach requires the assessment of the cloud vertical profile. This can be derived from instrumental measurements. Whenever these measurements are not available, the cloud attenuation may still be derived by using the values of the total columnar content of cloud liquid water (normalized to 0°C) provided by the reference figures of ITU recommendation in ITU-R, P.840 (ITU, 1999).

7.2.1.4 Rain fade

The growing demand of satellite communication services has congested the currently available radio spectrum in the popular C band to such an extent that a need has been felt for shifting to new higher spectral region, which can additionally provide larger bandwidth for broadband communication. So, new frequency band above 20 GHz is being explored for such purposes.

However, the advantages that we get from the higher band are not without a cost. The rain attenuates the signals in these higher frequencies. So, the signals suffer strong fade as the wavelength approaches the rain drop diameters. This effect is most conspicuous in tropical countries where the amount of rainfall is significantly high. Therefore, for satellite-based communication systems, it has become necessary to mitigate these effects such that the link can be sustained and the service can be effectively continued even in rainy conditions. For this purpose, it is required to find out a prediction model, for all ranges of rainfall rate as well as for all frequencies, particularly in the affected band.

Rain attenuation researches started from the latter half of the 20th century and since then, several theoretical and experimental studies were made in respect of rain attenuation to obtain better models.

The fundamental concepts in attenuation modelling comprise of mostly three aspects, viz. the approach of modelling, elements of the model and modelling algorithm. Depending upon the approach used for modelling, the extent of the usage of the empirical data varies. The approaches may be classified into the followings:

Empirical approach: The empirical approach develops the entire expression for attenuation directly from measured attenuation. Therefore, the model is based on the measured data and hence lacks the necessary physical foundation. The associated parameters thus do not relate directly to any actual rain parameters. Therefore, although the models are easy to apply, the approach leads to concerns on the range of applicability. The methods of Lin (1979) and the older models of International Radio Consultative Committee (CCIR) are examples of empirical models.

Semi-empirical approach: In this approach, the model is mostly developed on the physical basis, while the values of some of the parameters of the model are derived from the measurements. As far as the approach of modelling is concerned, almost all modern attenuation models are 'semi-empirical' in nature in that they employ attenuation data to derive the model parameters. The popular models in this approach are developed on the basis of either occurrence of a rain cell or of the rain profile, at a given location.

Rain cell basis: These are the models which were based on the occurrences of rain cells. They are based upon the idea of the random occurrences of 'rain cells'. Now from the statistics of the occurrence of the rain cells and the distribution of rain in the cells in the path of the signal, the rain attenuation is calculated. The rain cell models of Misme and Waldteufel (1980) and Lane and Stutzman (1980) are examples of this kind (Stutzman and Dishman, 1982).

Rain profile basis: Most of the modern models are based on the rain rate profile. They employ the temporal and spatial distribution of effective rain rate and derive the probabilistic rain fade for the given path through the rain. These models are typically simple in construction, easy to use and physically sound.

In such an approach of modelling, three different aspects are of importance (Fedi, 1981), viz. the relationship between specific attenuation and rain rate, the statistics of point rainfall intensity and the spatial distribution of rainfall and finally the statistical representation of the spatial distribution of rain intensity in terms of the point rain rate.

To understand the importance of these aspects in the development of the model, we shall consider here, just like our previous cases, that the attenuation can be expressed in terms of the specific attenuation and the effective length. Then the question zeroes down to—how to generate the specific attenuation and how to get the effective length. In the detailed discussion following, we shall learn to get the answer to these two questions and understand how different meteorological aspects come into play in such modelling.

Rain-induced specific attenuation

Rain-induced specific attenuation, γ, is the attenuation experienced by a satellite signal per unit distance travelled by it through the rain. It is generally expressed in dB/km and is a fundamental quantity in the calculation of rain attenuation statistics. The specific rain attenuation at any point in space depends upon the rain rate at that point through rain drops size distribution. The properties of the drops, like the rain drop shape, temperature and those of the incident wave such as the frequency, polarization and the relative orientation with the drop also plays role in it.

When an electromagnetic wave propagates through a medium containing raindrops, part of its energy is absorbed by the raindrops and dissipated as heat, while another part is scattered in all directions. Thus, through both these processes, the wave, incident upon the rain drops, loses power and attenuation of the wave is said to have occurred. Total attenuation experienced by a radio wave crossing a rainy path is given by the sum of the contributions of the individual rain drops that the wave encounter over the rainy path. Therefore, the wave propagates with monotonically decreasing power.

The specific rain attenuation is thus estimated theoretically by considering the total power lost by the propagating wave upon interaction with the raindrops over unit length, taken as 1 km. The value, therefore, depends upon how many rain drops the wave encounters per unit length and how much power the wave loses on each drop. The raindrops are considered here as equivolumetric spheres with a known distribution. Solutions for Mie scattering is then utilized to obtain the scattered power. Even more realistic shapes like oblate spheroids have been considered, however making the scattering solution more complex. Finally, the specific attenuation is calculated by integrating the effects of each individual raindrop for a given rain rate over the unit length.

Thus, to derive the specific attenuation, we have two distinct problems in hand, (a) to find out the numbers of rain drops of each size present in unit volume, (b) to find the amount of power the wave loses on a drop of given size. For (a), we need to have the size distribution of the rain drops. We have already learnt in the previous chapter about the different drop size distributions models which efficiently represent the rain in different regions across the globe. Few suitable out of these have been considered in the calculation of specific attenuation. These include the distributions of Laws and Parsons (1943), Marshall and Palmer (1948) and the models of Ajayi and Olsen (1985), etc.

Here, without considering any specific model, we shall take a generalized rain drop distribution and will gradually develop the specific rain attenuation term to understand how each of these elementary particles contributes to the rain attenuation process. Let the distribution be given by $N(D)$, where N represents the number of drops present per unit volume with diameter $D - D + dD$. We shall use this expression later for further explanation of the physical process behind the attenuation. We have also learnt how to get the DSD from the disdrometer counts, earlier in this chapter. Thus, we can assume that, using either the models or measurements, we can obtain the necessary distribution of the drops according to their sizes.

To find the amount of extinction in a drop of a particular size, we shall utilize theory of scattering that we have learnt in Chapter 3, for the case when an electromagnetic wave passes through a material medium. Recall that, while discussing Mie scattering theory, we derived quantitatively the amount of wave power scattered and absorbed from the electromagnetic wave by any particular raindrop size, in terms of its scattering cross section and extinction cross section. Here, we shall use these basic theoretical results of Mie scattering and expressions thereof and use them to achieve our objective. While developing the theory, we assumed that:

- The raindrops are spherical which cause attenuation with volume of water equal to that of the actual drop.
- Each drop's contributions are independent of the other drops, and the contributions of the drops are additive.
- The attenuation is due to both, energy absorbed in the rain drops and due to energy scattered by water droplets from the incident radio wave.

• The technique is independent of polarization and hence does not differentiate between horizontal and vertical polarizations.

As we are interested in knowing the total power lost in all directions, we shall use the integrated solution over all possible directions, represented by the total extinction cross section, to get the same. For convenience, we reiterate here that the scattering cross section in any direction is the ratio of the scattered power per unit solid angle in that direction to the total incident signal power flux density. This is shown in Fig. 7.4. This cross section, integrated over all directions, gives the total scattering cross section, which when multiplied with the incident signal power flux density gives the amount of total scattered power. A certain amount of the energy is absorbed, which results in a heating of the target. Similarly, the total absorption cross section is also defined as the total power absorbed by the drop per unit incident power flux density. The sum of the scattering and the absorption cross section gives the extinction cross section. Therefore, the extinction cross section is the area that, when multiplied by the power flux density (i.e., intensity) of electromagnetic waves incident on the drop, gives the total power scattered and absorbed by the drop, i.e. the total power lost by the wave upon incidence.

We can also readily recall here that the total scattering cross section was indicated as σ_s, while the total extinction cross section was indicated by σ_e. Each of these terms, normalized to the particle geometric cross section α ($\alpha = \pi D^2/4$), can be expressed as (Mätzler, 2002; Yushanov et al., 2013)

$$Q_s = \sigma_s/\alpha = \frac{2}{x^2}\Sigma(2n+1)\left[|a_n|^2 + |b_n|^2\right]$$

$$Q_e = \sigma_e/\alpha = \frac{2}{x^2}\Sigma(2n+1)\left[R_e(a_n + b_n)\right]$$

(7.22)

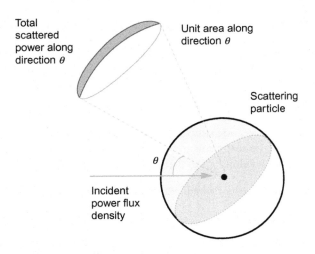

Total scattered power along direction θ

Unit area along direction θ

Scattering particle

θ

Incident power flux density

FIG. 7.4

Scattering cross section.

where, $x = (2\pi/\lambda) D/2$ and 'D' is the drop diameter and λ is the wavelength. n is the order of the multipole expansion, while a_n and b_n, obtained for the Mie solution, represent the corresponding coefficients for the magnetic and electric multipoles. The absorption cross section is just the difference of the extinction and the scattering cross sections.

From the above discussion, it is clear that the total extinction cross section considering all the drops of distinct discrete diameter D_i can be represented as

$$\sigma_T = \sum_i \sigma_e(D_i) N(D_i) \tag{7.23a}$$

If we consider the diameter of drops are in continuum, then it can be represented as

$$\sigma_T = \int \sigma_e(D) N(D) dD \tag{7.23b}$$

Now, let us find the total power loss by the signal over 1 km, where 1 km distance is taken as a convenient unit length through rain. For this, we first consider an infinitesimal distance dl along the direction of the propagation of the wave with a unit cross sectional area. The total power reduction in this space may be expressed as

$$-dP = \sigma_T \, dl \, P$$
$$\text{or, } P/P_0 = \exp(-\sigma_T l) \tag{7.24a}$$

where P_0 is the power at $l=0$. So, the expression P/P_0 over a length of 1 km may be obtained as

$$P/P_0 = \exp(-\sigma_T 1000) \tag{7.24b}$$

σ_T is generally expressed in cm^2. Then the above expression becomes

$$P/P_0 = \exp\left(-\sigma_T \times 10^{-4} \times 1000\right)$$
$$= \exp(-\sigma_T/10) \tag{7.24c}$$

So, the rain attenuation, in dB over 1 km is

$$\begin{aligned}
\gamma_R(\text{dB/km}) &= 10\log\left(\frac{P_0}{P}\right) \\
&= 10\left(\frac{\sigma_T}{10}\right) \log(e) \\
&= 0.4343 \sigma_T \\
&= 0.4343 \int \sigma_e(D) N(D) dD
\end{aligned} \tag{7.25}$$

where σ_e is given in cm^2 and N in numbers/m^3. Thus, the specific attenuation, γ, being developed from the scattering parameter, is related to the size of the drops and frequency of the wave, which are inherent to the term σ_e above and also on its distribution, $N(D)$. As we have considered the drops to be effectively spherical, the polarization does not remain a factor.

The temperature of the rain drops is another critical parameter. Although the assumed temperature has little effect for frequencies above Ka band, specific attenuation values otherwise vary with temperature of the raindrops. This is because, the refractive index of water, upon which the terms a_n and b_n of Eq. 7.22 depends, is a function of temperature. While 20°C is a reasonable assumption for terrestrial link attenuation prediction (Damosso et al., 1980), it is probably not representative of the temperature for earth-space paths (Thompson et al., 1980), especially for the tropical regions. Attempts have been made to include the temperature variation of specific attenuation with altitude, but this introduces a complexity into the calculation of attenuation while having little effect upon the results.

Power law relationship. The term rain-induced specific attenuation, commonly represented by γ_R (dB/km), is a fundamental quantity in calculating the rain attenuation or rain attenuation statistics. We have seen that the specific rain attenuation at any location over the earth depends upon the rain drop distribution at that point and also on the properties of the incident electromagnetic wave. The raindrop size distribution, in turn, is dependent upon the rain rate. Therefore, we should be able to represent γ_R in terms of the rain rate, R, which is a more convenient parameter of measurement and representation of rain.

The power law form of rain-specific attenuation in terms of R is very convenient and common in calculating rain attenuation statistics. It is expressed as

$$\gamma_R = kR^\alpha \, dB/km \tag{7.26}$$

where R is the rain rate in mm/h, and k and α are power law regression coefficients. Both k and α vary with frequency, while the term R takes into consideration the variation of the specific attenuation on rain drop size and its distribution. Thus, they commensurate with the variations of the individual terms of Eq. (7.25). Although many of the researchers have obtained the regression values k and α experimentally, it can be shown that the relation $\gamma_R = kR^\alpha$ is an approximation of a more generalized theoretical expression (Olsen et al., 1978). The empirical procedure to find k and α is based on equating the values for specific attenuation to the power law expression and deriving the regression coefficients out of it, given the corresponding rain rate R (mm/h). Thus, from both theoretical and numerical viewpoints, the relation reduces to the simpler form of the power law equation.

For the purpose of calculating the specific attenuation, it is adequate to assume the raindrops to be spherical in shape. This assumption makes k and α independent of polarization. However, practically, the attenuation in each drop also depends upon the plane of polarization of electromagnetic waves due to nonspherical nature of the drops. The polarization having component vertical to the local ground will be attenuated less than those having horizontal components of polarization. Attenuation values-computed assuming spherical drops generally lie between the extremes of values computed for vertical and horizontal linear polarization with distorted drop shapes. Nevertheless, the error incurred considering spherical shape of the drops is typically 10% or less in the frequency range of interest (Crane, 1977; Olsen et al., 1978). Therefore, specific attenuation values calculated with spherical assumption of spherical drops will produce considerably adequate results.

Point rainfall intensity and effective path rain
Once we have got the specific attenuation in terms of the rain rate, we need to determine the rain rate along the path to get the specific attenuation profile and hence to get the total attenuation by integrating them. If the rain rate profile, $R(l)$, were known along the extent of the propagation path, L, it would be a simple matter to calculate

the total attenuation by integrating the specific attenuation over the path. However, they are actually not known. Further, it is important to realize that the rain rate is not the same at all points along the signal path, neither the rain rate profiles can be expressed by any fixed mathematical expression. That is, the spatial occurrence of rain along any definite path is not deterministic. Rain events are rather localized compared to the total length of the signal path through the rain. Therefore, the signal experiences abrupt changes in spatial variations of the rain rate as it passes from the ground station to the satellite through the rain. As this variation does not comply to any specific mathematical representation, the models for spatial variations in rain need to be formulated based on local data (Ajayi and Olsen, 1985; Bandera et al., 1999). However, it is impractical to measure and introduce all the rain rate profile along the path in the model. Therefore, the point measurement of the rain rate is used and the spatial variation of the rain rate along the signal path is estimated from it. Now, given the point rain rate, the spatial profile of the rain is also not always the same. Therefore, we have to rely on the statistical nature of its spatial variation. The spatial inhomogeneity and its temporal variations are thus handled through statistical derivation of rain rate path profile averaged over a year. Thus, the use of statistical reckoning of rain rate spatial profile renders the corresponding model of attenuation long-term statistical in nature.

Since the path profile of the rain rate, obtained from point rainfall, using long-term statistics of spatial variation of the rates, varies for different values of the pint rainfall rate. Therefore, now the question that stands is, which value of the point rain rate is to be taken for the purpose. The answer depends upon what kind of temporal statistics we are interested in. For satellite communication purposes, it is the exceedance value of attenuation, which is of prime importance. For example, $A_{0.01}$ represents the 0.01% of exceedance value of rain attenuation. It gives that value of attenuation which will be exceeded only for 0.01% of time cumulatively over an entire year. In other words, for the rest 99.99% of time of the year, the attenuation will be below this value. Thus, if we know $A_{0.01}$ and keep this value as the rain margin in the satellite, we are sure that for 99.99% of time, the aforesaid margin will be able to compensate the fade. So, the link will be available for 99.99% of time with consequent outage of 0.01%. To obtain $A_{0.01}$, equiprobable rain rate, i.e. $R_{0.01}$ value, is necessary. Thus, most rain attenuation prediction models make use of the single value of exceedance rain rate $R_{0.01}$ (mm/h), i.e. the rain rate exceeded for 0.01% of the time over a year.

In the prediction techniques based on the use of point rain rate, probability density distribution of occurrence of rain rate over a year is estimated from measurements at the point of interest. From this distribution, the exceedance value is estimated as $R_{0.01}$, such that integrating these probability densities from this particular rain rate to the maximum possible rain rate will yield the cumulative probability value 0.01. As rain rate distributions may show considerable variability from year to year (Crane, 1977), care must be taken in estimating the exceedance. A simple way of dealing this is by using the probability density distribution at a site averaged over years. However, when adequate local data are not available, the distributions can be estimated from the rain climate region grid-based interpolation values at the point recommended by the ITU (ITU-R, 2012b).

The above discussion indicates that, we are able to represent a path profile of rainfall as a function of point measurements, R_0. Then, the total attenuation A can be expressed as the path integral of the spatial function of the corresponding specific attenuation, derived in terms of this representative rain rate R_0. Thus, the total attenuation is obtained by integrating over the geometric length L. Equivalently, it can be obtained as the product of the point specific attenuation $\gamma_{R0} = kR_0^\alpha$ and the effective rainy path, L_{eff}, which becomes a function of R_0.

L_{eff} can be obtained as the true geometric path multiplied by the reduction factor, r_{eff}. It is to be remembered that, not only for different values of rain rate, R, the effective path will be different, but even for a given rain rate, the spatial pattern will vary for different events of occurrence. This is because, depending upon the orientation of the rain cell profile, the signal path may be on the ascending side of the path variation or on the descending side of it for different events.

To derive the point-to-path rain relationship, typically exponential shapes of spatial profile are used (Valentin, 1977; Barsis and Samson, 1976) which agree with inferences from indirect measurements.

SAM model. To understand the approach of the rain attenuation modelling, we shall first discuss a rather simple attenuation prediction model proposed by Stutzman and Dishman. It is popularly known as 'Simple Atmospheric Model' (SAM) (Strutzman and Dishman, 1982) and is one of the extensively employed slant path attenuation prediction models.

The model utilizes elevation angle, specific attenuation, an empirical constant 'η' for the estimation. Like any other model, it first estimates the specific attenuation, γ_R, as

$$\gamma_R = kR^\alpha \; dB/km$$

k and α can be obtained from the ITU recommendation P-838 (ITU-R, 2005), while the original SAM model utilized the regression parameters proposed by Olsen et al. (1978) and is given as.

$$
\begin{aligned}
k &= 4.21 \times 10^{-5} f^{2.42} & 2.9 <= f < 54 \\
&= 4.09 \times 10^{-2} f^{0.699} & 54 <= f < 180 \\
\alpha &= 1.41 f^{-0.0779} & 8.5 <= f < 25 \\
&= 2.63 f^{-0.272} & 25 <= f < 164
\end{aligned}
\tag{7.27}
$$

Once the specific attenuation is obtained, the effective rainy path is required to be derived. The effective slant path length through rain depends on the rain height. Therefore, the geometric path taken by the signal, up from the receiver to the rain height, is first required to be known. In order to find the rain height, SAM utilizes its distinct assumption of the combined stratiform and convective forms of rainfall over the signal path. The rain height assumed in this model is considered to be a function of rain rate. It considers the effective rain height to be equal to the 0°C isotherm height for low rain rates, up to 10 mm/h, when it is assumed that the stratiform rain dominates. However, for higher rain rates over 10 mm/h, a rain rate-dependent term is added to the 0°C isotherm height to get the effective rain height. The model presupposes that under such conditions, the contribution of the convective rain increases and the effective starting height of precipitation gets raised. So, the overall effective height is expressed as

$$h_r = H_i \text{ km} \qquad \qquad \text{for } R \leq 10 \text{ mm/h}$$
$$h_r = H_i + \log\left(\frac{R}{10}\right) \text{ km} \quad \text{for } R > 10 \text{ mm/h} \qquad (7.28a)$$

where R is the point rain rate in mm/h and H_i is the 0°C isotherm height in kilometres. The breakpoint of 10 mm/h has been chosen because it corresponds to the maximum rain rate considered by the model for stratiform rain. The 0°C isotherm height H_i varies with latitude. In this model, this height is approximately expressed as (Crane, 1978)

$$H_i = 4.8 \text{ km} \qquad \qquad \text{for } |\varphi| < 30°$$
$$H_i = 7.8 - 0.1\varphi \text{ km} \quad \text{for } |\varphi| > 30° \qquad (7.28b)$$

where φ is the geographical latitude in degrees. The subsequent portion is the most important part of the model. Now that we know the path geometry and its limits, we now need to get the rainfall over it to get the total path attenuation. Now, this model chooses a rain rate profile, which again distinguishes between stratiform and convective rain. For stratiform type of rain with rain rate not more than 10 mm/h, the rain rate is considered to be spatially constant. For rains of mixed convective type, when the rate is more than 10 mm/h, the rain rate profile is considered to have an exponential variation, where the exponent depends upon the rain rate. The profile is given by the function,

$$R(z) = R \qquad \qquad \text{for } R \leq 10 \text{ mm/h}$$
$$R(z) = R \exp\left[-\eta \ln\left(\frac{R}{10}\right) z\right] \quad \text{for } R > 10 \text{ mm/h} \qquad (7.29a)$$

where, R is the point rainfall intensity in mm/h, z is the horizontal distance along the path and η is a parameter controlling the rate of decay of the profile with distance (Strutzman and Dishman, 1982). For $\eta < 0$, the rate gets enhanced with distance. Because the rain is assumed to be uniform in the vertical direction up to He, the rain profile $R(l)$ along the slant path l is same as the profile along the projection of this path on the ground. Thus, it can be derived using the simple trigonometric relationship $R(l) = R(z)$ where $z = l \cos \theta$, where θ is the elevation angle, giving

$$R(l) = R \qquad \qquad R \leq 10 \text{ mm/h}$$
$$R(l) = R \exp\left[-\eta \ln\left(\frac{R}{10}\right) l \cos\theta\right] \quad R > 10 \text{ mm/h} \qquad (7.29b)$$

This is valid until l remains within the rainy path below the rain height, i.e. for $l < L$, where $L = (h_r - h_s)/\sin \theta$ and h_s is the altitude of the earth station location. This expression is likely to be valid for elevation angles above about 10 degrees when the curvature of the earth does not play any significant role.

In this way, we get the rain path profile in terms of the point rain rate, R. Therefore, the specific attenuation at any distance l is given by

$$\gamma_R(l) = kR(l)^\alpha$$
$$= kR^\alpha \text{ db/km} \qquad \qquad \text{when } R \leq 10 \text{ mm/h}$$
$$= kR^\alpha \exp\left[-\eta\alpha \ln\left(\frac{R}{10}\right) l \cos\theta\right] \text{ db/km} \quad \text{when } R > 10 \text{ mm/h} \qquad (7.30)$$

Therefore, the total attenuation due to rain, A_R can be expressed in terms of the point rainfall R, obtained by integrating the spatial function of the specific attenuation over the total rainy path is given by

$$
\begin{aligned}
A_R(R) &= \int \gamma(l)dl \\
&= kR^\alpha L \text{ db} && \text{when } R <= 10\,\text{mm/h} && (7.31) \\
&= kR^\alpha \frac{1 - \exp\{-\eta\alpha \, \ln(R/10)L\cos\theta\}}{\eta\alpha\ln(R/10)\cos\theta} \text{ db} && \text{when } R > 10\,\text{mm/h}
\end{aligned}
$$

This Simple Attenuation Model (SAM) described above is a function of the point rainfall intensity R only with the rain attenuation value as the output. It does not require the spatiotemporal rain statistics like many other standard models. The decoupling of the attenuation model and the rain rate statistics allow evaluation of the model parameters independent of the long-term measurements of rain rates and of the errors associated with it.

Box 7.2 SAM Rain Attenuation Model

SAM is one of the simplest models to understand the role of each parameter in generating the rain fade. Run the MATLAB program SAM.m to generate the figure representing the total rain attenuation corresponding to the rain rates given in the text file 'rain.txt'. The corresponding figure is shown in the figure below.

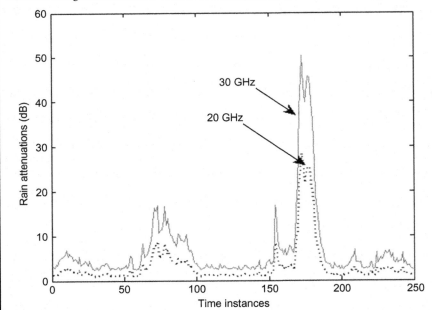

FIG. M7.2

Temporal variation of rain attenuation.

Change the different parameters and see the effects of the change in the plot. Change the frequency 'f', elevation angle 'T', and the rain rate profile parameter η represented by variable 'c' in the program.

Observe the sensitivity of each by carefully noting down how much the plot values change for changing one of the given parameters, viz. cc, a20, b20, a30, b30, rain height H_R and elevation angle T, with all others remaining constant.

ITU model. Attenuation due to rain, like any other attenuation discussed above, can be calculated as product of specific attenuation (dB/km) and effective path length L_e (km). ITU provides a model that represents rain attenuation exceedance value $A_R(p)$ for $p\%$ of time over a year in terms of $A_{0.01}$. $A_{0.01}$ can be expressed in product from as,

$$A_{0.01} = \gamma_R L_E \text{ dB} \tag{7.32}$$

This model enables calculation of long-term rain attenuation statistics from point rainfall rate, using the long-term statistics of rain occurrences. This ITU-R procedure provides estimates of the long-term exceedance values of the slant path rain attenuation at a location which is derived from the knowledge of the rain rate exceedance values for frequencies up to 55 GHz. This model utilizes the power law expression of the specific attenuation. The parameter $R_{0.01}$, representing the 0.01% exceedance values of the point rainfall rate for the location over an average year (mm/h), is derived and used in this expression. Once the specific attenuation is established, the effective path is obtained from the rain height relative to the earth station, the elevation and the location of the station. The frequency of the signal f and the effective earth radius R_e, taken as 8500 km, also are used for the purpose (Fig. 7.5).

The first step is to obtain the specific attenuation, γ_R. This is done using the frequency-dependent ITU coefficients given in Recommendation ITU-R P.838 (ITU-R, 2005) and the exceedance value of the rainfall rate for 0.01% of time, $R_{0.01}$, at the location of the ground station, using the relation

$$\gamma_R = k R_{0.01}{}^\alpha \text{ dB/km} \tag{7.33}$$

$R_{0.01}$ should be obtained from local measurements with an integration time of 1 min at the location of the ground station. As $R_{0.01}$ forms a mandatory parameter; so if there is no measurement done at the station or if it cannot be obtained from local data sources, an estimate can be obtained from the maps of rainfall rate given in Recommendation ITU-R P.837 (ITU-R, 2012b).

Once the specific attenuation is obtained, it is required to find the effective path through the rain. For this, first we determine the rain height, h_r, for the given location.

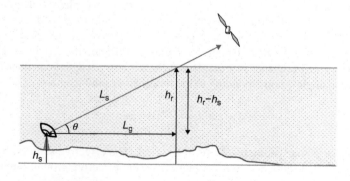

FIG. 7.5

Schematic of rain fade measurement scenario.

If local measurements for the same are not available, the height can be obtained from the ITU Recommendation ITU-R P.839 (ITU-R, 2013c), which gives the global variation of the rain height. Although the model is defined up to very low elevation angles, for brevity and for the sake of easy understanding, we shall keep the things simplified, by considering elevation angles $\theta > 5°$ only, such that the planar geometry holds good and earth's curvature does not affect the geometric estimations of the earth-satellite path.

Let the height of the station be h_s and θ be the elevation angle at the location as shown in Fig. 7.5. Then, if $\theta \geq 5°$, then the slant path length, L_s, below the rain height can be computed as

$$L_s = \frac{h_r - h_s}{\sin\theta} \text{ km} \tag{7.34}$$

Once the slant path is suitably obtained, the horizontal projection, L_G, of the slant path length can be calculated as

$$L_G = L_s \cos\theta \text{ km} \tag{7.35}$$

This is the ground trace of the signal path up to the rain height. Notice that this turns out to be $L_G = (h_r - h_s)/\tan\theta$, which is obvious for the planar geometry.

Had the rainfall be uniform all through in both time and space, then the calculations done up to this point would be enough for the required estimation. However, it is neither uniform, nor even constant. Therefore, some statistical averaging is required to be done over the length of the path through the rain. The following part of this section describes the method to get the same.

First, to get the effective rainy path from the geometric path, the reduction is done on the ground trace, L_G. The horizontal reduction factor, $r_{0.01}$, for 0.01% of the time is given by

$$r_{0.01} = \left[1 + 0.78\sqrt{\frac{L_G \gamma_R}{f}} - 0.38\left(1 - e^{-2L_G}\right)\right]^{-1} \tag{7.36}$$

Notice that the reduction factor $r_{0.01}$ decreases with $L_G \gamma_R$, the latter representing the attenuation corresponding to the total geometric ground trace, and increases with frequency. The reduction factor, thus, changes the effective ground trace from L_G to $L_g = L_G r_{0.01}$. The vertical height remaining the same, the effective elevation angle changes with the change in ground trace. So, it is now needed to calculate the modified elevation angle ζ corresponding to the modified ground length L_g, ζ is given by

$$\zeta = \tan^{-1}\frac{h_r - h_s}{L_G r_{0.01}} \text{ degrees} \tag{7.37}$$

Notice that for planar geometry, $\zeta > \theta$, if $r_{0.01} < 1$. The effective slant path through the rain, L_R, is then obtained in kilometers from the effective ground length as,

$$L_R = L_G r_{0.01} / \cos \theta \quad \text{if} \, \zeta > \theta$$
$$\quad\quad = (h_r - h_s) / \sin \theta \quad \text{otherwise} \tag{7.38}$$

From the above expression, it becomes evident that in the calculation of the effective slant height L_R through the rain, any reduction occurring in the original geometric path is accepted while any increment is disregarded and the slant path is kept unaltered. The vertical adjustment factor, $\nu_{0.01}$, for 0.01% of the time is then calculated as

$$\nu_{0.01} = \left[1 + \sqrt{\sin\theta} \left\{ 31 \left(1 - e^{-\frac{\theta}{1+\chi}} \right) \frac{\sqrt{L_R \gamma_R}}{f^2} - 0.45 \right\} \right]^{-1} \tag{7.39}$$

Where the term χ is dependent upon the geographical latitude λ as,

$$\chi = (36 - |\lambda|)^\circ \quad \text{when} \, |\lambda| < 36^\circ$$
$$\chi = 0^\circ \quad\quad\quad\quad \text{otherwise} \tag{7.40}$$

Then, the effective path length is obtained as the product of the effective rainy path L_R and the adjustment factor $\nu_{0.01}$, i.e.

$$L_E = L_R \nu_{0.01} \, \text{km} \tag{7.41}$$

Finally, the predicted attenuation exceeded for 0.01% of an average year is obtained as the product of the specific attenuation and the effective path length and is given by

$$A_{0.01} = \gamma_R L_E \, \text{dB} \tag{7.42}$$

Once the $A_{0.01}$ value is obtained, the exceedance values for other percentages of the year may be derived out of it using the following relation given in Eq. (7.43a). It is mostly dependent upon the attenuation $A_{0.01}$ and the required percentage, p. Additionally, latitude of the location and elevation angle also has a bearing upon it. For the range of 'p' between 0.001% and 5%, the expression is given as

$$A_R(p) = A_{0.01} (p/0.01)^{-\{0.655 + 0.033 \, \ln(p) - 0.045 \, \ln(A_{0.01}) - \beta(1-p)\sin\theta\}} \, \text{dB} \tag{7.43a}$$

Here the coefficient β is a function of the percentage 'p', latitude λ and elevation angle θ. The function is defined stepwise as

$$\beta = 0 \quad\quad\quad\quad\quad\quad\quad\quad\quad\quad \text{for} \, p \geq 1\% \, \text{or} \, |\lambda| \geq 36^\circ$$
$$\beta = -0.005(|\lambda| - 36) \quad\quad\quad\quad\quad \text{for} \, p < 1\% \, \text{and} \, |\lambda| < 36^\circ \, \text{and} \, \theta \geq 25^\circ \tag{7.43b}$$
$$\beta = -0.005(|\lambda| - 36) + 1.8 - 4.25 \sin\theta \quad \text{for all other cases}$$

This method provides an estimate of the long-term statistics of attenuation due to rain and can be used for estimating the static link margin while establishing a satellite link considering the availability term. This will be described in later sections in this chapter. When comparing measured statistics with the prediction, allowance should be given for the rather large year-to-year variability in rainfall rate statistics.

Box 7.3 ITU Rain Attenuation Model

Our objective in this box is to get acquainted with the ITU model for rain attenuation. In this section, we shall run the MATLAB program itu_rainattn.m to generate the figure representing the total attenuation due to the rain for a given satellite and receiver location and with known rain statistics in terms of $R_{0.01\%}$. The figure shows gradual decrement of the attenuation exceedance value with increasing time percentage. This means that as we are trying to know what is that rain rate value which is exceeded for a longer period of time of the year, the model shows lower threshold value of attenuation. This is expected. This way, if we ask for an attenuation value for a percentage of time that corresponds to the total rainy period of the year, the corresponding rain attenuation exceedance value will be zero, because positive attenuation occurs when there is rain.

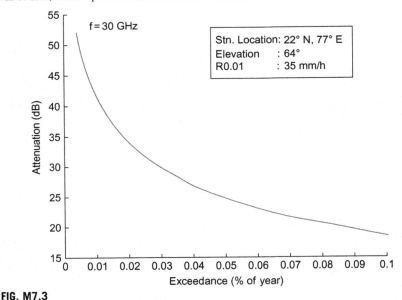

FIG. M7.3

Variation of attenuation exceedance with percentage of exceedance.

    Also observe the function is also sensitive to elevation and frequency. The program is reading the regression coefficients 'k' and 'α' from an external file, 'kalfa.txt'. Change the location values and $R_{0.01}$ value and see the variations in the curve.

7.2.1.5 Depolarization model

Frequency reuse in orthogonal polarizations is often employed to increase the capacity of satellite-based communication systems. This technique is restricted, however, by depolarization of the signal on atmospheric propagation paths. In Chapter 6, we have already illustrated the term depolarization and defined the quantitative metrics associated with it. Various depolarization mechanisms are prevalent in the troposphere. But the most common and important cause of depolarization is the rain.

    The total depolarization occurring in a signal is not only a function of its frequency, but also depends upon various other factors pertaining to rain, like the relative orientation between the signal polarization and the drops, the total numbers of the drops encountered by the signal etc. While the former may be represented as function of elevation angle, the wave polarization and the canting angle of the drops, this last factor may be surrogated by the total attenuation being experienced by the wave.

Under nominal propagation condition, the cross-polarization discrimination (XPD), which is the ratio of the copolar power to the cross polar power, should be large as the cross-polarization power is negligibly small under such circumstance. However, with depolarization occurring during the rain, the XPD deteriorates. We can consider that the contribution of each of the factors to the cross-polarization discrimination (XPD) is independent and hence can be represented and derived separately. We call as C_f, C_A, C_φ, C_θ and C_σ respectively, the contributions of the factors frequency, attenuation, polarization, elevation and canting angle to the XPD. We also mention below their respective expressions for frequencies $6 \le f \le 55$ GHz and for elevation angle $\theta \le 60$ degrees. The total expression for the depolarization is then just a combination of each of these individual factors.

The frequency-dependent term C_f can be calculated as

$$
\begin{aligned}
C_f &= 60.0 \log f - 28.3 \quad 6 <= f < 9 \text{GHz} \\
&= 26.0 \log f + 4.1 \quad 9 <= f < 36 \text{GHz} \\
&= 35.9 \log f - 11.3 \quad 36 <= f <= 55 \text{GHz}
\end{aligned} \tag{7.44}
$$

where f is the frequency of the signal in GHz. The basic observation from this part of the model is that the term C_f monotonically increases with frequency and is continuous at the break points. Thus, the term is piecewise linear with the logarithmic value of f.

The rain attenuation contribution to the XPD, given by the term C_A, can be expressed in terms of the copolar attenuation A_p, as

$$
C_A = -V(f) \log A_p
$$

Where,
$$
\begin{aligned}
V(f) &= 30.8 f^{-0.21} \quad 6 <= f < 9 \text{GHz} \\
&= 12.8 f^{0.19} \quad 9 <= f < 20 \text{GHz} \\
&= 22.6 \quad 20 <= f < 40 \text{GHz} \\
&= 13.0 f^{0.15} \quad 40 <= f <= 55 \text{GHz}
\end{aligned} \tag{7.45}
$$

Notice that, for a given frequency, the absolute value of the contribution term increases with increasing attenuation, A_p. Consequently, this term reduces XPD indicating larger depolarization with larger attenuation.

Similarly, we can find the polarization improvement factor C_φ as

$$
C_\varphi = -10 \log [1 - 0.484(1 + \cos 4\varphi)] \tag{7.46}
$$

The improvement factor $C_\varphi = 0$ for $\varphi = 45$ degrees and reaches a maximum value of 15 dB for $\varphi = 0$ or 90 degrees.

The elevation angle-dependent term C_θ is given by

$$
C_\theta = -40 \log (\cos \theta) \quad \text{for } \theta \le 60 \text{ degrees} \tag{7.47}
$$

As θ decreases, the total rainy path increases, and considering the negative sign, the contribution reduces the XPD. This is evident, as larger passage through rain will lead to larger depolarization.

Finally, the Canting angle-dependent term C_σ is given by

$$
C_\sigma = 0.0053 \sigma^2 \tag{7.48}
$$

σ is the effective standard deviation of the raindrop canting angle distribution, expressed in degrees. So, larger σ means more randomness in the drop orientation, and hence, it undoes the effect of depolarization, improving the XPD values.

Therefore, we get the value of XPD that does not exceed for $p\%$ of the time as

$$\text{XPD}_{\text{rain}} = C_f + C_A + C_\varphi + C_\theta + C_\sigma \, \text{dB} \qquad (7.49)$$

However, this model is not very useful for predicting XPD statistics for frequencies below 6 GHz.

7.2.1.6 Combining the effects

The net fade distribution due to tropospheric refractive effects, $A_T(p)$, is the combination of various component factors like the rain fade, cloud attenuation, gaseous absorption and other effects described above. For systems operating at frequencies above about 18 GHz, and especially those operating with low elevation angles and/or low margins, the effect of multiple sources of simultaneously occurring atmospheric attenuation must be considered very critically as the link quality and sustainability is sensitive to even smaller variations. The estimation of total attenuation due to multiple sources of simultaneously occurring atmospheric attenuation can be done by knowing the individual attenuation values for the required time percentage of occurrences.

Let, $A_G(p)$ be the gaseous attenuation due to water vapour and oxygen for a fixed exceedance probability of p expressed in dB. Also, let $A_R(p)$ be the attenuation due to rain, $A_C(p)$ be the attenuation due to clouds and $A_S(p)$ be the fade due to tropospheric scintillation, all estimated for the same fixed exceedance probability p. Then considering the mutual dependence of different factors, we get the combined attenuation as

$$A_T(p) = A_G(p) + \left[\{A_R(p) + A_c(p)\}^2 + A_s^2(p) \right]^{\frac{1}{2}} \qquad (7.50)$$

where p is the probability of the attenuation being exceeded in the range 50%–0.001%. For percentages of time lower than 1%, the cloud and the gaseous attenuation is capped to their 1% occurrence probability value.

7.2.1.7 Modelling of frequency scaling

In a practical satellite communication system, the uplink frequency is higher than the downlink frequency. Due to this difference, the attenuation value observed at the ground station over the downlink cannot be directly used for compensating the attenuation in the uplink. So to compensate the uplink fade in an open loop configuration, in a condition when the uplink fade information is not available to the ground terminal by any other mean, an efficient fade mitigation system needs precise estimation of the attenuation in uplink. This information is derived from the knowledge of the attenuation in downlink frequency. This is called frequency scaling of the attenuation. The scaling of the attenuation has an obvious dependence upon the frequencies in question. The knowledge of the relationship between attenuation values at different frequencies is needed, not only for carrying out the compensation activity with suitable FMT, but also for a broader use of evaluation of system performance based on measurements at other frequencies. The scaling relation is typically expressed by the Frequency Scaling Ratio, R_{fs}, which is the ratio between attenuation values, A_1 and A_2, expressed in dB, for two different frequencies, f_1 and f_2.

Two different frequency scaling approaches can be considered, viz. the long-term frequency scaling and instantaneous frequency scaling (IFS). The long-term frequency scaling gives the ratio of the attenuation values at two different frequencies for the same probability of occurrences. It is thus a statistically derived number which allows the calculation of the long-term cumulative distribution function of attenuation at one frequency, when the same for the other frequency is known or measured. Expressed mathematically, R_{fs} becomes,

$$R_{fs}(p) = \frac{A_2(p)}{A_1(p)} \tag{7.51}$$

The total long-term availability and performance of a system operating at one frequency can be predicted from attenuation measurements made at another frequency, using this ratio. Since attenuation is, in principle, the combination of contributions from different causes, like rain, gaseous absorption, clouds and scintillation, these contributions depend on frequency in different ways. So, any particular value of attenuation in one frequency may be arrived at by different contributions from the individual factors of the attenuation. This is the main cause why the same attenuation in one frequency does not always map to the same attenuation in the other. This leads to the variations in the long-term frequency scaling ratios. It is thus intuitive that the frequency scaling technique is most satisfactory when one cause predominates.

Long-term frequency scaling has been applied in the past to extrapolate rain attenuation mainly from Ku to Ka band. In Ku band, the nonrainy contributions to attenuation are generally negligible and the frequency scaling technique works particularly well. Further, the path independent nature of the ratio makes the frequency scaling more stable. One of the long-term frequency scaling models has been recommended by the ITU-R (ITU-R, 2015a) and is given below

$$\frac{A_2}{A_1} = \frac{g(A_2)}{g(A)} = \left(\frac{\varphi_2}{\varphi_1}\right)^q \tag{7.52}$$

Here

$$\varphi(f) = \frac{f^2}{1 + 10^{-4} f^2} \quad \text{and} \quad q = 1 - 0.00112 \sqrt{\frac{\varphi_2}{\varphi_1}} (\varphi_1 A_1)^{0.55}$$

The term 'A_1' represents the attenuation in frequency f_1, while A_2 represents equiprobable value of attenuation in the frequency f_2. Both f_1 and f_2 are expressed in GHz. The term φ increases with frequency and therefore the ratio $(\varphi_2/\varphi_1) > 1$ when $f_2 > f_1$. $(f_1 A)^{0.55}$ increases with A_1 for a given f_1. Thus, the frequency scaling ratio, $R_{fs} = A_2/A_1$, changes conditioned on the value of A_1.

The IFS, on the other hand, relates simultaneous attenuation values at different frequencies. Thus, it provides a one to one mapping between the two attenuations at any instant of time. Modelling of the IFS is needed for real time adaptive FMT. That is, the timely activation of a fade countermeasure in the uplink relies on the attenuation measured on the downlink and then conversion to the higher uplink frequency. Such estimation is done at every instant and continued over the period of the attenuation. We shall learn about the particular fade mitigation techniques (FMT) later in this chapter.

The IFS ratio of rain is a stochastic quantity and has been empirically characterized for the variability around its mean values (Sweeny and Bostian, 1992; Laster and Stutzman, 1995). IFS ratio R_{fs} can be considered to be a log-normal variable about the mean, as long as rain dominates the attenuation phenomenon. The conditional mean, 'μ_{IFS}' conditioned on the attenuation value experienced at the lower of the two frequencies, is very close to the long-term ratio, evaluated on an equiprobability basis. It also fits well with a power law expression (Matricciani and Paraboni, 1985)

$$(\mu_{IFS}|A_1) = aA_1^{-b} \tag{7.53}$$

where a and b are the frequency-dependent regression parameters. The standard deviation σ_{IFS}, however, is found to be almost independent of the mentioned attenuation and its value is given by:

$$\sigma_{IFS} = 0.13 \tag{7.54}$$

IFS fluctuations are essentially due to the natural variability of the various phenomena. As the natural variations of the individual contributors are slow, they do not change abruptly over shorter interval of time.

7.2.1.8 Fade compensation using static margin

For proper planning of any satellite-based system, it is necessary to have appropriate design that can cater to the needs of the system during constrained conditions and sustain the link under severe impairments. For the purpose, it is required to estimate the amount of fade that can occur over the period exceeding the designed outage time. Therefore, to protect the link under such constrained condition and to provide the designed availability figure, a portion of the total amount of fade is required to be compensated. This is typically done in case of a satellite system by adding some static power margin. During the design time, long-term statistical attenuation data is used to find out the minimum value of fade that needs to be compensated to provide the designed availability. In this approach, the statistical fade distribution over an average year within the area of service is determined from long-term data. From this distribution, the total exceedance fade value, $A_T(p)$ for the probability of $p\%$ of time over the year is determined. This amount of power loss is to be compensated by the system and hence is added to the system as a fixed margin. This ensures that any attenuation that exceeds the value A_p cumulatively happens only for a percent of time less than the designed outage period.

7.3 FADE MITIGATION TECHNIQUES (FMT)

In applying the static margin to the satellite link to mitigate the fade, a fixed and predetermined amount of attenuation is compensated. However, for high frequency signals and for regions where the rain is ample and intense, the amount of fade is enormous and cannot be compensated just by adding fixed margin. This is because there is a limitation on the amount of additional power that can be provided to the link. Further, this would also lead to drainage of power when the atmospheric condition is favourable and fadeless.

Therefore, an efficient way to tackle this pragmatic situation is to resort to the real time adaptive FMT. Adaptive FMT means, these mitigation techniques are invoked only when there is a large fade in excess to the static margin that needs compensation. The action taken to combat the impairment is in accordance with the amount of real-time fading experienced by the signal. The link resources, in such cases are adaptively allocated so that the data transfer remains unaffected as much as possible. Fade Mitigation, therefore, can be formally defined as an arrangement through which a satellite link is sustained against possible atmospheric fades, occurring to a signal by judicious and optimal redistribution of available resources, thus compensating or avoiding the impairment (Castanet et al., 2009). Consequently, this implementation results in increased availability of satellite link and continuity during the adverse atmospheric conditions. We shall learn about such real time adaptive FMTs in the next section. Few techniques will also be discussed in Chapter 9.

The typical approach of a FMT arrangement is to achieve the mitigation objective by performing a chain of four major activities, which are, Measure, Predict, Decide and Act.

For a FMT implementation, the typical arrangement consists of a control loop which is constituted by three interdependent modules, viz. prediction module, decision module and the implementation module. The input to this control module is the measured values of the fade, while the output is the definite FMT command. As the name suggests, the prediction for the forthcoming attenuation is done from the input measurements in the Prediction module. It is in the decision and implementation module that the kind of resource rearrangement required for the type of fade being experienced is actually determined and methods of implementing them are executed with necessary resource management. Fig. 7.6 shows the schematic of such a typical control loop for a FMT.

To mitigate the tropospheric attenuation using any standard FMT, it is essential to have a-priori knowledge about the forthcoming level of attenuation. A good mitigation technique should have minimum drainage of resources and lowest probability of lock of loss in the receiver. So, the prediction should not fall below the actual

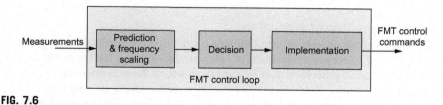

FIG. 7.6

Schematic of FMT control loop.

attenuation and the amount of overestimation should be minimum. This has, therefore, obvious dependence upon the individual performances of the measurements, predictions and choice of implementation scheme. FMT implementation relies on the principles of EIRP control techniques, adaptive transmission techniques and diversity protection schemes which are described in the subsequent sections.

7.3.1 FMT CONCEPTS

7.3.1.1 Measurements

To take a decision on the FMT necessary for keeping the link alive, it is first required to estimate the upcoming fade so that adequate protection may be taken against it within time. For the same, it is required to measure the fades, such that from the time series of this ongoing fade, the upcoming fade values may be predicted. We have already learnt about the different measurement techniques at the beginning of this chapter. The most important feature necessary for the measuring system are accuracy and continuity.

7.3.1.2 Predictions

The prediction of the fade is one of the most important components in the control loop. It predicts the upcoming fade values from the current time series of the fade. There are different methods of predictions, each utilizing the past time series measurement values till the current instant to achieve the end.

In making attenuation prediction for this purpose, any under-prediction will lead to an under-compensation of fade and may result in a loss of link-availability, whereas over-prediction of attenuation will lead to drainage of power or resources. Therefore, two basic features of the prediction algorithm are necessary, first the process should provide no under-prediction of attenuation and second, the instances of over-predictions should be minimum. Further, the controllers require a small but finite reaction time for activating the mitigation technique. So, the prediction should be made over interval which can provide a comfortable time margin in advance so that the implementation of compensation may take place simultaneously with the fade.

Different types of prediction techniques have been adopted for the purpose. These may be broadly categorized into the following, but nonexhaustive, classes (Roy et al., 2012).

Statistical approach of prediction

Using statistical parameters of rain fade has been a popular method for prediction. The log-normal stationary distribution of rain attenuation along with appropriate and predefined dynamic properties may be used from which the rain attenuation may be obtained (Masseng and Bakken, 1981). Few other models use a weighted average of fade slopes over the past samples and uses it for the prediction of the samples ahead (Kastamonitis et al., 2003). A near real-time fade prediction algorithm based on the two sample model (Bolea-Alamanac et al., 2003; Van de Kamp, 2002) has also been found to be performing well.

Regression-based predictor

Regression is another sought out technique for prediction. First-order linear regression (Dossi, 1990) has been used to calculate the fade slope over previous samples with increasing added bias in favour of the most recent sample, which are then utilized for predicting fade. Prediction based on Auto Regressive Moving Average (ARMA) equations with a moving window of samples of recent past is also popular (Gremont et al., 1999). An algorithm that switches between Integrated Auto Regressive Integrated Moving Average and Generalized Autoregressive Conditional Heteroscedasticity (ARIMA/GARCH) model for short-term fade prediction (De Montera et al., 2008) is also used for prediction.

Artificial neural network (ANN) method

ANN has been extensively used to predict tropospheric attenuation and was deployed for the decision making block of the controller (Malygin et al., 2002). Adaptive form of the neural network has been popularly used in both linear (Chambers and Otung, 2005) and nonlinear form (Roy et al., 2012) for prediction. Few hybrid prediction techniques combining both the ARMA and Adaptive Linear Neuron (ADALINE) algorithm have also been proposed (Chambers and Otung, 2008).

7.3.1.3 Decision

The Decision block typically finds the answers to three key questions in the FMT process. These are:

   i. Whether any FMT is needed or not at any instant of time
   ii. If FMT is needed, then how much fade compensation is required?
   iii. Which type of FMT can be most effectively used?

The FMT block finds answers to these questions based on the predictions of the forthcoming expected attenuation, provided by its preceding block. The basic lookouts while deciding in regard to the above questions are the availability of the resources, effective gain provided by the techniques, expected duration of the FMT, etc. Although we treat here the decision block separately, it is *typically* not a separate entity but is typically implemented as a logical part of the implementation module.

7.3.1.4 Mitigation techniques

Various standard FMTs are applied depending on the extent of attenuation (Castanet et al., 2002) and are usually actuated by the control system. These controllers have the ability of detecting and predicting the actual level of total attenuation in satellite link. Depending upon the predicted attenuation depth, the control system takes a decision and activates a suitable FMT to compensate the probable loss.

Making use of FMTs involves in real time the redistribution of resources to meet the link budget to sustain the link with minimum required performance even under most constrained propagation conditions. Resources and parameters which are adaptively modified in the link are the link power, link signal structure components like modulation, coding, etc., link origin and destination location, time, etc. Broadly, the FMTs can be divided into two different categories on the basis of its approach of mitigation. These are,

a. Compensation type FMT: Those techniques which attempt to compensate the effects due to fade that the link is experiencing by modifying the link and signal parameters.
b. Avoidance type FMT: Those techniques which attempt to avoid the effects of the fade by rerouting the link through change in link path, frequency, time of transmission, etc.

However, these can be categorized on different other basis, too. The performance of a satellite system depends upon the quality of the received signal and more precisely on the term E_b/N_0, i.e. the ratio of the energy contained in a demodulated data bit to the noise spectral density. This term is again dependent upon the term C/N_0, the latter indicating the signal quality.

The E_b/N_0 at the receiver may be expressed as

$$E_b/N_0 = (\text{EIRP} - L_{\text{fs}} - L_A + G_R/T + G_c - k) - R \text{ dBHz} \tag{7.55}$$

where EIRP is the Equivalent Isotropic Radiated Power transmitted by the source, L_{fs} is the free space path loss, L_A is the atmospheric loss, G_R/T is the ratio of the receiving antenna gain to the noise temperature and is the receiving station figure of merit. G_c is the FEC coding gain available. R is the rate of the data. The final objective is to maintain a certain threshold value of E_b/N_0. So when large L_A attempts to pull down the E_b/N_0 value, the restoration of the same can be done by modifying any of the controllable parameter to the right hand side of the equation. We shall see in the following few subsections, how that is done for each of these parameters.

7.3.2 COMPENSATION TYPE OF FMT

To keep the link alive and working with desired performance, the C/N_0 of the signal, or more precisely, the E_b/N_0 has to be retained above a certain threshold value. Therefore, the propagation loss over the static link margin has to be compensated by changing some terms in the right hand side of the Eq. (7.55). Each of the individual components present in the equation can be modified to make the final E_b/N_0 above the required threshold so that the required performance is met. A compensation type FMT works on the concept of reallocating the link resources to compensate the fade due to the propagation. Changing the first term in Eq. 7.55, i.e. the EIRP, is the simplest of such techniques and will be described first. Other techniques like Adaptive Modulation (ADMOD), Adaptice Coding (ADCOD), etc. are also compensatory, but they are achieved at the cost of compromised performances. These techniques will be described subsequently.

7.3.2.1 EIRP control

We have seen in Eq. (7.55) that EIRP is directly related to the E_b/N_0 at the receiver. This EIRP is the most convenient parameter to modify to meet the fade encountered by the signal during propagation. Thus, EIRP control consists of varying either the carrier power or the transmission antenna gain in order to compensate for the power losses due to propagation effects. However, the adjustment of the antenna gain carried out on-board the satellite, a technique referred to as spot beam shaping (SBS), may be viewed as a separate type of satellite EIRP control. EIRP control techniques, viz. Uplink Power Control and Downlink power control, are described below.

Uplink power control (ULPC)

In ULPC, the output power of a transmitting Earth station is enhanced to compensate the fade in the uplink signal. It is, therefore, an Earth Station EIRP control technique. ULPC aims to keep a constant level of all the carriers at the input of the transponder, even during the time of the fade. The schematic for ULPC is shown in Fig. 7.7A. The transmitter power is increased during the fade conditions to counteract it and again decreased when more favourable propagation conditions are recovered. However, only minimum amount of power variation is implemented. This limits the excess drainage of power and also limits the possibility of driving the transponder amplifier to nonlinearity and also any interference in clear sky conditions.

This FMT is simple to implement since it requires only the knowledge of the factor of power increment in the earth station power amplifier within its limits. It is convenient to implement as this technique can play on the granularity, i.e. the minimum resolution of the power increment, which is not possible so easily with other FMT.

Generally, power control may be implemented in two ways such as open loop power control system and closed loop power control system (Ippolito, 1986). In case of open loop ULPC system, the transmitted power is modified based on the measurements of recently received power in the downlink, either from a pilot/beacon signal or from the information channel itself. On the other hand, for closed loop power control system, the transmitted power is adjusted based on reported power measurements over the channel. The transmitter (earth station) decides whether or not to vary the output power after receiving feedback information from the ground receiver at the other end and not based on estimates of the attenuation only. This theoretically results in a much more comprehensive control system. In practice, however, when applied to satellite systems the closed loop system must cope with propagation delays of the round trip time between the communicating elements, especially during the most aggravating atmospheric phenomena like the deep rain fades, scintillations, etc. which have short durations.

We know that a satellite transponder serves different channels which are placed adjacently in frequency or time, separated by guard bands or time. A possible problem caused by ULPC is adjacent channel interference. When the power of one

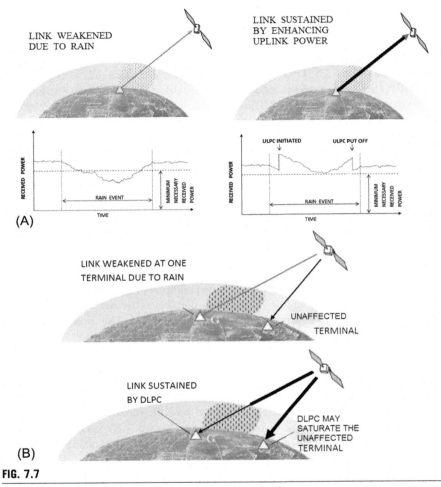

FIG. 7.7

(A) Uplink power control and (B) downlink power control.

channel is sufficiently increased compared to the other, the latter being not under fade condition, a part of the energy of the affected channel falls into adjacent channels and may lead to adjacent channel interference.

Further, an increase of an earth station transmit power may also impair the operation of adjacent satellite systems causing adjacent satellite interference, when the inter satellite separation is narrow.

On the other hand, ULPC requires adjustments only in the earth station resource which can provide some extra power support. Therefore, it can be implemented in a very flexible manner. Despite the drawbacks mentioned above, ULPC constitutes an effective countermeasure against signal fading and is a very popular technique used today.

Downlink power control (DLPC)

In DLPC, the on-board channel output power is adjusted to the magnitude of downlink attenuation. DLPC aims to allocate a limited additional power available on-board in order to compensate a possible degradation in terms of downlink C/N_0 due to propagation fade on a particular region.

However, DLPC is implemented with limited power levels for enhancement. This is due to the fact that the satellites have limited power resource. Also, the satellite footprint is usually large, and only a part of it is typically affected by the rain, while for the rest of the region, the signal remains unaffected. Therefore, enhancing the power to large extent to cater the need of the affected signals may lead to overdriving the signals in unaffected receivers into saturation and consequently leading to nonlinear effects.

So, apart from the possible adjacent channel interference, DLPC may also create intermodulation interference, a type of interference caused by intermodulation products generated by the nonlinear amplification of multiple carriers. The activation of DLPC, therefore, may possibly lead to a higher interference degradation to the link budget.

On-board beam shaping (OBBS)

The major drawback of DLPC is the fact that, since the satellite power resource is limited, the downlink power cannot be increased beyond a certain marginal limit. Moreover, when the satellite covers a wide area as footprint of a single antenna beam, the rain, being a localized phenomenon, occurs only in a small region of it. However, the DLPC simultaneously pulls the signal level of those receivers also which are not affected by rain fade and drives the receiver amplifier to saturation leading to nonlinearity.

It is possible to overcome both these two impediments by another adaptive technique called 'On-board Beam Shaping'. In this method, the total area of service is subdivided into many smaller regions and each of them is served by a smaller spot beam.

On-board beam shaping technique is based on active antennas and allows these spot beam gains to be adapted to propagation conditions. Under fade conditions, the antenna beam width gets more concentrated on the affected area and thus increasing the antenna gain. This in turn enhances the EIRP only for the affected region and so takes care of the signal fade. The OBBS technique, therefore consists of appropriately shaping the satellite antenna pattern so that the power received by the ground terminals remains nearly constant, even under rain fade conditions (Panagopoulos et al., 2004).

A major advantage of OBBS is that individual estimates of the attenuation across the receivers are not needed since compensation is carried out over the entire area under rain rather than for a single terminal site. Still, as the compensation is achieved by shaping the antenna pattern in accordance to the need and not by gross power enhancement, wastage of power is highly reduced. Further, since the power back-off is not compromised with, the undesirable effects of intermodulation interference due to amplifier nonlinearity are minimized.

Although current high frequency, high throughput satellites are utilizing this technique, further maturity in beam shaping technology for implementing adaptive antennas and the control of spot beam distribution can be expected.

7.3.2.2 Adaptive waveform techniques

Techniques belonging to this category focus on modifying the parameters of the signal structure used by the system, whenever the link quality is degraded. They can be further classified into three categories: Adaptive coding (ADCOD), Adaptive modulation (ADMOD) and Data Rate Adaptation (DRA).

Adaptive coding (ADCOD)

The communication links are always enforced with the Forward Error Correction techniques. This is done by adding redundant bits to the information bits, which are actually used at the recipient's end to detect and correct any error occurred while demodulating and detecting the data bits. Now, due to the propagation impairments causing large fading to the signal, the probability of erroneous detection increases. During this time, the FEC may be made more stringent by adding large numbers of redundant bit. Again, this can be relaxed when the fade vanishes. So, ADCOD consists in implementing a variable coding rate matched to impairments originating from propagation conditions (Proakis and Salehi, 2008). In Eq. (7.55), we have seen that the E_b/N_0 improves with larger coding gain G_c. Therefore, during large fades, FEC with high gains can be utilized. However, it leads to the consequent increase in the overhead and its associated consequences.

ADCOD can be implemented in two ways. On the one hand, it can be done with accompanied increment in bandwidth thus maintaining the data rate. Otherwise, and more often it is implemented with constant bandwidth, sacrificing the data for the overhead FEC bits.

However, for fade mitigation, it is important to correct burst errors rather than random errors. This is because the most dominant impeding conditions, like the rain, are continuous over finite interval of time causing the signal to degrade in spells and errors to occur in bursts. Therefore, the availability of a satellite link can be preserved by using the variants of the codes more resistant to bursts. A technique known as interleaving is effective at minimizing the effect of burst errors. In this technique, the sequence of data bits is first rearranged in a matrix form of definite dimensions. Then the constituent bits are restructured in a different sequence. The resultant bits are transmitted in the new sequence to which burst errors take place. When this sequence is reorganized at the receiver, to get back its original order, the burst errors get dispersed randomly over the different places over the entire length of the sequence. So, after reorganizing the bits to actual original sequence, the errors are spread and can be considered independent. However, interleaving proves efficient only against very short fades, particularly against scintillation.

Whenever the link suffers from severe propagation impairments, more efficient coding schemes that may be employed within the scope of ADCOD may originate from concatenated codes, i.e. combinations of block codes with convolution codes or by using Turbo codes. For the rest of the time, a less complex coding scheme may be used. Despite the decoding delay and computational complexity involved, the significant coding gain produced by turbo codes has drawn the interest of satellite communications, even leading to the standardization of these codes in small terminal satellite networks (Panagopoulos et al., 2004).

Adaptive modulation (ADMOD)

In Eq. (7.55), we have seen that the link parameters are adjusted to maintain the minimum necessary E_b/N_0. The E_b/N_0 is dependent upon the type of modulation. The ADMOD is the technique in which we reduce this minimum necessary E_b/N_0 by changing the modulation type. Reduced value of E_b/N_0 is only obtained when the bandwidth efficiency of the modulation is less, i.e. there are lesser numbers of bits that can be transmitted for a given bandwidth. So, when the C/N_0 at the input of the demodulator decreases due to propagation effects, ADMOD maintains the link by decreasing the necessary E_b/N_0 values for a certain required BER level by reducing the spectral efficiency (in bps/Hz). Hence, it is done by sacrificing the effective throughput for a given bandwidth.

During heavy rainfall events, ADMOD techniques exchange spectral efficiency for power requirements. The most commonly used modulation scheme in satellite communications is PSK (phase shift keying) modulation. To obtain higher spectral efficiency, that is, to transmit more bits per second without requiring more RF bandwidth, ADMOD employs higher-order PSK schemes such as QPSK, 8-PSK, 16 and 64-APSK or QAM, etc., having higher bandwidth efficiencies, respectively (Livieratos and Cottis, 2001).

In general, as the order M of an M-array PSK or M-array QAM scheme increases, the spectral efficiency of the communication link becomes higher. On the other hand, since the separation between adjacent amplitude and/or phase states is reduced, they become difficult to distinguish under noisy state and hence these higher-order modulation schemes become more susceptible to errors. As a result, an ADMOD system utilizes highly efficient modulation schemes such as 16-PSK, 64-PSK or 256-QAM under clear sky conditions and more robust modulation schemes such as BPSK or QPSK under the condition of heavy meteorological impairments.

ADCOD and ADMOD are preferably implemented as a closed loop FMT where the receiving earth station communicates with the transmitting station through a terrestrial return link in low transmission rates. In packet format, the information regarding the coding and modulation type to be adapted is provided in the header. In some implementations, both the coding and the modulation are simultaneously changed to obtain higher gain. It is hence termed as Adaptive Coding and Modulation (CODMOD). Such implementations are described in Chapter 8.

Data Rate Adaptation (DRA)

There exists another technique, in which the information data rate is adaptively modified to maintain a constant BER even during the fade. The technique is called data rate adaptation (DRA). Considering back the Eq. (7.55), we have seen that the received E_b/N_0 is varies inversely to the data rate R, while other factors remain constant. So, for a given minimum required E_b/N_0, the available value, represented by the right hand side of the equation, may reduce due to the increasing values of the attenuation L_A, and to keep it unaffected, the data rate R may be reduced by equal factor.

So, here, user data rates should be matched to the propagation conditions. Nominal data rates are used under clear sky conditions with no degradation of the service quality with respect to the system margin, whereas reductions of data rates are introduced according to fade levels. However, it eventually leads to reduction of information exchange rate.

7.3.3 AVOIDANCE TYPE FMT

Avoidance type FMT are countermeasures oriented against rain fades or other types of impairments in which the effective link is modified or duplicated in such a manner that the effect of the impairment on the overall link is avoided. A common technique to avoid is to shift the link in time, space or in frequency. The action is not to modify the link budget but instead to establish a new one, such that the impairments, which are basically temporally and spatially bound and frequency-dependent, does not influence the modified link thus established. One of the methods of implementing the avoidance type of FMT is to have diversity. In Eq. (7.55), this technique leads to effective reduction in L_A.

7.3.3.1 Diversity

The literal meaning of diversity is to be in the state of having variety. In a satellite system, variety in the links may be established by having different locations of transmission, different locations of reception, different frequency to establish the link or even by having different times to establish the link. The objective of these techniques is to re-route information in the network so that the atmospheric impairments may be avoided. Consequently, the minimum performance of the service may be maintained. As such, they constitute the most efficient FMTs. Four types of diversity techniques can be considered:

a. Site diversity (SD)
b. Orbital diversity (OD)/satellite diversity (SatD)
c. Frequency diversity (FD)
d. Time diversity (TD)

The first two techniques, viz. SD and OD take advantage of the spatial structure of the rainfall or the impeding medium, whereas FD and TD are based on their spectral and the temporal dependence respectively. There are two factors widely used to describe diversity performance, viz. (a) diversity gain and (b) diversity improvement.

We shall define the terms after discussing the site diversity (SD).

For very low fade values, the joint statistics will be almost similar to that of the single link. However, with increasing fade values, the probability of occurrence of any fade will be lesser jointly, i.e. with diversity than for a single station. It is actually the statistical distributions and correlation of the features of the impeding factors and the corresponding fades, upon which the diversity technique stands. Therefore, before moving into the details of the individual diversity systems, it is important to know two basic statistical features of fade (ITU-R, 2015b). Here, we shall consider rain fade, in particular.

a. Spatial correlation: The joint probability of occurrence of rainy conditions at different locations is reduced by several per cent for two separated stations, up to about 600 km. Beyond this range, the occurrences become more or less independent and hence no further improvement can be expected.
b. Frequency correlation: For the same signal path, the frequency correlation of the impairments decreases as the frequency separation increases.

Site diversity (SD)

High attenuations are due to regions of heavy rain, which are typically concentrated over a small geographical extent. Therefore, two earth stations at two distinct locations can establish links with the satellite which at given time suffer separate attenuations. SD, utilizing this idea, is the diversity technique in which two or more ground stations are used with geographically separated locations and hence with duplicated link along two different paths. This is shown in Fig. 7.8. The diversity advantage is achieved by changing the network routes, between these two stations, to avoid the severity of adverse propagation effect in one and thereby sustaining the performance of the link. Therefore, it applies only for the Fixed Satellite Service. SD thus actually takes advantage of the fact that two fades experienced by two Earth Stations separated by a distance higher than the size of a convective rain cell (more than 5 km) are statistically independent. Therefore, the joint probability of fade on these two links is considerably low compared to any of the individual link. In other words, when large fade is affecting one link, then, due to the separation, it is most unlikely that the other link will also be equally affected or affected at all. However, SD is an expensive option to implement.

The SD can adopt different algorithms for implementation like (Ippolito, 1986)

a. Primary predominant: In this type, the primary site is used as long as its signal level remains above the threshold. The diversity site is only used when the C/N_0 at the primary site falls below a certain predetermined threshold.
b. Dual active: In this both the sites are always active. The Earth station affected by a weaker event among the two and therefore having better C/N_0 in the received signal is used to route the information to the destination.
c. Combination: Here, signals from both the sites are additively combined to get a better level. No switching is necessary as both the sites are continuously used.

While the diversity system is able to re-route traffic through suitably selected alternatives, they are classified as 'Balanced' if the attenuation threshold and other configurations on both the links are equal and 'Unbalanced' if they are not. At frequencies above 20 GHz, path impairments other than rain can also affect the SD performances. The most important factor affecting the diversity gain offered by SD systems is the separation distance (D) between the earth stations. Larger values of D result in surpassing intense rain cells producing diversity gains up to around 20 dB.

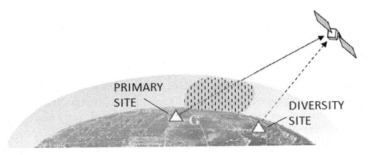

FIG. 7.8

Site diversity.

SD techniques are more efficient if the atmosphere exhibits inhomogeneity, so that alternative paths less affected by rain attenuation exist (Panagopoulos et al., 2004) when one is heavily impeded. This corresponds to the convective type of precipitation that occurs only for low time percentages. For SD, the main limitations come from cost consideration due to the additional earth station and the implementation of a terrestrial link to further process the jointly received signals.

We shall here define the terms diversity gain and diversity improvement, to understand the implication of each, and then shall mention the mathematical expressions to estimate them.

In a space diversity, there exist two different ground stations which are geographically apart. Therefore, it is unlikely that both the links will experience the same rain rate profile along their distinctly separate paths. Hence, the two stations will not experience the same attenuation at the same time. So, in SD, switching over to the nonfading route avoids the fade and reduces the overall probability of occurrence. Similar statement is valid for other kinds of diversity, also.

This indicates that, the joint probability of occurrence of a given value of fade is much lower than that for any single station. So, given a fixed total attenuation value A, the reduction in the probability of occurrence of this attenuation in diversity mode compared to the single mode of operation of a station is called the '*diversity improvement*'.

Similarly, for a given percentage of time, say $p\%$, let the exceedance value be As during the operation with a single station. Then with diversity, since the instances of occurrences of any value of attenuation on the effective link reduce, the probability density of occurrence consequently get reduced over the whole range of nontrivial attenuation values, the exceedance value of attenuation, corresponding to $p\%$ also reduces. So, for a given percentage $p\%$, the difference in the exceedance attenuation value between the single station operation and diversity mode is called the diversity gain. In brief, diversity gain is defined as the difference between the single site attenuation (without diversity reception) and the joint attenuation (with diversity reception) both expressed in dB for the same exceedance probability level. Mathematically, this can be written as

$$G_D = A_s(p) - A_D(p) \qquad (7.56)$$

G_D is considered as gain in the system because the result is effectively the same as having a gain in the system with single station. So, if a single site needs margin A_s, to maintain a link with necessary performance, the same can be achieved by having two sites with margin A_D which are separated to have a diversity gain of $A_s - A_D$. Fig. 7.9, showing exceedance plots with and without diversity illustrates the definitions of diversity gain and diversity improvement. The x axis in this figure represents the exceedance value for attenuation, while the y axis represents the corresponding probability of occurrence. So, obviously the diversity curve remains below the curve for a single station. For a definite value of probability on y axis, the horizontal difference between the curves, representing the attenuation difference for the same probability of occurrence, with and without diversity is the diversity gain.

Site diversity gain. The diversity gain, G (dB), between pairs of sites can be calculated with the empirical expression (ITU-R, 2015a), which can be used for site separations of less than 20 km. The total gain can be considered to be composed of independent contributions from different factors. The net diversity gain G may be assumed to be a function of spatial separation, frequency, elevation angle and baseline which combine multiplicatively to give the total diversity gain.

The gain contributed by the spatial separation 'd' is given by

$$G_d = a\left(1 - e^{-bd}\right) \qquad (7.57a)$$

Parameters 'a' and 'b' are dependent upon rain characteristics, or alternatively, on the single station rain attenuation A in dB, as

$$a = 0.78A - 1.94\left(1 - e^{-0.11A}\right) \text{ and}$$
$$b = 0.59\left(1 - e^{-0.1A}\right) \qquad (7.57b)$$

The frequency-dependent factor G_f is of the form

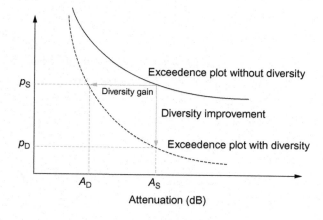

FIG. 7.9

Diversity gain and diversity improvement.

$$G_f = e^{-0.025f} \qquad (7.58)$$

where f is the frequency of the signal in GHz. Similarly, the gain dependent on elevation angle θ takes the form

$$G_q = 1 + 0.006\theta \qquad (7.59)$$

where θ is the elevation angle in degrees. It is clear from the above that, given a spatial distance 'd' between the diversity sites, since for higher elevation the effective separation between the paths is more, larger gain is achieved through this factor.

There is also a gain contribution in terms of the baseline orientation angle ψ. The baseline orientation angle ψ is the acute angle that the base line joining the two diversity stations on the ground makes with the surface projection of the propagation path towards the satellite. The expression for the corresponding gain is

$$G_\psi = 1 + 0.002\psi \qquad (7.60)$$

This is evident as the largest separation of the signal paths occurs when ψ is $\pi/2$, while for $\psi = 0$, the two paths almost coincide and hence this term does not provide any additional gain.

Finally, the net diversity gain is given as

$$G = G_d \cdot G_f \cdot G_\theta \cdot G_\psi \, \text{dB} \qquad (7.61)$$

Focus 7.1 Estimation of Diversity Gain

Two diversity sites are located at a distance of 5 km both looking at an elevation of 62 degrees exactly Southwards for a Ka band satellite system with downlink at 20.2 GHz. The rain attenuation in 20.2 GHz that needs to be supported for 99.95% availability is 14.5 dB. The objective here is to find the diversity gain for these two stations.

The attenuation for the mentioned availability at the given frequency is 14.5 GHz. Therefore, this value of attenuation has the exceedance probability percentage $p = 0.05\%$

With the given values, we get

$$a = 9.7636 \quad \text{and}$$
$$b = 0.4516$$

With these values of 'a' and 'b', the gain contributed by spatial separation is

$$G_d = 9.7636 \times (1 - \exp(-0.4516 \times 5))$$
$$= 8.7427$$

The frequency-dependent gain is,

$$G_f = e^{-0.025f}$$
$$= 0.6035$$

The estimate of the elevation-dependent gain is

$$G_\theta = 1 + 0.006\theta$$
$$= 1.3720$$

Further, since each of the stations is looking exactly Southwards, the baseline orientation angle ψ is 90 degrees and hence the corresponding gain term is

$$G_\psi = 1 + 0.002\psi$$
$$= 1.18$$

Finally, the overall gain for this diversity arrangement becomes

$$G_D = 8.7427 \times 0.6035 \times 1.3720 \times 1.18$$
$$= 8.5420$$

This means, with site diversity in place, the same availability of 99.95% may be maintained with only 14.5–8.5 = 6 dB margin.

Frequency diversity (FD)

We have seen in earlier part of the chapter that the propagation impairments, like the rain fade, cloud attenuation, gaseous attenuation, the scintillation, etc. all depend upon frequency. Broadly, higher frequency signals experience larger tropospheric impairments including fade. Frequency diversity is a technique to mitigate the fade incurred in a signal by the virtue of this fact. In this avoidance type of FMT, the information during the time of the fade event is transmitted on a carrier of different frequency that is less prone to these effects.

To implement such FMT, it is required to have payloads and ground stations equipped with two different frequencies. These frequencies must be wide apart to obtain considerable gain out of the technique. When a fade is occurring, links are re-routed using the lowest frequency band available in the payload, which are less sensitive to the continuing atmospheric propagation impairments. For example, this technique may employ the use of high frequency bands, like Ka, during normal operation and switches over to spare channels at lower frequency bands like the C band when the attenuation due to rain exceeds a certain threshold.

Apart from the dual frequency requirement, another drawback in the use of FD is associated with capacity allocation. The lower bands are unable to offer equally large bandwidths to the system and the throughput needs to suffer.

Orbital diversity (OD)

Satellite diversity is just the reverse architectural concept of space diversity. In this avoidance technique of FMT, there are two different satellites, placed at two different locations which transmit the same information to one earth station. The satellites are widely separated to provide an effective diverse path. Fig. 7.10 explains the architecture of OD in a simplified form.

Under the constrained condition of rain, OD allows earth stations to choose between two different satellites. Similar to SD, OD also adopts a re-route strategy for the network and, therefore, can be applied only for fixed satellite services (Matricciani et al., 1987). If more than two satellites are available for the purpose, it allows the Earth Stations to choose between various satellites. Finally, it may lock to one for which the most favourable link with respect to the propagation conditions exists.

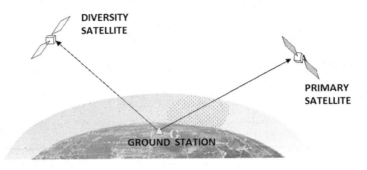

FIG. 7.10

Orbital diversity.

As the tropospheric fades are encountered at the lower flank of the link, no considerable gain will be availed unless the satellites are apart widely enough, so that the signal passes through adequately separated paths. There will be still some statistical correlation nearest to the ground as the different paths tends to converge to the earth station situated in the rain.

Furthermore, with OD, an interruption of service may occur during the shift from one satellite to the other. The duration of this interruption could be minimized employing sophisticated tracking software or if active beam antennas are used for the ground segment. Switching between two antennas is also another option.

Time diversity (TD)

The main factor causing the attenuation is the rain. However, rain is time-limited and TD takes this feature in its advantage. TD FMT utilizes the idea of delayed transmission of the data to avoid the fade. That is, the information, which was supposed to be transmitted during the fade event, is stored and repeated or retransmitted after a delay comparable with the rain interval. Therefore, TD can be implemented for services that can withstand data latency and in which real time operation is not essential. Hence, the performance of such system, in terms of throughput, actually depends upon the fade duration which determines the retransmission interval. It is also dependent upon the time intervals between fades which determines the window of retransmission. Video on demand, multimedia, and data applications can be services of such category (Ismail and Watson, 2000). However, some researchers do not consider TD as any real adaptive FMT.

7.3.4 JOINT FMT

For satellite systems operating at Ka and Q/V-band, each individual FMT is more or less adapted to a specific range of availability. Such methods implemented individually yield small gain and hence can only compensate relatively small magnitudes of propagation impairments. However, these fade mitigation methods are quite complementary and can be implemented simultaneously as joint FMTs to extend the depth of attenuation which it can compensate. It is possible to improve the performance of the mitigation to a large extent, by carrying out such combination of different kinds of FMTs when there is a need of larger fade compensation. Therefore, advanced satellite communication systems use such combined techniques to form a more sophisticated and powerful fade compensation scheme. For example, systems may compensate total attenuation (uplink plus downlink) over a link by ULPC combined with MODCOD or DRA. Joint FMT is a very promising solution to improve the performance of a Satellite Communication System (Castanet et al., 2009). However, the design of these FMT is strongly dependent on the system requirements, e.g. service availability, user minimum data rate, system capacity or interference reduction.

7.3.4.1 Implementation of a simple FMT

There are different approaches for implementing any particular FMT or their combination. It primarily depends upon the satellite application. To understand and appreciate how exactly these techniques are actually implemented in a satellite system, let us take a very simple case for a networked satellite system for communications. Here we shall discuss an ULPC technique, based on closed loop approach. Implementation of such power control technique may improve the performance of existing network by avoiding wastage and underutilization of power.

Let us consider that the network is consisting of a centralized HUB and numbers of remote terminals connected in a star topology through the hub via the satellite. This is shown in Fig. 7.11. The forward link, that is from the hub to the terminal, is a continuous TDM broadcast link, received at all the remote terminals. The return link from the terminals to the hub is a narrow band satellite link.

Closed loop power control consists of monitoring and control of transmit power, in real time, of all remote terminals to mitigate fading, maintain functional links and

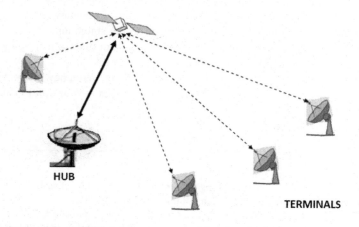

FIG. 7.11

Architecture of a closed loop FMT.

prevent generation of intermodulation products in transponder and interference by other satellites. The monitoring and control processor can be made centralized and conveniently implemented in the Network Monitoring System (NMS) located at the HUB station.

The hub polls all the terminals one by one to get measured C/N_0 readings at each of the terminal end. The terminals measure the receive C/N_0 from the received signal transmitted from the hub and send the value back to it through the return link in response to the request during the polling. Along with C/N_0, the current transmitted power by the terminal is also sent. The hub also measures the downlink rain attenuation of each of the return link by itself. Based on these information, the centralized NMS calculates the actual fading in both uplink and downlink, and accordingly, finds out the required transmit power for the terminals. This information regarding the required transmit power is then sent to the terminal for reconfiguration. The NMS also calculate the required uplink power at HUB side and adjust accordingly at regular interval. However, the transmitted power at the hub is increased only to a level which does not drive the satellite transponder to saturation causing intermodulation products. If this enhanced power at the hub does not mitigate fade at the worst affected terminal, then coding gain is increased and modulation is changed to mitigate the excess fading.

CONCEPTUAL QUESTIONS

1. From the given expressions in Eq. (7.22) for the total extinction and scattering cross section, derive the total absorption cross section using the 'a' and 'b' coefficients.
2. In a time interval of 1 min, 100 drops were measured in a disdrometer for each class of rain drop. Does this mean the drop density distribution is uniform?
3. Keeping uplink frequency lower than the downlink frequency is sometimes beneficial during rainy condition—comment on the statement.
4. We can operate a satellite link even with insignificant margin just by increasing the numbers of diversity sites—is the statement correct?
5. What are the advantages and disadvantages of using frequency diversity?
6. Time diversity is not at all a diversity technique. What is your opinion about this statement?

REFERENCES

Ajayi, G.O., Olsen, R.L., 1985. Modelling of a tropical raindrop size distribution for microwave and millimetre wave applications. Radio Sci. 20, 193–202.
Bandera, J., Papatsoris, A.D., Watson, P.A., Tozer, T.C., Tan, J., Goddard, J.W., 1999. Vertical path reduction factor for high elevation communications systems. Electron. Lett. 35 (18), 1584–1585.

Barsis, A.P., Samson, C.A., 1976. Performance estimation for 15-GHz microwave links as a function of rain attenuation. IEEE Trans. Commun. COM-24 (4), 462–470.

Bolea-Alamanac, A., Bousquet, M., Castanet, L., et al., 2003. Implementation of short term prediction models in fade mitigation techniques control loops. In: Joint COST 272/280 Workshop, ESA/ESTEC, Noordwijk, The Netherlands, 26–28 May, 2003.

Castanet, L., et al., 2002. Interference and fade mitigation techniques for Ka and Q/V band satellite communications systems. In: Proceedings of COST 280 Workshop: Propagation Impairments Mitigation for Millimeter Wave Radio Systems, Noordwijk, The Netherlands, 2003.

Castanet, L., Bolea-Alamañac, A., Bousquet, M., Claverotte, L., Gutierrez-Galvan, R., 2009. Performance assessment of Fade Mitigation Techniques for the GEOCAST IST project with transparent and OBP architectures. Int. J. Space Commun. 22 (1), 1–12.

Chambers, A.P., Otung, I.E., 2005. Neural network approach to short-term fade prediction on satellite links. Electron. Lett. 41 (23), 1290–1292.

Chambers, A.P., Otung, I.E., 2008. Near-optimum short-term fade prediction on satellite links at Ka and V-bands. Int. J. Satell. Commun. Netw. 26 (1), 31–43.

COST Action 255, 2002. Radiowave Propagation Modelling for Satcom Services at Ku Band and Above. ESA Publications, Noordwijk.

Crane, R.K., 1977. Prediction of the effects of rain on satellite communication systems. Proc. IEEE 65, 456–474.

Crane, R.K., 1978. A global model for rain attenuation prediction. In: Paper Presented at Electronics and Aerospace Systems Convention, Arlington, USA.

Damosso, E., DeRenzis, G., Fedi, F., Migliorini, P., 1980. A systematic comparison of rain attenuation prediction methods for terrestrial paths. In: Proceedings of URSI Open Symposium on Effects of the Lower Atmosphere on Radio Frequencies Above 1 GHz, Lennoxville.

De Montera, L., Mallet, C., Barthe's, L., et al., 2008. Short-term prediction of rain attenuation level and volatility in Earth-to-Satellite links at EHF band. Nonlinear Process. Geophys. 15 (4), 631–643.

Dossi, L., 1990. Real time prediction of attenuation for applications to fade countermeasures in satellite communications. Electron. Lett. 26 (4), 250–251.

Fedi, F., 1981. Rain Attenuation on Earth-Satellite links: a prediction method based on joint rainfall intensity. Ann. Telecommun. 36, 73–77.

Gremont, B., Filip, M., Gallois, P., et al., 1999. Comparative analysis and performance of two predictive fade detection schemes for Ka-band fade countermeasures. IEEE Sel. Areas Commun. 17 (2), 180–192.

Gunn, R., Kinzer, G.D., 1949. The terminal velocity of fall for water droplets in stagnant air. J. Meteorol. 6, 243–248.

Ippolito Jr., L.J., 1986. Radio Wave Propagation in Satellite Communication. Van Nostrand Reinhold Company, USA.

Ismail, A.F., Watson, P.A., 2000. Characteristics of fading and fade countermeasures on a satellite Earth link operating in an equatorial climate, with reference to broadcast applications. IEEE Trans. Microwave. Antennas Propag. 147 (5), 369–373.

ITU-R P.840, 1999. Recommendation ITU-R P.840-3: Attenuation Due to Clouds and Fog. ITU, Geneva.

ITU-R P.838, 2005. Recommendation ITU-R P.838-3: Specific Attenuation Model for Rain for Use in Prediction Methods. ITU, Geneva.

ITU-R P.835, 2012a. Recommendation ITU-R P.835-5: Reference Standard Atmospheres. ITU, Geneva.

ITU-R P.837, 2012b. Recommendation ITU-R P.837-6: Characteristics of Precipitation for Propagation Modelling. ITU, Geneva.

ITU-R P.676, 2013a. Recommendation ITU-R P.676-10: Attenuation by Atmospheric Gases. ITU, Geneva.

ITU-R P.836, 2013b. Recommendation ITU-R P.836-5: Water Vapour: Surface Density and Total Columnar Content. ITU, Geneva.

ITU-R P.839, 2013c. Recommendation ITU-R P. 839-4, Rain Height Model for Prediction Methods. ITU, Geneva.

ITU-R P.618, 2015a. Recommendation ITU-R P.618-12: Propagation Data and Prediction Methods Required for the Design of Earth-Space Telecommunication Systems. ITU, Geneva.

ITU-R P.679, 2015b. Recommendation ITU-R P.679-4: Propagation Data Required for the Design of Broadcasting-Satellite Systems. ITU, Geneva.

Karasawa, Y., Yamada, M., Allnutt, J.E., 1998. A new prediction method for tropospheric scintillation on earth-space paths. IEEE Trans. Antennas Propag. 36, 1608–1614.

Kastamonitis, K., Gremont, B., Filip, M., 2003. Short-term prediction of rain attenuation based on fade slope. Electron. Lett. 39 (8), 687–689.

Lane, S.O., Stutzman, W.L., 1980. Spatial rain rate distribution modeling for earth-space link propagation calculations. In: URSI/IEEE AP-S International Symposium, Quebec.

Laster, J.D., Stutzman, W.L., 1995. Frequency scaling of rain attenuation for satellite communication links. IEEE Trans. Antennas Propag. 43 (11), 1207–1216.

Laws, J.O., Parsons, D.A., 1943. The relation of raindrop size to intensity. Eos. Trans. AGU 24, 452–460.

Lin, S.H., 1979. Empirical Rain Attenuation Model for Earth-Satellite paths. IEEE Trans. Commun. COM-27, 812–817.

Livieratos, S.N., Cottis, P.G., 2001. Availability and performance of single multiple site diversity satellite systems under rain fades. Eur. Trans. Telecommun. 12 (1), 55–56.

Malygin, A., Filip, M., Vilar, E., 2002. NN implementation of a Fade Countermeasure controller in a VSAT link. Int. J. Satell. Commun. Netw. 20 (2), 79–95.

Marshall, J.S., Palmer, W.M., 1948. The distribution of rain drops with size. J. Meteorol. 5, 165–166.

Masseng, T., Bakken, P.M., 1981. A stochastic model for rain attenuation. IEEE Trans. Commun. 29 (5), 660–669.

Matricciani, E., Paraboni, A., 1985. Instantaneous frequency scaling of rain attenuation at 11.6-17.8 GHz with SIRIO data. IEEE Trans. Antennas Propag. AP-33 (3), 335–337.

Matricciani, E., Mauri, M., Paraboni, A., 1987. Dynamic characteristics of rain attenuation: duration and rate of change. Alta Freq. 56, 33–45.

Mätzler, C., 2002. MATLAB Functions for Mie Scattering and Absorption: Research Report No. 2002-08 June 2002. Institute of Applied Physics, University of Bern, Bern.

Misme, P., Waldteufel, P., 1980. A model for attenuation by precipitation on a microwave earth space link. Radio Sci. 15 (3), 655–665.

Olsen, R.L., et al., 1978. The aR^b relation in the calculation of Rain Attenuation. IEEE Trans. Antennas Propag AP-26 (2), IE3 TAP, 318–329.

Otung, I.E., 1996. Prediction of tropospheric amplitude scintillation on a satellite link. EEE Trans. Antennas Propag. 44 (12), 1600–1608.

Panagopoulos, A.D., Arapoglou, P.D.M., Cottis, P.G., 2004. Satellite communications at KU, KA, and V bands: propagation impairments and mitigation techniques. IEEE Commun. Surv. Tutorials 6 (3), 2–14.

Proakis, J.H., Salehi, M., 2008. Digital Communications, fifth ed. Mc Graw-Hill, New York, USA.

Roy, B., Acharya, R., Sivaraman, M.R., 2012. Attenuation prediction for fade mitigation using neural network with in situ learning algorithm. Adv. Space Res. 49, 336–350.

Strutzman, W.L., Dishman, W.K., 1982. A simple model for the estimation of rain-induced attenuation along earth-space paths at millimetre wavelengths. Radio Sci. 17 (6), 1465–1476.

Sweeny, D.G., Bostian, C., 1992. The dynamics of rain induced fades. IEEE Trans. Antennas Propag. 40, 275–278.

Thompson, P.T., Dissanayake, A.W., Watson, P.A., 1980. The frequency dependence of microwave propagation through rainfall. In: Proceedings of AGARD Conference, 284, pp. 5.1–5.9.

Valentin, R., 1977. Attenuation and depolarization caused by rain at frequencies above 10 GHz. In: Proceedings of URSI, Commission—F, Open Symposium, La Baule, France.

Van de Kamp, M., 2002. Rain attenuation as a Markov process, using two samples. In: Eighth Ka-band Utilization Conference, Baveno, Italy, 25–27 September 2002.

Van de Kamp, M.M.J.L., Riva, C., Tervonen, J.K., Salonen, E.T., 1999. Frequency dependence of amplitude scintillation. IEEE Trans. Antennas Propag. 47 (1), 77–85.

Yushanov, S., Crompton, J.S., Koppenhoefer, K.C., 2013. Mie scattering of electromagnetic waves. In: Proceedings of COMSOL, Boston.

FURTHER READING

Allnutt, J.E., 2011. Satellite to Ground Radio Wave Propagation, second ed. The Institution of Engineering and Technology, United Kingdom.

IEEE, 1990. IEEE Standard Definitions of Terms for Radio Wave Propagation. IEEE Std. 211-1990, The Institute of Electrical and Electronics Engineers, USA.

Matricciani, E., 1997. Prediction of fade durations due to rain in satellite communication systems. Radio Sci. 32 (3), 935–941.

Noise, multipath and interference

8.1 NOISE

It is very important to introduce at this point the concept of 'Noise'. In this chapter, we shall learn about noise, its definition, sources and characteristics and the effects it creates on the satellite communication systems. In communication systems, noise is an undesired random electrical disturbance added to the useful information signal in a communication channel.

The signal, passing through the medium and through different units of the receiver, gets added with some uncorrelated random electrical variations, which is referred to as noise. The noise gets algebraically added with the signal and hence is additive in nature. Therefore, the signal which is received at the receiver is the original signal plus the added noise. This added noise being random has no correlation of its present value with those of the past. Hence, these added values cannot be definitely known and neither can be predicted. However, over a long period of time, they generally follow Gaussian amplitude probability distribution. Further, when these noise values are added up over a considerable time, the positive and the negative terms add up to equal magnitudes and hence the sum becomes zero. This means, the noise has a zero mean. Yet, the square of these values when averaged over time gives a definite variance σ^2 of the noise. In short, the noise which gets added up with the signal is generally an additive, zero mean, Gaussian, random signal. We shall also learn about its spectral characteristics later in this chapter.

Since the signal variables change due to this noise addition, it affects the measurements done in the receiver and results in error in identifying the correct information out of it. Thus, the overall performance of the signal communication gets deteriorated. This is analogous to the condition when we want to hear a speaker, speaking in a crowded room. He is audible and intelligible to all when the audience is silent. Now, suppose many people in the audience themselves start talking. This creates noise in the background which now makes it difficult for the listeners to realize what the speaker is actually saying. More the people talk across, more the noise is created, making understanding more difficult. If the noise continues to increase, then at some point, the speaker becomes unintelligible, if he still speaks with the same intensity (Acharya, 2014).

Similarly, in a satellite communication system, the received signal power is meaningless unless it is compared with the power received from unwanted noise sources over the same bandwidth. Such noise sources include thermal radiation from

Satellite Signal Propagation, Impairments and Mitigation. http://dx.doi.org/10.1016/B978-0-12-809732-8.00008-9

the earth and sky, cosmic background radiation and random thermal processes in the receiving system.

The noise is a summation of unwanted energy from natural and sometimes man-made sources. It is, however, typically distinguished from interference like crosstalk, deliberate jamming or other unwanted electromagnetic interference from specific transmitters, which may also lead to similar degradation of the signal.

The amount of noise that is added with a certain signal determines the received signal quality in terms of carrier-to-noise ratio (CNR) at the receiver input. This, in turn, leads to a definite signal-to-noise ratio in the demodulated signal. In a digital communications system, this also yields a certain E_b/N_0 (information bit energy to noise density ratio) and results in a definite bit error rate (BER). This BER defines the performance of the system.

8.1.1 SOURCES OF NOISE

The signals may be considered as systematic electrical variations. As the signal travels from the transmitter to the receiver, it propagates through the medium and through the electronic and other hardware components at the transmitter and receiver, as well. In the whole process, it picks up some uncorrelated random electrical variations, generated by natural or man-made sources, which are within the band of the signal. Such random electrical variations, generated by any means, get added up as a noise to a meaningful signal. These kinds of unwanted noise are generated in all space and at all time and across all frequencies. Correspondingly, anything that can generate such random electrical signal acts as a source of noise. For a propagating wave, there are two main sources of noise. Depending upon the origin, the noise, which affects the satellite signals, can be broadly categorized into the following.

a. Antenna noise comprising of Sky noise and Atmospheric noise
b. Receiver Thermal noise

To understand the reason and the characteristics of the noise generated by these sources, we shall appreciate the fact that any energetic charged particles will exhibit finite kinetic energy and hence temperature. As a result, these particles will be accelerated and radiate energy. Particles present everywhere, at finite temperature, thus exhibit emission of random uncorrelated, incoherent radiations due to motion of their inherent charges. This is the basic and the most simplified background of the black body radiation, which we shall learn next.

8.1.2 BLACK BODY
8.1.2.1 Planck's law
At the outset, it is important to have a general view of the noise sources and about their characteristics. For that, it is suggestive to get a very brief idea about the black body. A black body is any material that can act as a hypothetical perfect absorber of

energy, with no reflecting power regardless of frequency or angle of incidence. A black body is also a perfect radiator over all possible frequencies, and hence, in thermal equilibrium emits electromagnetic radiation in equal amount as it absorbs. This maintains the balance of heat energy of the body. Therefore, at a certain definite temperature, the amount of heat absorbed by the body is equal to the energy radiated by it in the form of radiation. This radiation is emitted according to Planck's law.

Planck's law describes the energy distribution of the electromagnetic radiation emitted by a black body in thermal equilibrium, at a definite temperature, T. The law was proposed by Max Planck in 1900 and is a pioneering result of modern physics and quantum theory. As Feynman has mentioned, ".. it was the starting of the end of the classical physics" (Feynman et al., 1992). According to this law, the available energy at temperature T is not equally distributed to all the particles, but the distribution is exponentially reducing with increasing energy. Using this distribution, the spectral radiance of a body, $B(f)$, describing the amount of power emitted due to the radiation per unit area of the body, per unit solid angle, per unit frequency range at absolute temperature T is given by (Ghosh, 1990)

$$B(f) = \frac{2hf^3}{c^2} \frac{1}{e^{hf/kT} - 1} \tag{8.1}$$

where, h is the Planck constant and its value is 6.626069×10^{-34}, k is Boltzmann's constant with value 1.38×10^{-23} J/K. (However, some authors represents Planck's law in terms of energy density). The variation of the spectral radiance for different

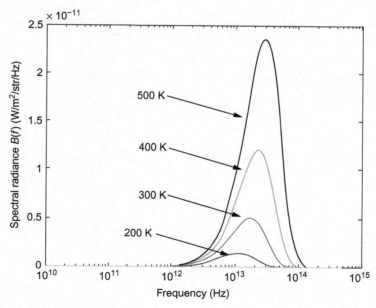

FIG. 8.1

Spectral radiance distribution according to Planck's law for black body radiation.

frequencies is shown in the Fig. 8.1. In SI unit, B has the unit of Watt/Steradian/metre^2/Hz. From this, the total spectral irradiance, i.e. the total power radiated by the body per unit area per unit frequency at temperature T is obtained by integrating $B(f)$ over the range of solid angles and is given as

$$I(f) = \frac{8\pi h f^3}{c^2} \frac{1}{e^{hf/kT} - 1}$$

$$= \frac{I_0(f)}{e^{hf/kT} - 1}$$

(8.2)

Note that, in the expression, except the variable f, the only parametric term is the temperature T, while all other terms are constants. Thus, the radiation spectrum has a definite pattern depending upon the value of the temperature. In other words, the radiation will exhibit a distribution in energy, which is characteristic of the temperature T and does not depend upon any other properties of the medium. Therefore, at a definite temperature maintained by the body, the radiation has a definite spectrum that is determined by the temperature alone and it also characterizes the total radiation. This makes it possible to derive the temperature of the body from

---

Box 8.1 Low Frequency Approximation of Planck's Law

The objective in this box is to see the behaviour of the radiance for lower frequency regime. It is the area where our concern lies since the satellite signals are in GHz range only. Run the Matlab programme Planck.m to obtain the plots for radiance for different absolute temperatures starting from 200 to 500 K. The variation is shown in part (A) in the Fig. M8.1 below. It represents the Rayleigh-Jeans approximation of Planck's law for lower frequencies.

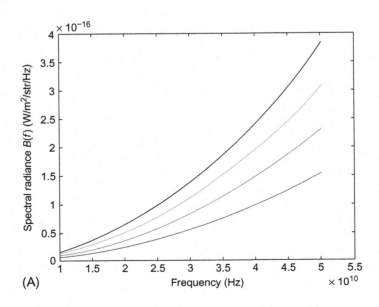

(A)

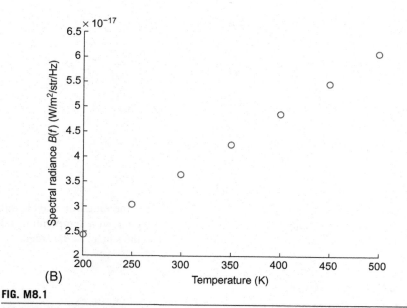

FIG. M8.1

(A) Rayleigh Jeans Approximation of Planck's radiation law, (B) variation of the radiance with temperature at a given constant frequency.

Observe that, although the radiance varies in a parabolic fashion with frequency, it is evident from part (B) of Fig. M8.1 that for a given frequency, the variation with temperature is linear. However, the constant of linearity, i.e. the gradient of the line is a function of frequency.

Vary the frequency range to much higher values, say a few orders higher, and see how the above variations change.

the measurement of its radiation over the frequency. In turn, considering the equilibrium condition, the amount of absorption by the body at that frequency can also be derived out of it.

Real materials emit energy at a constant fraction of black body energy levels. This fraction is a property of the material of the medium and is called the emissivity, ε. Thus, by definition, a black body in thermal equilibrium has an emissivity given by $\varepsilon = 1.0$. Materials with emissivity less than 1 are also called Grey bodies.

8.1.2.2 Radiation temperature

Our range on interest lies only over a small portion of the entire emission range of a hot black body and for lower values of the frequencies. For the lower frequencies, the radiation formula gets modified into Raleigh–Jeans formula (Ghosh, 1990), which is actually a low frequency approximation of the Planck's formula. This is expressed as

$$\begin{aligned}
B(f) &= \frac{2hf^3}{c^2} \frac{1}{e^{hf/kT} - 1}\bigg|_{f \ll} \\
&= \frac{2hf^3}{c^2} \frac{1}{\{1 + hf/(kT) - 1\}} \\
&= \frac{2hf^3}{c^2} \frac{kT}{hf} \\
&= 2\left(\frac{f}{c}\right)^2 kT \\
&= \frac{2}{\lambda^2} kT \\
&= g(\lambda)T
\end{aligned} \tag{8.3}$$

where k is the Boltzmann constant and $g = 2/\lambda^2 \, k$ and λ is the radiation wavelength. Therefore, the spectral radiance becomes linearly varying with temperature and inversely to the square of the wavelength, i.e. directly to the square of the frequency.

8.1.3 ANTENNA NOISE

8.1.3.1 Sky noise temperature

Every particle at finite temperature emits incoherent radiations. We have learnt this in our previous section. For the given temperature, these radiations can extend over a large bandwidth with a definite spectral profile. These incoherent radiations are picked up by the receiver antenna and it gets added up with the signal as noise.

For a given frequency, 'f', or wavelength, 'λ', since the incoherent radiations generated by the particles of the clear astronomical sky and received by the system as noise are only a function of the temperature 'T', this noise can be represented in terms of the corresponding noise temperature, and is referred to here as the 'Sky Temperature', T_S.

At this point, it is to be remembered that, the spectral radiance in the band appropriate to the satellite signals is represented by the Rayleigh–Jeans formula, which clearly shows frequency-dependent variation. The expression in Eq. (8.3) is only for irradiance and not for the total power 'seen' by the receiver antenna. When the antenna actually observes a large black body in form of extensive astronomical 'sky', the amount of radiation that is directed towards it is proportional to the area it 'sees' within its main lobe. Using the definition of spectral radiance, $B(f, \theta, \varphi)$, the power received by an antenna of area A_e from a section of the radiating surface of area S at a distance r, per unit frequency range is given by,

$$\begin{aligned}
P_r(f, \theta, \varphi) &= B(f, \theta, \varphi) S \cdot \frac{\hat{r}}{r^2} A_e(f, \theta, \varphi) \\
&= B(f, \theta, \varphi) d\omega A_e
\end{aligned} \tag{8.4}$$

In the above expression, $d\omega = S \cdot \hat{r}/r^2$ is the solid angle subtend by the radiating surface at the antenna point. The total power that the antenna receives within the solid angle $d\omega$ is:

$$P_r(f, \theta, \varphi) = \frac{1}{2} \times \frac{2kT(f, \theta, \varphi)}{\lambda^2} A_e(f, \theta, \varphi) \, d\omega \, df \qquad (8.5)$$

where df is the infinitesimal frequency range around the frequency f at which the radiation is considered. Notice the factor $\frac{1}{2}$ in the above expression, which results because the total radiation from the black body is in all possible polarizations. But, since the antenna can receive only one of the possible two components of the polarization, the actual received power is one half of this total amount. The polarization of the receiving antenna therefore causes it to ignore or reject half of the available thermal power. If we now replace the effective area by the gain expression

$$G(f, \theta, \varphi) = \frac{4\pi}{\lambda^2} A_e(f, \theta, \varphi)$$

we get (MIT, 2003):

$$P_r(f, \theta, \varphi) = \frac{kT(f, \theta, \varphi)}{\lambda^2} df \times \lambda^2 \frac{G(f, \theta, \varphi)}{4\pi} d\omega$$
$$= kT(f, \theta, \varphi) df \frac{G(f, \theta, \varphi)}{4\pi} d\omega \qquad (8.6)$$

To obtain the total noise power directed towards the antenna, we integrate the above expression over all the possible solid angles over which the antenna can receive, and assuming that T is same over the entire range, we get

$$P_{r,\,total} = kT \int df \frac{1}{4\pi} \int G(f, \theta, \varphi) \, d\omega \qquad (8.7a)$$

$$= kT \int df \frac{1}{4\pi} 4\pi$$
$$= kTB \qquad (8.7b)$$

This expression assumes that $hf \ll kT$ and that signals arriving from different directions have their powers superimposed. The scenario is illustrated in Fig. 8.2. Notice that, this equation establishes the fact that, the total noise power received is only dependent upon the temperature and bandwidth and nothing else. Thus, temperature can be used as surrogates for received noise power from the mentioned 'sky' per unit bandwidth and is called as Sky Noise Temperature, T_S. However, if the antenna look angle is large enough that the temperature varies over the total scene such that T cannot be taken to be constant, then we have to integrate equation (Eq. 8.7a) with the spatial profile of the temperature. In such condition, the effective noise temperature corresponding to the received noise becomes

$$T_S = \frac{\int T(\theta\varphi) G(f, \theta, \varphi) \, d\omega}{\int G(f, \theta, \varphi) \, d\omega} \qquad (8.8)$$

It is clear from the above derivation in Eq. (8.7b) that, over the bandwidth B, the total power reaching the antenna from the black body surface can be written as

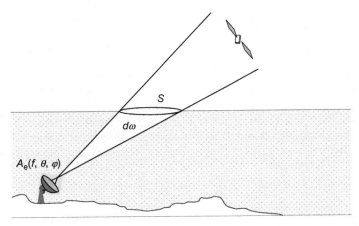

FIG. 8.2

Atmospheric noise received by an antenna.

$$N_S = P_{r,\,total} = kT_S B$$

So, the effective spectral whiteness originates from two effects. The antenna beam width, defining the solid angle within which it can receive the radiated power, narrows proportionately to the square of the frequency, while the power radiated by the black body increases with the square of the frequency by equal proportion. So, when we change to a lower frequency, the surface brightness falls by a definite factor, as the irradiance is inversely proportional to λ^2. But, the solid angle from which the antenna collects the noise power also increases and more surfaces come into view. Since this increment is directly proportional to λ^2, it increases by the same factor by which the irradiance falls. Hence, the received intensity per unit bandwidth remains uniform even with change in frequency. In short, it is the combined effect of the antenna gain, the source temperature and the collection region which determines the total cumulative noise power picked up in the antenna.

8.1.3.2 Atmospheric Noise Temperature

The receiving antenna will see a volume of the atmosphere within its beam width. The particles present there and having a finite temperature will absorb as well as emit radiation. The latter will be received by the receiving antenna. A receiver receiving signals from the atmosphere will also pick up the background sky noise in its field of view. Therefore, the total noise power reaching the antenna can be represented in terms of the noise temperature contributions from the total volume of the atmosphere visible to the antenna plus the astronomical background.

At the outset, we recall that in the Rayleigh–Jeans region of the spectrum, the spectral radiance is given by

$$B(f) = \frac{2}{\lambda^2} kT$$
$$= g(\lambda)T$$

where, $g = 2 k/\lambda^2$ is a function of wavelength λ and becomes constant for a given fixed value of λ. Now, assume that a hot source is radiating noise energy, which is passing through the atmosphere. The energy can be represented by the corresponding noise temperature T. A fraction of the radiation is absorbed along the path of propagation in a medium, here the atmosphere. The energy traversing through the atmosphere thus gets reduced. The Beer–Lambert law, also referred to as the Beer–Lambert–Bouguer law, governs the reduction (Sportisse, 2009). It states that the amount of energy lost by the noise while passing through a horizontal slice of atmosphere per unit length is proportional to the energy it carries, i.e. proportional to its current temperature. This absorbed energy determines the temperature of the layer and at equilibrium is equally reradiated. We shall derive here, in a very simplistic fashion, the amount of noise that reaches the receiver, after being originated from a section of the atmosphere due to this reradiation. For the same, let us assume that a noise, radiated by a source with initial radiation energy E_0 or with equivalent noise temperature T_0, subsequently has a variation $T(h)$ as it passes through the atmosphere. The energy lost by the radiation in terms of equivalent noise temperature reduction is

$$dT = -\alpha T \, dh$$
$$\text{or,} \quad T(h) = T_0 \exp(-\alpha h) \tag{8.9a}$$

$$\text{or,} \quad E(h) = E_0 \exp(-\alpha h) \tag{8.9b}$$

Here h is the depth of the atmosphere and α is the absorption coefficient. T_0 is the noise temperature before entering the atmosphere, i.e. at height where we consider the depth $h = 0$. Therefore, from the solution, it is evident that the effective noise temperature of the atmosphere exponentially decreases upon traversing distance h downwards. Now, that we have got the temperature profile, we can use it for the following purposes, (a) to obtain the total sky noise power that reaches the ground passing through the whole atmospheric block and (b) to find out the total absorptive attenuation that any signal would experience on travelling through it and (c) to find the received noise components in terms of this attenuation.

First, consider that the energy E_g reaching the ground antenna can be derived from Eq. (8.9b) as

$$E_g = E_0 \exp(-\tau)$$
$$\text{or,} \quad E_0/E_g = \exp(\tau) \tag{8.10}$$
$$= A$$

where $\tau = \alpha h$ and A is the attenuation factor. So, the attenuation A is thus expressed in terms of E_0 and E_g in decibels as

$$A(dB) = 10 \log(E_0/E_g)$$
$$= 10 \log[\exp(\tau)] \tag{8.11}$$

The above Eq. (8.9a) again shows that the temperature that reaches the receiver at the ground through the atmosphere is $\exp(-\alpha h)$ part of the temperature T_0 at height h. Putting this relation in the relation in Eq. (8.11), we get attenuation A, in dB as

$$A(dB) = 10 \log (T_0/T_g) \tag{8.12}$$

where T_g is the noise temperature received at the ground. So, if we identify T_0 as the sky noise temperature, the part of this noise temperature received at the ground due to this component can be expressed as

$$T_g = T_S \times 10^{-A(dB)/10} \tag{8.13}$$

The total power received by the ground antenna due to the radiation from the entire volume within its visible region is the integral of the radiation from each of the layers of width dh. Using our knowledge that at equilibrium, the amount of energy absorbed is equal to that radiated and the fact that only an exponential part of the total radiated noise power reaches the ground, the total power received at the antenna, due to the atmospheric brightness, per unit bandwidth becomes

$$N_B = \int_0^{hm} k\alpha T(h) \exp(-\alpha h) dh$$
$$= k\alpha \int_0^{hm} T(h) \exp(-\alpha h) dh$$

Here, we are using the term 'k' instead of $g(\lambda)$ due to the reason explained in Section 8.1.3.1. Readers may also notice that, α may be identified with the emissivity, ε. At this point, if we consider for simplicity that the atmosphere is a homogeneous layer of depth, h, with mean radiation temperature, T_m, then

$$N_B = k\alpha T_m \int \exp(-\alpha/g)h|_{h_m}^0$$
$$= k\alpha T_m (1/\alpha)\{1 - \exp(-\alpha h_m)\} \tag{8.14}$$
$$= kT_m[1 - \exp(-\tau_m)]$$

where $\tau_m = (\alpha h_m)$ is the optical thickness of the atmosphere visible to the antenna. This indicates that the total radiation energy received is dependent upon the term α and hence it depends upon how much translucent the atmosphere is. The corresponding noise temperature, $T_b = T_m[1 - \exp(-\tau_m)]$, is called the atmospheric noise temperature.

Now we consider the fact that the receiver antenna sees radiation from any sources behind the Grey Body. Therefore, the antenna will simultaneously collect some noise from astronomical sources in the field of view. Even if the atmosphere was perfectly transparent and there were no bright stars and galaxies in the field of view, the receiver would still pick up cosmic background radiation. The noise temperature contribution from the extra-terrestrial sources, primarily cosmic, while looking towards the far sky will be

$$N_S = kT_S \exp(-\tau_m) \tag{8.15}$$

T_S is the effective sky noise temperature of the far objects in the field of view, dominated by cosmic noise, when no other significant source is in view of the antenna. The total noise contributions at the antenna, also known as the Antenna noise, N_A, can therefore be represented as,

$$N_A = kT_S \exp(-\tau_m) + kT_m[1 - \exp(-\tau_m)] \tag{8.16}$$

In terms of noise temperature, the total antenna noise temperature can be written as

$$T_A = T_S \exp(-\tau_m) + T_m[1 - \exp(-\tau_m)]$$

$$= \frac{T_S}{A} + T_m\left(1 - \frac{1}{A}\right) \tag{8.17a}$$

Remember, here we considered unit bandwidth all through the derivation and hence have defined the received noise power as $N_A = kT_A$. Note that, here we've also ignored any scattering of the power by the atmospheric elements and also 'man-made' noise like the radiation from radars, etc.

The actual atmosphere and sky noise contributions will depend upon the amount of atmospheric volume observed by the antenna, and hence, in turn will be dependent upon the zenith angle. Looking vertically up to the zenith, the antenna will be facing the smallest possible thickness of atmosphere and hence minimum noise will be experienced. As the zenith angle increases, the antenna signal path has to pass through a much longer path through the atmosphere. The noise will then increase accordingly. One can also explain this increment considering that for larger zenith angles, the attenuation A increases leading to larger atmospheric noise. Further, note that, for no atmospheric attenuation, i.e. $A = 1$, only the sky noise term prevails. When a is very large, with $A \rightarrow \infty$, the sky noise term vanishes and the antenna temperature is only determined by the atmospheric contribution.

Using the above relations and following the same argument used to derive Eq. (8.13), the relation between the atmospheric noise temperature, sky noise temperature and the expected attenuation may be expressed using attenuation in dB as (Ippolito, 1986)

$$T_A = T_S \times 10^{-A(dB)/10} + T_m\left(1 - 10^{-A(dB)/10}\right) \; K \tag{8.17b}$$

where, A is the total attenuation experienced at the ground receiver in dB. Here, T_A is the temperature corresponding to the total antenna noise at the ground station antenna and T_m is the atmospheric mean radiating temperature in Kelvin.

Recall that in Chapter 7 we have read about the radiometer. The radiometer measures this total noise. If the total temperature measured by the radiometer be T_A, then we can say using the above equation,

$$T_A = T_S \exp(-\tau_m) + T_m(1 - \exp(-\tau_m))$$

$$\text{Or, } \; T_A - T_m = T_S \exp(-\tau_m) - T_m \exp(-\tau_m)$$

$$= (T_S - T_m) \exp(-\tau_m)$$

$$\text{Or, } \; \frac{T_A - T_m}{T_S - T_m} = \exp(-\tau_m) \tag{8.18}$$

$$= 10^{-A(dB)/10}$$

$$\text{Or, } \; A(dB) = 10 \log\left(\frac{T_m - T_S}{T_m - T_A}\right)$$

This is the way the absorptive attenuation from the radiometric measurements is derived.

8.1.4 THERMAL NOISE

The noise that the antenna picks up from the atmosphere is in addition to similar noise generated inside the receiver itself. Thermal noise is generated by the random thermal motion of charge carriers, usually electrons, inside an electrical conductor. The characteristics of this noise are not different from those of the radiating black body. The noise power thus created is independent of the applied voltage on the conductor and is only dependent upon the temperature of the conductor.

Thermal noise is approximately white, meaning that its power spectral density is nearly equal throughout the frequency spectrum. Therefore, in the time domain the noise is completely uncorrelated. The amplitude of the signal has a Gaussian probability density function. A communication system affected by thermal noise is often modelled as a zero mean, additive white Gaussian noise (AWGN) channel. The root mean square (RMS) voltage due to thermal noise, generated over bandwidth B (Hz), is given by

$$N = kTB \tag{8.19}$$

where k is Boltzmann's constant in Joules/Kelvin and T is the resistor's noise temperature in Kelvin and may not be the actual physical temperature of the body.

Noise figures. The signal which is received at the receiver is already contaminated with noise. Therefore, it has certain finite value of the signal-to-noise ratio (SNR). This term, SNR, represents the relative amount of signal power over that of the noise and is an index of the signal purity. When such a signal is passed through an amplifier, both the signal and the noise get equally amplified. Hence, under ideal condition, there should not be any change in the SNR. However, for all pragmatic cases, the amplifier itself adds some noise to the signal that it amplifies. Therefore, at the output of it, the SNR deteriorates, compared to that at the input.

The Noise figure of an amplifier is the ratio of the SNR at the input to the SNR at the output of the amplifier. So,

$$F = \text{SNR}_{\text{in}}/\text{SNR}_{\text{out}} \tag{8.20}$$

If S and N be the signal and noise power, respectively, at the input, G be the gain of the amplifier and N_a be the additional noise contribution of the amplifier, then

$$\begin{aligned} \text{SNR}_{\text{in}} &= S/N \quad \text{and} \\ \text{SNR}_{\text{out}} &= \frac{GS}{GN + N_a} \end{aligned} \tag{8.21}$$

Therefore, the Noise figure F may be defined as

$$\begin{aligned} F &= \frac{S/N}{[GS/(GN + N_a)]} \\ &= \frac{GN + N_a}{GN} \end{aligned} \tag{8.22}$$

We can readily identify the numerator as the total noise power at the output. Thus, the noise figure F can also be defined as the actual total noise power at the output of the amplifier to the noise power that would be present there, only due to the input noise.

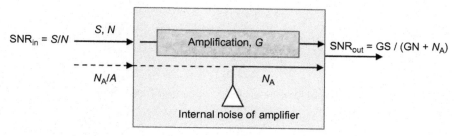

FIG. 8.3

Noise figure.

Again, from N_a, we can derive a hypothetical equivalent noise at the input of the amplifier, defined as N_{a_in}, so that this noise, when amplified by an ideal amplifier of the same gain, G, would give us the amount of noise N_a. Therefore,

$$N_{a\_in} = N_a/G \qquad (8.23)$$

So, using Eq. (8.23) in Eq. (8.22), we get

$$F = \frac{GN + N_a}{GN}$$

$$= 1 + \frac{N_a}{GN} \qquad (8.24)$$

$$= 1 + \frac{N_{a\_in}}{N}$$

From the definition of the noise figure, it is evident that the noise figure value is always greater than unity. When the amplifier is ideal, it does not add any excess noise to the amplifying signal. Then, $N_{a\_in} = 0$ and the noise figure term becomes $F = 1$. As N_a increases, F deviates from unity to even higher values. Typically, the noise is expressed in dB rather than as a factor. The schematic for noise figure is shown in Fig. 8.3.

It is required to note here the presence of the term N. This is the original noise content of the source signal, which is a variable term in general. Thus, the noise figure, by this definition, remains relative to the noise content of the input signal.

Focus 8.1 Relative Values of Noise Figure

In this Focus, we wish to compare between the F for two different signals at input. Suppose there is an amplifier which adds noise equivalent to T_a Kelvin to the signal. Now, if a signal with noise equivalent to T_1 K incorporated in it appears at the input, the Noise figure becomes

$$F = 1 + kT_a/kT_1$$

$$= 1 + T_a/T_1$$

If T_A be 50 K and T_1 be 75 K, then the Noise figure becomes

$$F = 1 + 2/3$$
$$= 1.67$$
$$= 10 \log (1.67) \, \text{dB}$$
$$= 2.23 \, \text{dB}$$

Now, if the incident signal has noise equivalent to temperature 200 K, then

$$F = 1 + 50/200$$
$$= 1 + \tfrac{1}{4}$$
$$= 1.25$$
$$= 10 \log (1.25) \, \text{dB}$$
$$= 0.9691 \, \text{dB}$$

Thus, we see that, for the same noise of the amplifier, the noise figure value is less when the original input signal has comparatively high noise added with it. Else, when the input signal has low noise added, the figure appears high.

However, for the same noise added, this dependence upon the input noise is quite deceptive. Therefore, there must be a fixed standard noise reference with respect to which the noise figure is to be represented.

Therefore, this definition of the noise figure, with a variable term, is not appropriate to be used to describe the noise-adding characteristics of the amplifier. Whenever an amplifier is to be demarcated by its noise figure, it should be definite and unambiguous. Therefore, a standard reference term is required to be used here. So, the noise figure, F, of an amplifier is represented with respect to a standard noise at the input. This reference noise is taken as that noise which is generated by a body at the standard noise temperature $T_0 = 290$ K. Thus, we can define the noise figure as

$$\begin{aligned} F &= 1 + \frac{N_{a\_in}}{N} \\ &= 1 + \frac{kT_{a\_in}}{kT_0} \\ &= 1 + \frac{T_{a\_in}}{T_0} \end{aligned} \tag{8.25}$$

It follows from the above that the noise contribution of an amplifier equivalently converted as its input noise is

$$T_{a\_in} = (F - 1)T_0 \tag{8.26}$$

So, the total effective noise temperature at the input of the amplifier is $(F - 1)T_0$ making the corresponding noise power equal to $k(F - 1)T_0$. This makes the total effective SNR at the amplifier input, considering the noise added by the amplifier, as

$$\frac{S}{kT_i + (F - 1)kT_0} \tag{8.27}$$

where, T_i was the noise temperature at the input of the amplifier due to the noise actually present in the signal.

Extending this idea to all the subsequent amplifiers in the receiver chain, the effect of the noise added by the next amplifier with noise figure F at its own input is $k(F_1 - 1)T_0$. But in order to bring all the noise contributions to the same reference, we calculate its equivalent noise at the input of the prefatory amplifier, which is basically a LNA. The equivalent noise due to the first amplifier following the LNA with Noise figure F_1 and amplification A_1 at the input of this LNA is $(F_1 - 1)kT_0/A_{LNA}$. So, we see, the effect of the noise added by the next amplifier drastically gets reduced by the factor of the LNA gain, A_{LNA}, when compared with the signal. Similarly, that of the next amplifier with F_2 and A_2 as Noise figure and gain will be $(F_2 - 1)$ $kT_0/(A_1 A_{LNA})$. The effects thus get less conspicuous for amplifiers of later stages. Considering all such contributions, the effective noise becomes

$$N_{eff} = N_a + (F - 1)kT_0 + \frac{(F_1 - 1)kT_0}{A_{LNA}}$$
$$+ \frac{(F_2 - 1)kT_0}{A_1 A_{LNA}} + \frac{(F_3 - 1)kT_0}{A_2 A_1 A_{LNA}} \quad (8.28)$$
$$+ \dots$$

It is obvious from the above equation that for amplifiers following the LNA, the noise injection to the actual signal is quite low as they are effectively reduced by the gain of all preceding amplifiers. Therefore, the gain of the LNA is kept considerably high. Further, as the LNA noise is the only component that is conspicuously added, N_{a_in} or F is kept as low as possible.

8.1.5 SYSTEM NOISE

We have already come across the different types of noise that the signal is affected with in course of its travel to the receiver. The theoretical part being explained, we shall now concentrate on a practical receiver and find out how much noise is incorporated by the antenna system and the adjacent hardware. In this we shall consider each of the noise components discussed above. We shall start with the atmospheric noise.

Follow Fig. 8.4 to identify the noise addition along the signal transmission paths. The signals transmitted to or from the satellites is nothing but the continuous flow of electromagnetic energy. It is the first task of the receiver to receive this energy and convert it into electric parameters of current and voltages, which can be utilized to derive the information therein.

8.1.5.1 Noise by the atmosphere

As the ground antenna receives signal from a part of the sky for a definite orientation of the receiver, it also picks up the noise coming from there. Not only the atmosphere that prevails in the path of the propagation of the signal emits noise, but also the background contributes to it. We have derived the expression for the noise temperature picked up by the antenna. It is also worth mentioning here that, while a ground

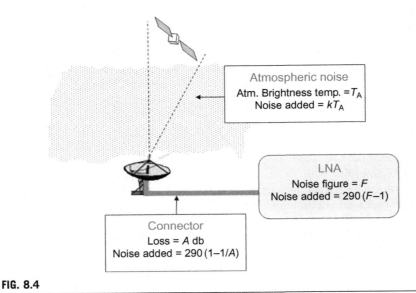

FIG. 8.4

Noise addition along the signal transmission path.

receiver sees the cool background of the cosmic space with temperature of about 2.7 K, the ground looking antennas in the satellite look towards the comparatively hotter background of the earth. For a sky looking antenna, if the atmospheric attenuation at any point be A, then using the equation (Eq. 8.17b), the noise temperature picked up by it can be written as

$$T_A = T_m \left(1 - 10^{-A/10}\right) + T_c \times 10^{-A/10} \, \text{K}$$

$$= T_m \left(1 - 10^{-A/10}\right) + 2.7 \times 10^{-A/10} \, \text{K}$$

(8.29)

For attenuation values less than 6 dB and frequencies below 50 GHz, T_m can be approximated, with a weak dependence on frequency. T_m value can be chosen as 265 K in the 10–15 GHz range and as 270 K in the 20–30 GHz range. When surface temperature is known, a first estimate of T_m can also be obtained by multiplying the absolute surface temperature by 0.95 in the 20 GHz band and by 0.94 for the 30 GHz window (ITU-R, 1997).

8.1.5.2 Noise by connectors from feed to LNA
As the signal is transmitted to the next unit, i.e. the LNA, through the connector, some noise gets added up during this process. The amount of the noise power added depends upon the loss of the connector. Simultaneously, the sky noise and the signal get attenuated while it passes through this connector. This composite noise as well as the received signal is further treated in the next section of the receiver, i.e. the LNA. Therefore, if A_c be the attenuation offered by the connector with temperature T_c, and

if T_A be the antenna temperature, then the total noise temperature, T_T, coming to the other end of the connector can be expressed using Eq. (8.17a) as

$$T_T = \frac{T_A}{A_c} + T_c\left(1 - \frac{1}{A_c}\right) \tag{8.30}$$

Therefore, as A_c increases, the component of the antenna sky temperature reaching the input of the next stage through the connector reduces, while the contribution of the connector to the noise increases.

8.1.5.3 Cascaded amplifiers and effective gain

The antenna is merely a transducer that converts the radiative electromagnetic energy of the signal to electric energy. Therefore, the signal strength that is available is very less and needs to be amplified. This amplification is done by the LNA. Low noise is necessary for the first amplification due to the reason we have already discussed in previous section. So, if F be the noise figure of the LNA, A_{LNA} is the amplification, then the total noise that is available at the output of the LNA is

$$N = (F - 1)T_0 \times A_{LNA} \tag{8.31}$$

8.1.5.4 Addition of noise to signals

A noise signal is typically considered as a linear addition of random electrical variation to a useful information signal. The noise level in an electronic system is typically measured as an electrical power N in watts or dBm, a root mean square (RMS) voltage (identical to the noise standard deviation for zero mean noise). However, it is always represented in relative terms of the signal to which it is attached. So, the typical signal quality metric involving noise is signal-to-noise ratio (SNR or S/N). For a RF carrier with the information modulated in it, the signal quality is generally represented as the carrier to noise ratio (CNR). For an amplifier or similar other hardware cascaded in the path of the signal, the noise characteristics of the hardware are described in terms of the additional noise introduced by the hardware and are represented as the noise figure. So, if S be the signal power received at the antenna, T_A be the atmospheric noise, A_c be the attenuation of the connector and F be the noise figure of the LNA, then the total SNR that is available at the output of the LNA is

$$SNR = \frac{S}{Bk\{T_A + (1 - 1/A_c)T_0 + (F - 1)T_0\}} \tag{8.32}$$

Focus 8.2 Total Noise Added to a Signal

The objective of this Focus is to have a hands-on estimate on the noise which typically gets added to a signal as it passes from the transmitter to the receiver. Assuming the source is noise-free, the first incorporation of noise takes place in the form of atmospheric noise. Let the sky noise temperature be contributed by the cosmic noise only and is taken as 2.7 K. The mean radiation temperature of the medium can be typically taken as 276 K. Then, if we consider rain-free condition with absorptive attenuation of 2 dB at 30 GHz, the contribution of the atmospheric noise in terms of temperature is

$$T_{N_1} = 276 \left[1 - 10^{-2/10} \right]$$
$$= 276(1 - 0.631)$$
$$= 276(0.369)$$
$$= 101.84$$

The temperature due to the cosmic noise is

$$T_{N2} = 2.7x\,10^{-2/10}$$
$$= 1.703\mathrm{K}$$

Therefore, the total noise picked up by the antenna is

$$T_A = T_{N1} + T_{N2} = 101.84 + 1.703$$
$$= 103.543\mathrm{K}$$

Now, the next contributor is the next element in the channel path, that is the connector from the antenna to the LNA. The attenuation of the connector can be typically considered as 0.5 dB. The noise contributed by this element is hence

$$T_{N_3} = 290 \times (1 - 1/A)$$
$$= 290 \times \left(1 - 10^{-0.5/10} \right)$$
$$= 31.53\mathrm{K}$$

However, this also reduces the antenna noise by a factor A, and hence, the effective T_A becomes

$$T_{A\,\mathrm{eff}} = 103.543/10^{0.5/10}$$
$$= 92.283\ \mathrm{K}$$

Thus, the total noise temperature till this point is

$$T_x = 92.283 + 31.53$$
$$= 123.813\mathrm{K}$$

Finally, the LNA has a typical Noise figure of 2 dB. This means figure in factor form is

$$F = 1.58$$

Therefore, the noise temperature that the LNA adds at its input is

$$T_{\mathrm{LNA}} = 290 \times 0.58\mathrm{K}$$
$$= 168.2\mathrm{K}$$

Thus, the total noise temperature overall is

$$T_{\mathrm{Total}} = 123.813 + 168.2\mathrm{K}$$
$$= 292.013\mathrm{K}$$

The corresponding noise power density is

$$N_0 = 10\log(292.013) - 228.6$$
$$= 24.654 - 228.6$$
$$= -203.946\mathrm{dB/Hz}$$

For a typical satellite signal of EIRP 24 dBw and path loss of −206 dB and received antenna gain of 40 dB, the total signal power received is

$$S = 24 - 206 + 40$$
$$= -142\,\text{dB}$$

Then, the received signal-to-noise density ratio is

$$S/N_0 = -142 - (-203.946)$$
$$= 61.946\,\text{dBHz}$$

We can also find out the ratio of the antenna gain G to the antenna temperature T, i.e. G/T of the received antenna as

$$G/T = 40 - 24.654$$
$$= 15.346\,\text{dB}$$

8.1.6 SPECTRAL CHARACTERISTICS OF NOISE

Noise is a random process, characterized by stochastic properties such as its variance, distribution, etc. Before closing our discussion on noise in this chapter, for the sake of logical completion we need to have a discussion of the characteristics of the noise spectrum.

Let us consider a sequence, which is a time series of continuous or discrete values representing a physical process. Such a sequence can be classified as deterministic and random. Noise is such a random variation which is unpredictable such that the future values of the sequence cannot be predicted with certainty even if the past values are known. Though unpredictable, it has statistical regularity.

The power spectral density of a signal is the distribution of the power carried by the signal at different frequencies per unit frequency range. Noise may also be characterized by its power spectral density, more popularly known as the noise spectral density $N_0(f)$ in watts per hertz. We have already seen that the noise, which gets added up with the satellite signal, is present in all frequency and have almost the same power content in all frequencies in the range of our interest. It is called 'white noise' due to such feature. Therefore, their spectral amplitude is same in all frequencies within the band of interest. However, the different frequency component of the noise have mutually different phases which vary randomly. Noise addition takes place mainly when the signal is modulated around the centre frequency f_c of the carrier and within the passband of the system. So, the random noise can be expressed in a band pass form as (Lathi, 1968)

$$n(t) = n_r \cos\left(2\pi f_c t + \varphi_n\right) \tag{8.33a}$$

where, f_c is the carrier frequency and n_r and φ_n are the amplitude and phase of the added noise, respectively. This expression can be resolved into

$$n(t) = n_c(t) \cos\left(2\pi f_c t\right) + n_s(t) \sin\left(2\pi f_c t\right) \tag{8.33b}$$

where, n_s and n_c are the amplitudes of low frequency noise components such that $n_c = n_r \cos \varphi_n$ and $n_s = n_r \sin \varphi_n$. From Eq. (8.33b), we can also consider that the low pass noise components $n_c(t)$ and $n_s(t)$ are frequency shifted by the carrier frequency f_c to form this passband noise. In the above expression, the noise components are actually bandlimited around dc. Both n_c and n_s are slowly varying random components. In other words, the noise components vary slowly but in a random fashion. Further, since both n_r and φ_n are unbiased random, the mean noise power in each of the components are the same, i.e. $\overline{n_c}^2 = \overline{n_s}^2$.

8.1.7 COMBATING THE NOISE

There are many different ways by which the noise can be tackled. Since the atmospheric noise cannot be avoided, the basic approach taken is to reduce the thermal noise at the source and receiver and increase the received signal power with respect to the noise. This improves the overall C/N_0 of the received signal and in turn improves the performance. The implementation of the same may be done through the following ways.

8.1.7.1 Good antenna figure of merit

The signal received at the receiver is expressed as

$$S = \frac{\text{EIRP}}{L_P \times L_A} G_R \tag{8.34}$$

where EIRP is the EIRP of the transmitter, L_P is the path loss, L_A is the additional atmospheric loss of the signal power during propagation and G_R is the received antenna gain. Similarly, the received antenna noise power is $N = kT_A$, where T_A is the antenna temperature. Therefore, the signal-to-noise ratio becomes,

$$S/N = \frac{\text{EIRP}}{L_P \times L_A} \times \frac{1}{k} \times \frac{G_R}{T} \tag{8.35}$$

So, the relevant parameter of the receiving station determining the received signal quality is the term G_R/T, which is a figure of merit of the receiving station. The receiving system noise temperature, as we have learnt, is the summation of the antenna noise temperature and the RF chain noise temperature from the antenna terminals to the receiver output. The antenna noise temperature, being mostly dependent upon the sky noise temperature, is difficult to abate. Therefore, the thermal noise added by the RF chain is targeted for reduction. In the process, the antenna feed to LNA connector is kept short in length, while LNA Noise figure is kept low and the physical temperature is controlled. The components are also chosen with very low insertion loss to achieve the end.

8.1.7.2 Preinstallation noise survey

The noise received by the antenna constitutes mainly of the atmospheric noise. However, it also sometimes picks up the terrestrial noise or other man-made noise through its side lobes. In order to reduce such chances of receiving avoidable additional noise, the common exercise followed is to have a preinstallation noise survey. This assesses for the presence of any excess noise of terrestrial or similar origin at the location of the receiver such that for cases where there is an option to select the receiving location, the most benign location, in terms of noise, is selected for the purpose.

8.1.7.3 Standard communication techniques

Finally, to achieve the required performance even in presence of the anticipated noise, different standard mitigation techniques are used like employing the Forward Error Correction Coding or by boosting the power with respect to the noise by standard power control techniques.

8.2 MULTIPATH

Until now, we have considered that the signal passing from the transmitter traverses a single definite path to reach the receiver. However, this may not be true in all cases. At certain times, the signals may traverse through multiple paths to reach the receiver. When such a phenomenon happens, multipath is said to occur.

8.2.1 MULTIPATH DEFINITION AND TYPES

By definition, multipath is the propagation phenomenon that results in radio signals reaching the receiving antenna by two or more paths caused by atmospheric effects like reflection, refraction, diffraction, scattering or by similar effects from other natural or man-made terrestrial objects.

Multipath causes multipath interference that includes both constructive and destructive interference and phase shifting of the signal. Destructive interference may also cause extended depreciation of the received signal, leading to fading. The overall effect on the signal is thus dependent upon the amplitude and the relative phase of each individual component of the multipath signal. To understand the phenomena of multipath, it is necessary to know the reasons why the signals traverse through different paths. But, before we explore the causes, it will be helpful to classify the multipath into two broad divisions. Although this distinction is not universal, it will help us in understanding the physical facts. The received signal may be constituted by the direct line of sight component added with another component picked up by the antenna. This additional component may reach the receiver from a different path coming through reflection, refraction or diffraction of the original signal. So, here we divide the multipath into two types, viz. the refractive multipath and the reflective multipath.

8.2.1.1 Refractive multipath

When the different paths followed by the direct and the multipath component are due to different bending of the signal along its traversed path due to different refractive indices, which ultimately reach the same receiver, refractive multipath is said to occur.

For example, when satellite navigation signals propagate via the ionosphere, it is possible for the energy to be propagated from the transmitter to the receiver via many different paths. The profile of the electron density of the ionosphere is not smooth, and as a result, any signal entering the ionosphere will be refracted accordingly and will take a numbers of paths to eventually reach the receiver. As a result, multipath will occur. With changes in the ionosphere with time, the effective paths through which the signals reach the receiver will also change with consequent changes in their relative phase. Therefore, the overall received signal, produced by the vector sum of the individual components at the receiver, will also abruptly change with time.

Signals using frequencies at VHF and above are affected by the troposphere. The signal is refracted as a result of the gradual changes in refractive index (RI) profile, especially within the first few kilometres above the ground. Here, the water vapour variation strongly influence the effective RI. However, relatively abrupt changes in refractive index can also occur as a result of turbulence or other weather conditions. Under such conditions, there can be a number of effective paths through which signals travel to reach the antenna leading to the multipath effect.

Although, the refractive multipath effects are more severe for terrestrial communication systems, it is also observed for satellite communication for very low elevation angle and the signal has to traverse a long path through the atmosphere.

8.2.1.2 Reflective multipath

When the signal from the transmitter reaches the receiver not only via the direct path, but also as a result of reflections from objects such as buildings, hills, ground, water bodies, etc. that are adjacent to the main path, reflective multipath is said to occur. The different path followed by the multipath component will have a phase, different from that of the direct component due to the extra path traversed by this signal and also due to the arbitrary phase change occurring at the time of reflection. Although, the reflected signal may have reduced amplitude, it may still be significant depending upon the reflecting material. In other words, for reflective signals, the magnitude of reflected signal received and its phase, which depend upon the reflection coefficient of the reflector, determines the effect of such multipath.

The schematic of the two kinds of multipath phenomena is shown in the Fig. 8.5.

8.2.2 MULTIPATH FADING

The effective signal received as a result of the multipath experiences variations in its intensity due to interference of the direct and multipath components of the signal. This leads to the multipath fading. The fading also varies in time due to the variation in the overall scenario of the multipath. This may include variations in the transmitter or receiver position, or in the characteristics of the propagation path that give rise to the multipath. The fading can often be relatively deep, i.e. the signals fade completely fade away, whereas at other times the fading may not cause the signal to fall below a usable strength.

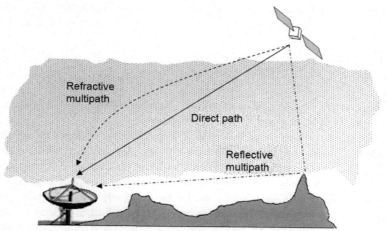

FIG. 8.5

Different multipath scenario.

The overall signal at the receiver being the vector sum of the multipath components may both enhance or depreciate the signal. As an example, considering a very simplified situation, let us take two components of the same signal, S, reaching the receiver through different paths and hence with different phase, as

$$S_1 = A \cos (\omega t + \varphi_1)$$
$$S_2 = A \cos (\omega t + \varphi_2) \tag{8.36}$$

S_1 is the direct signal, while S_2 is the multipath signal. Considering both having the same amplitude, A, at the receiver, when they add up, the combined signal becomes,

$$S_c = A \cos (\omega t + \varphi_1) + A \cos (\omega t + \varphi_2)$$
$$= 2A \cos \left(\frac{\varphi_1 - \varphi_2}{2} \right) \cos \left\{ \omega t + \left(\frac{\varphi_1 + \varphi_2}{2} \right) \right\} \tag{8.37}$$

Thus, we can see that, due to this combination of the signals as a result of the multipath, the resultant signal thus produced is dependent upon the relative strength and phase of the multipath signal. Now, the channel characteristic is not strictly time invariant and may change with time, changing φ_1 and φ_2. This makes the phase and the amplitudes of the constituent signals, which are now trigonometric functions of φ_1 and φ_2, to vary accordingly. This changes the characteristics of the effective received signal with time and consequently leads to different issues at the receiver. In our simplified example, as both φ_1 and φ_2 are random, it may happen that for prolonged time the value of the term $(\varphi_2 - \varphi_1)/2$ remains more than $\pi/3$. Therefore, $|\cos (\varphi_1 - \varphi_2)/2| < 1/2$ and hence the amplitude remains below 'A' for the period and hence will exhibit extended depreciation of signal strength.

In general, there may be multiple path of propagation. Associated with each path, there will be a corresponding random phase, φ, and an amplitude attenuation factor,

α. Both the random phase and the attenuation factors are time variant due to the varying nature of the medium. The received signal, thus, may be expressed as (Proakis and Salehi, 2008)

$$S(t) = \sum_i \alpha_i A \exp\left[j(\omega t - \varphi_i)\right]$$
$$= \sum_i \alpha_i e^{-j\varphi_i} A \exp\left(j\omega t\right) \tag{8.38}$$

Thus, the received signal consists of sum of a number of time variant phasors, with the ith component having amplitude α_i and phase φ_i.

We already have mentioned before that, large dynamic changes in the medium are needed to change the values of α by a considerable amount, while φ changes by 2π radians whenever the path variation is by an amount of λ. For small λ of the satellite signals, this amount of variation in effective path can be readily observed. Consequently, φ changes by 2π by relatively small variation of the medium. As each path variation is independent and random, for large numbers of paths, the central limit theorem may be applied and thus the overall phase variations may be modelled as Gaussian.

Due to the perpetual presence of the direct signal, for which $\varphi = \varphi_0 = 0$, the resultant received signal amplitude cannot be zero mean, and in such cases, the effective amplitude variation can be represented by Rician distribution.

Multipath fading can affect radio communications channels in two main ways, flat fading and selective fading.

Flat fading. This form of multipath fading affects all the frequencies across a given channel either equally or (in a more practical sense) almost equally.
Selective fading. Selective fading occurs when the multipath fading affects different frequencies across the channel to different degrees. The RI for refractive multipath and where reflection is involved, the reflectivity $|r|$, are mostly frequency-dependent, making the phase and amplitude variations during the reflection, dispersive. So, selective multipath fading occurs because, due to refraction, reflection, etc., even though the physical path length changes by the same length, the phase change as a result of these effects will be different across the bandwidth used. It will mean that the phases and amplitudes of the signal will vary differently across the channel.

Box 8.2 Multipath Effects on Signal

We intend to understand in this box the effect of multipath on a signal received by a receiver from satellite. For this, we shall run the Matlab programme Multipath.m. It obtains the signal upon combining different component signals reaching the receiver through multipath. There is only one direct signal path, represented by S and 'n' numbers of multipath components, represented by S_m. The multipath signal differs from the direct one by random phase difference that varies with time. The programme then plots the resultant signal across time. The resultant variation of the received power with respect to the expected direct signal power is shown in the below figure.

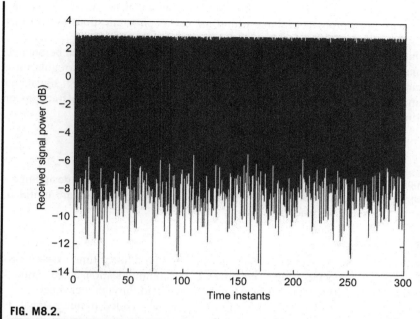

FIG. M8.2.

Effect of multipath on received signal power.

It can be observed that, for $n = 1$, although the signal can get boosted up to 3 dB, with the given reduction factor 0.4 for the amplitude of the multipath signal, under constructive combination, it can go down by about 8 dB typically causing multipath fading, during the destructive combination conditions.

Change the values of the number of multipath signals 'n' and the multipath signal amplitude attenuation factor 'alf' to see the resultant variations.

In certain applications, the time of arrival of a definite phase of the signal is of importance. For example, for navigation signals, the time of arrival of a definite phase of the signal is used to find the range of the transmitting signal from which the position of the user is determined. In such application, the multipath poses a major threat. This is because, sometimes under constraint conditions, the multipath signal gets comparable strength as the direct signal. Then the receiver can lock onto the multipath signal rather than the direct one. As the multipath signal traverses different equivalent distance from the transmitter than its direct counterpart and also experiences abrupt phase change, the derived range from these signals gives erroneous results for the satellite range, thus deteriorating the position accuracy of the system, which is derived from the former.

Multipath fading may also cause distortion to the radio signal. As the various paths that can be taken by the signals vary in length, the signal transmitted at a particular instance will arrive at the receiver over a spread of times. This can cause problems with phase distortion when data transmissions are made. As a result, it may be necessary to incorporate features within the radio communications system that enables the effects of these problems to be minimized.

Scintillation can be treated as a kind of multipath. We have discussed before that the signals also scatter or diffract from the atmospheric elements with irregularities. The scintillated signal is formed as a result of the combination of a direct signal and other multipath signal coming through different paths, due to such diffractions or scattering at the irregularities.

8.2.3 MULTIPATH MITIGATION

Multipath fading is a feature that needs to be taken into account when designing or developing a radio communications system. The following methods are useful in mitigating the multipath effects.

8.2.3.1 Equalizers

Multipath fading can be viewed as transmission through a linear time varying system. This may be corrected using an equalizer. In telecommunication, equalization is the reversal of distortion incurred by a signal transmitted through a channel.

When equalizers are used, first a known signal, also called training signal, is transmitted. The equalizer estimates the channel response as a function of time, which is possible since the receiver knows both the expected signal and that actually received in distorted form. Then, assuming that the channel remains invariant, it compensates the channel effects on the signal by introducing reverse effects while the signal passes through it. The actual waveform of the transmitted signal is thus preserved and any additional group delay and phase delay between different frequency components, introduced by the channel are cancelled out. Effectively, equalizers render the frequency response flat over the total bandwidth. Since the multipath scenario changes with time, it is effective to use adaptive equalizers for mitigating the multipath effects more efficiently (Chakrabary and Datta, 2007).

8.2.3.2 Diversity

Multipath being a local phenomenon, and time-dependent too in some cases, use of diversity technique helps in mitigating the multipath effect. Both space diversity and time diversity are effective in this case. Since we have already learnt about these diversity techniques in the last chapter, these are not elaborated here. Effectually, the diversity techniques can provide significant performance improvement for multipath.

8.2.3.3 Rake receiver

The multipath is nothing but the same signal arriving to the receiver through different paths and hence with different delays or phase shifts. The delay profile of the channel provides multiple versions of the transmitted signal at the receiver.

A rake receiver is a radio receiver designed to counter the effects of multipath. It does this by using several 'subreceivers' called fingers. Each finger is nothing but a correlator, each separated by definite clock delays, i.e. with different reference phases and hence assigned to a different multipath component. Each finger, correlating independently to an increasingly delayed multipath signal, thus decodes a particular single multipath component. The receiver estimates the delay and also the amplitude of the delay profile. Finally, the contributions of all fingers are combined in order to reverse the effect of the delay and use the signals on each transmission path to its best. This could very well result in higher signal-to-noise ratio in a multipath environment. In short, since each component contains the original information, the magnitude and time-of-arrival of each component is computed at the receiver for channel estimation and then all the components are added coherently to improve the final signal quality.

8.3 INTERFERENCE
8.3.1 DEFINITION

Interference is the disruption of a signal travelling along a medium, due to another signal or effect that can alter the signal characteristics in the definite frequency spectrum. For satellite signals which are basically electromagnetic waves, interference is known as electromagnetic interference (EMI) or radio-frequency interference (RFI). Eventually, it affects the communication system by distorting the signal or degrading its SNR. In the process, it deteriorates the performance or even lead to malfunctioning or unavailability of the system.

Interference is formally defined as "The effect of unwanted energy due to one or a combination of emissions, radiations, or inductions upon reception in a radio-communication system, manifested by any performance degradation, misinterpretation, or loss of information which could be extracted in the absence of such unwanted energy" (ITU-RR, 2009).

8.3.2 SOURCES AND TYPES OF INTERFERENCE

Interference can occur from varied sources. Although not conventional, it is possible to classify interference depending upon the source of the interfering signal into the followings.

Interference may occur from the same signal in question and can also be initiated by different signals in the same satellite. Therefore, one way of classifying interference may be into following two broad classes,

a. *Self-sourced interference.* It is the interference occurring to a constituent component of a signal due to another component in the same signal.
b. *External source interference.* It is the interference that happens due to other signals in the same satellite or from other satellites and sources.

Inter symbol interference, caused due to limited bandwidth, may be considered as a self sourced interference. Same frequency may be used by many satellites, mostly for the same service. For example, the L1 frequency 1.5 GHz is used for GNSS purposes, by many systems. Therefore, a certain amount of interference is very much probable in such a scenario from a separate source but in the same frequency. Depending upon the magnitude of the effect, interference is classified as follows (ITU-RR, 2009).

a. *Permissible interference.* This is the interference which occurs within a predefined level and for predefined interval of time. The worst level of such permissible interference is tolerable for shorter durations of time, while feebler levels for comparatively longer durations.
b. *Acceptable interference.* Acceptable interference occurs when the interference exceeds the predefined permissible threshold in terms of levels or time, but still does not disrupt the affected system considerably enough to degrade its services.
c. *Harmful interference.* Harmful interference occurs when the interference is deep and/or long enough to deteriorate the services of the affected systems.

Again depending upon the reason of the origin of the source, the man-made interference may also be divided into

a. *Intentional interferences.* Intentional interference may be in the form of radio jamming, which introduces large noise in the communication band of interest and thus degrades the system to a level of no use. It is used as a tool in electronic warfare. These are typically long-term interferences.
b. *Nonintentional interferences.* Nonintentional interferences are basically short-term interferences with accidental type of origin. For example, mispointing of antenna, incorrect gain setting, crossing over of line of signt for two satellites in different orbits, etc. lead to such interferences.

Further, on the basis of the nature of the spectrum of the interference, the electromagnetic interfering signal can be categorized as follows:

a. *Narrowband interference.* This is said to occur when the interfering signal has a narrow spectrum. This type of interference typically emanates from intended transmissions, such as radio and TV stations or cell phones.
b. *Broadband interference.* Broadband interference, on the other hand, is said to have occurred when the interfering signal has a broader band of component frequencies. It is typically unintentional radiation from sources such as electric power transmission lines.

8.3.3 INTERFERENCE EFFECTS

Although we have categorized the interference types on the basis of their characteristics and sources, it is the effect on the affected system that defines the kind of interference and its nomenclature. Therefore, here we discuss the different interferences depending upon their effects.

Interferences may be identified depending upon the kind of effects that the interference is leading to. Most of these interferences have origin in the hardware at the transmitter and that of the receiver, rather than in the propagation path. Therefore, here we shall only briefly mention about those interferences as they have implication in the designing of the satellite links, the subjects being treated in the next chapter.

Intersymbol interference (ISI). Whenever the signal passes through a channel with limited bandwidth such that a part of the signal spectrum is either severed or distorted, it results in distortion in the shape of the data bits or pulses in the time domain. Therefore, a succession of discrete pulses transmitted through the channel are smeared to the extent that they are no longer distinguishable from one another and then Intersymbol interference is said to have occurred. The Nyquist criterion is to be followed between the bandwidth and the bit rate in order to avoid ISI.

Intermodulation interference (IMI). The total satellite signal may contain a number of individual modulated signals, say S_1, S_2, S_3, etc., with different subcarrier frequencies, say f_1, f_2, f_3, etc., respectively. IMI is caused due to generation of new frequency components in a multifrequency signal as a result of amplitude and phase nonlinearity in amplifiers. Whenever, there is nonlinearity present in a device, with signals S_1 and S_2 at the input, the resultant output signal will be of the form

$$S_o = a_0 + a_1(S_1 + S_2) + a_2(S_1 + S_2)^2 + \ldots$$
$$= a_0 + a_1 S_1 + a_1 S_2 + a_2 S_1^2 + a_2 S_2^2 + 2a_2 S_1 S_2 + \ldots \quad (8.39)$$

Due to the product terms like $S_1{}^2, S_2{}^2, S_1 S_2$, etc., new signals with frequencies equal to the sum and differences of the frequencies of the individual signals will appear. For example, $S_1{}^2$ will produce a new frequency component of $2f_1$, while $S_1 S_2$ will produce component of frequencies, $(f_1 + f_2)$ and $(f_1 - f_2)$. For higher order nonlinearity, these frequency components, called the intermodulation products, having frequencies which are sum of either positive or negative integral multiples of the original signal frequencies will be generated. Some of these signal components will lie in the passband of the system and will interfere with original signal to produce what is called IMI.

When amplifiers in the channel have nonlinearities and FDMA-type signal is fed to it, the channel simultaneously amplifies several carriers at different frequencies. So, if n sinusoidal signals with frequency $f_1, f_2, f_3 \ldots f_n$ are fed to the amplifier, the output will not only contain the n frequencies, but their intermodulation products with frequencies

$$f_{\text{im}} = m_1 f_1 + m_2 f_2 + m_3 f_3 + \ldots m_n f_n \quad (8.40)$$

where m_j may be positive or negative integers. The order of IMI is defined as the sum of the order of contribution of each component, i.e. $X = |m_1| + |m_2| + \ldots$ (Maral and Bousquet, 2006). However, the amplitude of the intermodulation products reduces with increasing order.

When the centre frequency of the passband is large compared to the bandwidth, then all positive summations of individual order 'm_i' including all even order difference products in the above equations generate a frequency largely away from the centre and thus outside the bandwidth. Only some odd orders with positive and negative m_j components, i.e. odd ordered difference products, say $m_1 = 3$ and $m_2 = 2$ or $m_1 = 2$ and $m_2 = 1$, produce the products whose frequencies lie within the bandwidth. Even the higher order odd components also do not interfere.

When the carriers are modulated, the IMI products are no longer spectral lines since their power is dispersed over a spectrum which is extended over a band of frequencies. If the number of carriers is significantly high, superposition of the spectra of the IMI products leads to a spectral density distribution, which is nearly constant over the whole of the amplifier bandwidth and this justifies considering the IMI products as white noise with psd denoted by N_{0im}. Its value depends upon the transfer characteristics of the amplifier and also on the numbers of carriers being amplified. The carrier power to the IMI noise power spectral density ratio (C/N_{0im}) can be associated to each carrier of the amplifier at the output since it affects each carrier.

IMI is avoided by operating the amplifier in the linear region by reducing or backing off the input drive. Reduced input drive leads to reduced output power and results in a downlink power-limited system that is supposed to operate at a reduced capacity.

Co channel interference (CoCI). In electronics, crosstalk is any phenomenon by which a signal transmitted on one circuit or channel of a transmission system creates an undesired effect in another circuit or channel. Crosstalk is usually caused by undesired capacitive, inductive, or conductive coupling from one circuit, part of a circuit, or channel, to another. A Co channel interference occurs between two signals that are on the same frequency channel. CoCI is avoided by proper frequency planning and designing of the channel with adequate spectral separation between the services.

Adjacent channel interference (AdCI). Adjacent channel interference occurs when the transponder is simultaneously shared by multiple carriers having closely spaced centre frequencies. Different carriers are filtered by the receiver so that each of the stations gets its intended signal. Filtering would have been easier to realize, had there been a large guard band between adjacent channels. But, this is not practically feasible as that would lead to insufficient use of transponders bandwidth. The result is that a part of power of the carriers, adjacent to the desired one, also gets captured by the receiver due to partial overlapping of the channel filter characteristics. So, this becomes a source of noise caused by the very narrow guard band, wide roll off by filters and nonlinearity. Trade off between economic use and technical limitations and performance needs to be done.

Cross-polarization interference. Cross-polarization interference occurs in satellite systems which reuse the same frequency in two orthogonal polarizations. The interference occurs due to coupling of energy from signal in one polarization to the signal in orthogonally polarized state. It can also happen in communication systems with orthogonal circular polarization. Cross-polarization interference occurs due to coupling between two polarization states due to the depolarization effects during the propagation. We have already seen that the rain and ice crystals can cause considerable depolarization, which effectively reduces the XPD for each polarization. Cross-polarization interference can be avoided by designing the system such that orthogonal polarizations are in different frequencies or by implementing the systems with large initial XPD.

Adjacent satellites interference. This kind of interference is caused by the presence of side lobes in the radiation pattern of the earth station antenna in addition to the desired main lobes. If the angular separation between two adjacent satellite systems is not too large, it is quite possible that the power radiated through the side lobes of the antenna for one satellite is directed towards the antenna direction of the adjacent satellite. Similarly, transmission from an adjacent satellite can interfere with the reception of earth station through the side lobes of the ground receiving antenna to cause this type of interference. Moreover, for small earth station antennas, the main beam lobe is large for lower frequencies. This will cause the antenna to see adjacent satellites within its main lobe at one of its ends and with considerable gain. This will cause interference by adjacent satellites.

For applications like satellite navigation, the ground user receiver antenna is nearly omnidirectional to accommodate all visible satellites of the same system. If under such condition, two different satellites from two different systems transmit in the same or nearby frequencies, simultaneous reception of signals in both frequencies is unavoidable and this may lead to intersystem interference.

During the interference, the power of one channel entering into the other due to this effect may be treated similar to a noise reducing the C/N_0 and degrading the signal quality. Thus, the effective C/N_0 may be expressed as:

$$\frac{C}{N+I} = \left\{ (C/N)^{-1} + (C/I)^{-1} \right\}^{-1} \tag{8.41}$$

CONCEPTUAL QUESTIONS

1. The antenna temperature will only be determined by the contributions from internal hardware, like the connector and LNA, when it is not looking towards the sky—Comment on this statement.
2. Is it correct to multiply the sky noise power kT by the antenna gain G, since the noise is also coming through the antenna like the signal?
3. Why the distribution of amplitude variation for a multipath-affected signal in decibel scale is not even about zero in figure M8.2?
4. How can we distinguish between a multipath fade and a source amplifier jittery variation due to malfunctioning?

REFERENCES

Acharya, R., 2014. Understanding Satellite Navigation. Elsevier Inc., Waltham, MA, USA.
Chakrabarty, N.B., Datta, A.K., 2007. An Introduction to the Principles of Digital Communications. New Age International Publishers, Kolkata.

Feynman, R.P., Leighton, R.B., Sands, M., 1992. Feynman Lectures on Physics. Narosa Publishing House, New Delhi, India.

Ghosh, A., 1990. Tapgatitattya (Thermodynamics). Paschimbanga Rajya Pushtak Parshad, Kolkata, India.

Ippolito Jr., L.J., 1986. Radiowave Propagation in Satellite Communication. Van Nostrand Reinhold Company, New York, USA.

ITU-R, 1997. Recommendation ITU-R P.1322-1: Radiometric Estimation of Atmospheric Attenuation. ITU, Geneva.

ITU-RR, 2009. Radio Regulations Article-1. Terms and Definitions. http://life.itu.int/radio club/rr/art01.htm (Accessed 13 October 2016).

Lathi, B.P., 1968. Communication Systems. John Wiley and Sons, New Delhi, India.

Maral, G., Bousquet, M., 2006. Satellite Communications Systems, fourth ed. John Wiley & Sons Ltd, West Sussex, England.

MIT, 2003. MIT Open courseware. https://ocw.mit.edu/courses/electrical-engineering-and-computer-science/6-661-receivers-antennas-and-signals-spring-2003/readings/ch3new.pdf (Accessed 22 December 2016).

Proakis, J.H., Salehi, M., 2008. Digital Communications, fifth ed. Mc Graw-Hill, New York, USA.

Sportisse, B., 2009. Fundamentals in Air Pollution: From Processes to Modelling. Springer Science & Business Media B.V., Dordrecht, The Netherlands.

Satellite link performance

In our previous chapters, we have seen the impairing effects that satellite signals experience while traversing through the atmosphere. We found that the ionospheric delays, scintillation or the tropospheric attenuation, depolarization, etc. can cause severe degradation to the signal quality and impede the services. We have also come to know about the techniques which can be used to mitigate these effects. It is logical now to learn how to design a satellite link amid these impairments with best use of resources and how to use the mitigation techniques to alleviate the deteriorating effects of propagation. We have already had an overview of the relevant parameters and expressions in Chapter 1 and there we asserted to deal with these aspects in details at a later part of the book. We shall do it here, in this chapter. Continuing with our previous discussion on the factors which quantify the system performances that we had in Chapter 1, here we shall learn them with more mathematical substantiation.

9.1 LINK DESIGN ASPECTS

The basic objective of designing a satellite system is to ensure that it performs the desired functionalities well to achieve the application goals under all conditions. The satellites have some primary application, like communication, remote sensing of the earth parameters, probing into the deep space or collecting the features of the sun or any other planet or merely global telecommunications. All the information or data collected or reaching the satellite are only relevant when they are transmitted to the earth for analysis and use. Thus, the basic common need is the proper communication of the information, although the required performance goals can be different for respective pertinent applications. So, communication of data is a major component of the total functionality, if not the only objective, as in telecommunication services. To achieve the same, communication signals need to qualify the desired performance figures. These performance figures, in one hand, need to cater the application requirements and have to be pragmatic enough to consider the most stringent impairment, on the other. To warrant the performance under all possible conditions, we need to utilize the resources efficiently so that it meets the objectives mitigating the possible detriments as well.

Satellite Signal Propagation, Impairments and Mitigation. http://dx.doi.org/10.1016/B978-0-12-809732-8.00009-0

Here, we shall briefly mention the different aspects of the system designing with our major emphasis on the considerations of the propagation effects in it. The required performance metrics, to achieve faithful and reliable communication include accuracy, availability, continuity, etc. over all operational conditions. To assess the performances quantitatively, we need to know the signal parameters that affect these metrics. Therefore, in designing the link, it is necessary to have appropriate knowledge of the system requirements, understanding of the propagation behaviour and environments and comprehensive information about the transmission and the receiving systems, at the same time.

The link design of any satellite communication system is done with the following common broad objectives:

a. To faithfully and efficiently (both cost- and resource-wise) carry the information through the available channels.
b. To guarantee availability and continuity of the link for a specified percentage of the time.

The receiver acquires the signal and demodulates it to identify the correct sequence of these binary bits. So, for the receiver to perform in a manner so that the bits are correctly decoded, its ability to distinguish between the contrasting levels or states of the signal, representing two or more binary bits, must be ensured and enhanced as required.

We have already learnt in Chapter 1 that, in order to clearly distinguish the signal states, for proper decoding of the bits at the receiver, it requires that the total energy contained in a bit be sufficiently higher than the average noise power. In other words, a threshold minimum E_b/N_0 ratio at the receiver input for a given percentage of time is necessary for maintaining the required minimum bit error rate (BER). We have already defined the terms E_b/N_0 and BER there. This term E_b/N_0, obtained after demodulation, is again dependent upon the signal quality, the latter being defined by the carrier power to the noise spectral density ratio, C/N_0 in the modulated signal. While the value of C/N_0 determines the accuracy, the percentage of time that this signal quality is maintained determines the availability of the system.

The design is principally done to ensure that the threshold is maintained above the required value for a particular data rate and for the required period of time. However, the propagation and other impairments tend to tow them down below the sufficient levels. Insuring the link against all these odd effects requires fitting transmission power in terms of the equivalent isotropically radiated power (EIRP), modulation scheme, coding type and data rate.

The targetted performance of a satellite system is therefore ensured and achieved through proper allocation of the mentioned satellite resources to establish such a link. The link design includes the process of selecting the suitable signal parameters like frequency, polarization, modulation, EIRP, etc., such that at the receiver the data is received with necessary data rate and accuracy, meeting the objective of the application. Thus, the main factors related to satellite system design are:

• Choice of frequency band and polarization
• Choice of modulation and multiple access technique

- Selection of EIRP
- Consideration of atmospheric propagation effects and propagation noise
- Choice of antenna system
- Choice of demodulation technique at the receiver

9.1.1 CHOICE OF OPERATING FREQUENCY

The performance of a signal is dependent upon its frequency in many respects. The frequency determines the bandwidth which in turn limits the bit rate and hence the information rate that the signal can carry. Higher data rate needs larger bandwidth which is available for higher carrier frequency. As a rule of thumb, only 5% of the centre frequency may be obtained within the workable bandwidth. Therefore, for a carrier frequency 'f' Hz, the bandwidth available is 0.05 f Hz. For example, a 20 GHz signal will approximately allow a bandwidth of 1 GHz which can carry data bits at the rate depending upon the the modulation but of the order of 1 Gbps. Therefore, high throughput satellites need high centre frequencies.

Satellite links may use from VHF up to Ka or even higher bands as per the requirement. The frequency yielding the better link performance, in a given band for the specific applications, is appropriately chosen, while the choice of the band of frequency of operation is governed by the kind of application that it will be used for. ITU has allocated different frequency bands for different applications. The allocations have been mentioned in Chapter 1 and in Table 1.1. The allocation by the ITU is on the basis of coexistence with other services, minimum interference and considering the propagation factors. The frequency bands are also reallocated for fixed services considering the spatial diversity. For such allocation, total globe is divided into three regions, viz. (ITU, 2017):

Region 1—consisting of Europe, Africa and parts of northern Asia.
Region 2—consisting of North and South America and adjacent countries.
Region 3—Indian subcontinent, Australia, New Zealand and parts of Asia not included in above regions.

Each band may be allocated to users in each region mentioned above for one or more services, with equal or different rights. There are two categories of service, namely Primary and Secondary. The allotment of a frequency channel within an allocated band of service in a definite region may be exclusive for one user or on share basis provided to different users.

9.1.2 ATMOSPHERIC PROPAGATION EFFECTS

Any kind of communication by satellite needs propagation considerations. The propagation conditions can be entirely different for different frequencies and for different regions. The choice of the exact frequency in the recommended band is dependent upon these propagation factors, also. Further, the conditions experienced by the broadcasting type of services are not comparable with those occurring in the fixed-satellite service. For fixed services, it is easy to have the numbers as the concerned link conditions are localized to one place. Broadcast type of services, on the other hand, has to cater large geographic areas, with considerably stringent quality of service. Different portions of the service area may be affected differently by the propagation effects. Maps of these impairments for large areas are required. The temporal and spatial variation may lead to consideration of statistical behaviour of these impairments. Therefore, first it is necessary to fix the kind of service and the specific frequency. Then we need to consider the effects of the propagation on that definite frequency.

Over the satellite signal paths, several propagation effects may require consideration, depending upon the frequency of the signal. For higher frequencies of operation, like for high throughput satellites (HTS), the most important impairments come from the tropospheric effects, including gaseous absorption, attenuation and depolarization by clouds, rain and other hydrometeors. For comparatively lower frequencies, as for example in L band used in navigation applications, the ionospheric effects such as propagation delay, scintillation and Faraday rotation are important. Further, local environmental effects, including attenuation multipath, etc. by buildings and vegetation, may also affect under certain circumstances.

9.1.2.1 Tropospheric effects

As far as satellite-based communication is concerned, higher data rates are preferred as it can cater to the increasing demands of broadband communications. Higher bit rates need higher frequency bandwidth. Further, it also allows to reduce the size and weight of the trans-receiver systems. Therefore, the global trend is to shift towards the higher frequency bands like Ku and Ka. However, these advantages are not without a cost. With higher frequency bands are associated larger tropospheric propagation impairments.

We have seen that the performance of a satellite link depends upon the signal carrier to noise density ratio, C/N_0. Therefore, anything that reduces the carrier power C or increases the noise density N_0 effectually deteriorates this parameter and in turn degrades the performance. Signal impairments caused by the troposphere are not considerable enough for frequencies below X band when they have path elevation angles exceeding 15°. As elevation decreases and/or frequency increases, these impairments become more and more severe. Fluctuations of signal amplitude and angle of arrival become more significant. The effects are of particular importance for high latitude service areas where the elevation angles to the geostationary satellites are particularly low. We have learnt about these factors in Chapters 6 and 7. Here, we shall list out those factors and their extent of influence and recollect the salient features:

i. Attenuation by atmospheric gases—gaseous attenuation is of minor importance in comparison to rain attenuation. In the 22 GHz band, however, water vapour absorption can be quite large.

ii. Cloud attenuation—Attenuation due to cloud will not be serious for frequencies below 20 GHz and is accounted for in the rain attenuation prediction method. Cloud attenuation may be estimated if the liquid-water content is known.

iii. Rain attenuation—For satellite applications, the rain attenuation exceedance for 0.1% of the time of the year is usually significant and of greatest concern. The value depends upon the rain rate at the location.

iv. Scintillation—Small-scale irregularities in the tropospheric refractive index originating from inhomogeneous mixing of gases of various densities or due to presence of raindrops can induce rapid fluctuations in signal amplitude which are significant for frequencies above 10 GHz or/and path elevation angles below 10°. It is of importance, especially for small-margin links.

v. Depolarization—Hydrometeors, principally concentrations of rain drops and ice crystals, can cause statistically significant depolarization of signals at frequencies above 2 GHz.

vi. Sky-noise and other factors: Sky Noise received by the antenna and thermal noise in the hardware are the principal contributors to the noise added to the signal. Sky noise increases with the rain and will further reduce C/N_0 of the received signal in addition to the signal attenuation due to it. In addition, snow and ice accumulations on reflector and feed surfaces of the antenna can seriously degrade antenna pointing, gain and cross polar characteristics for significant portions of a year deteriorating the C/N_0.

9.1.2.2 Ionospheric effects

At frequencies below 4 GHz, ionospheric effects are important. However, their influence varies depending upon the paths, time and locations. For general engineering use, estimated maximum values of ionospheric effects (ITU-R P.531) are summarized in Table 9.1 for various frequencies. Typical concerns for ionosphere being the fluctuations in the angle of arrival, signal scintillation and Faraday rotation for polarized signals, delay is important for navigation and other applications where the time of travel of the signal is an essential parameter.

Table 9.1 Indicative values for ionospheric effects for low elevation angles and one-way traversal

Effect	1 GHz	10 GHz
Faraday rotation	~100°	~1°
Propagation delay	225 ns	2.25 ns
Variation in the direction of arrival (r.m.s. value)	12″	0.12″
Scintillation	>10 dB peak-to-peak	≈2 dB peak-to-peak

From ITU, 2013. Recommendation ITU-R P.531-12: Ionospheric Propagation Data and Prediction Methods Required for the Design of Satellite Services and Systems, International Telecommunication Union, Geneva.

9.1.2.3 Effects of local environment

The satellite receivers on the ground are expected to be outdoor in the open. However, in specific receiving locations, effects of local relief features, vegetations or man-made structures may become important for consideration. This effect is predominant particularly in navigation systems where the satellites are moving across the sky.

The loss of signal power may result from entry into building, entry into vehicle or the receiver passing under natural vegetation cover or through constrained geophysical terrains. It can also get degraded due to the reflection, scattering and shadowing by buildings or geophysical features or simply from the ground causing multipath effect. In many cities, due to the presence of a large numbers of high buildings around, the satellites become inaccessible, the scenario being called as urban canyon effect. However, these losses are to be characterized locally.

9.1.3 TYPICAL COMMUNICATION SYSTEMS

With the selected frequency and consideration of the atmospheric impairments affecting the frequency during propagation, it is now necessary to estimate certain system parameters which affect the signal quality at the receiver. These include the EIRP at the transmission side and the antenna parameters at the receiver's side. This estimation is governed by the link budget. A 'link budget' is accounting of all of the gains and losses from the transmitter, through the medium to the receiver in a telecommunication system to assess the signal condition. The link budget allows to have the estimate of the expected received signal quality given the transmission power and propagation losses. Conversely, it allows finding the transmission power for a predefined signal quality across the transmission path.

The principal elements of the link budget are

- EIRP
- Signal power flux density
- Received power
- System noise temperature
- Carrier to noise power ratio

We have already mentioned the link budget equations in Chapter 1. Here, we shall once again go through the exercise to establish the equation and describing the aspects related to the link designing along with it.

9.1.3.1 EIRP

The received power at the receiver is dependent upon the power radiated by the transmitter. Let us consider that the transmitting antenna radiates a power P_T. For an isotropic antenna, this power would have radiated equally on all directions. However, typical antenna has finite directive gain, G_T. This gain represents the ratio of the power density in the direction of the maximum gain of the directive antenna to the power density of an isotropic antenna for the same total radiated power. The Effective Isotropic Radiated Power (EIRP) of the antenna is given as the product of the radiated power and the directive gain of the antenna.

$$\text{EIRP} = P_T \times G_T \tag{9.1}$$

For transmission from the satellite, the radiated power is limited by the dc power generated by the satellite while the antenna directivity cannot be increased beyond a limit as large antenna providing higher directivities are also heavier in weight and not convenient to be borne by a satellite. Therefore, these constraints in establishing a downlink signal is to be compensated by adequate resource in the ground, which we shall see later.

However, for uplink signal, the radiated power and the antenna gain are generally not a concern since large earth stations generally do not have much limitations in these regards. They can provide EIRP as large as 100 dBw or even more. However, exceptions are like mobile or a small terminals.

9.1.3.2 Signal power flux density

Signal power flux density is the amount of signal power that crosses a unit surface area where the surface is perpendicular to the direction of the propagation of the signal. As the signal traverses larger distance from the transmitter, the power diverges out in space. As the same power is dispersed over a larger cross sectional area, the flux density value diminishes and it becomes an inverse square function of distance from the transmitter. Considering the transmitter to be approximately a point source, the power flux density at distance R from it may be estimated by considering a hollow sphere of radius R with the source at the centre. The total power transmitted being P_T at the transmitter, the power flux, i.e. the signal power that would pass through a unit area for an isotropic antenna, is $\varphi_i = P_T/(4\pi R^2)$. Since the antenna in the link is directive with gain G_T, the effective power flux density Φ_a, due to the directivity, will be

$$\Phi_a = \frac{G_T P_T}{4\pi R^2} = \frac{\text{EIRP}}{4\pi R^2} \tag{9.2}$$

There are two important points to be remembered here. First, for services, mainly with secondary allocation, there is a constraint on the maximum power flux that can be used, such that it does not affect or interfere with the primary service there. So, the power cannot be increased over a finite threshold value. This has a bearing on the performance of the services available.

Second, for uplink, due to the use of higher frequency, the beam is more concentrated, i.e., large gain and hence higher EIRP. So, for the same transmitting antenna size, it provides larger power flux for a given transmission power, so that even a smaller satellite antenna can have enough received power.

9.1.3.3 Received signal power

The flux density of the power being known, the signal power received at the receiver is proportional to the area with which the receiving antenna effectively collects the signal power. We have seen in Chapter 1 that, this area is directly related to the gain of the antenna and so the total power received may be given by,

$$P_R = \frac{\text{EIRP}}{(4\pi/\lambda)^2 R^2} G_R \tag{9.3}$$

Thus, the received power is dependent upon the range and wavelength or frequency, in addition to the power transmitted. For higher frequencies, the received power is less than that of the lower frequencies given the same EIRP and received antenna gain, G_R. This reduction can be compensated either by increasing the receive antenna gain G_R or by increasing EIRP. In logarithmic (decibel) scale, it is represented as,

$$P_R = \text{EIRP} - L_{FS} + G_R \text{ in dB} \tag{9.4}$$

Here, $L_{FS} = (4\pi/\lambda \times R)^2$ is the free space loss. In addition to the free space loss occurring due to the obvious spreading of the power with distance, there is also associated, some loss due to the passage of the signal through the atmosphere. It is called the atmospheric propagation loss, L_A. Considering the latter, the received power becomes

$$P_R = \text{EIRP} - L_{FS} - L_A + G_R \text{ in dB} \tag{9.5}$$

9.1.3.4 System noise temperature

We have learnt in our last chapter that the receiving antenna picks up additive noise along with the signal, which is inevitable at a finite temperature. This noise is received from different sources, primarily from the atmospheric radiation of incoherent emissions. Noise is also added by the front end LNA of the antenna system. We worked out there that if the total noise temperature is T, the total noise incorporated in the system and added to the signal per unit bandwidth of the signal is $N_0 = kT$. Therefore, the ratio of the carrier power to the noise spectral density may be expressed as

$$\frac{C}{N_0} = \text{EIRP} - L_{FS} - L_A + G_R - kT \text{ in dB}$$
$$= \text{EIRP} - L_{FS} - L_A + \frac{G_R}{T} - k \text{ in dB} \tag{9.6}$$

So, the C/N_0 is a function of the EIRP, the propagation losses L_{FS} and L_A and the receiver figure of merit, G_R/T.

Now, before we move forward, it is suggestive to recall the theoretical relation between C/N_0 and the BER it results in. We know that, the BER, or more precisely, BEP, is a function of E_b/N_0 and the nature of the function depends upon the modulation type. We have also defined E_b as the energy contained in a bit, i.e. the energy accumulated from the carrier over the duration of a bit period. Thus, if only a single channel of information is contained in a carrier, then

$$\frac{E_b}{N_0} = \frac{C \times T_b}{N_0} = \frac{C}{N_0} \times \frac{1}{R_b} \tag{9.7a}$$

Where T_b is the bit duration and R_b is the information bit rate. If there are more information channel in the same carrier, the total carrier power is accordingly apportioned. Typically, this division of power is accredited to individual channel at the time of transmission in terms of EIRP. E_b/N_0 of each channel is then estimated with its respective EIRP and can be written as

$$\left(\frac{E_b}{N_0}\right)_j = EIRP_j - L_{FS} - L_A + \frac{G_R}{T} - k - R_j \text{ in dB} \tag{9.7b}$$

It is to be noted that, here we consider the bits are not coded with any FEC.

9.2 SATELLITE LINKS

A satellite system should always qualify for a definite application by its performance. Once the required performance of a system is defined, the corresponding Bit Error Rate (BER) is quantified. The BER in turn demands a definite value of E_b/N_0 at the receiver. The E_b/N_0 required for a given BER is again dependent upon the type of modulation. The following plot shows the values of the E_b/N_0 necessary for different BER and for different modulations (Fig. 9.1).

Therefore, if we need to ensure a specified performance, we have to have a threshold value of E_b/N_0 at the receiver as suggested by Fig. 9.1. Therefore, the required E_b/N_0 is fixed from the performance requirement and modulation used.

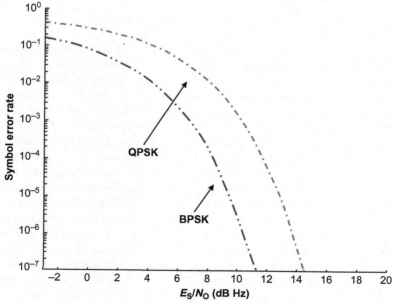

FIG. 9.1

Variation of bit error rate with E_b/N_0 for different modulations.

The necessary bit rate, R is also defined for the system. Further, for the frequency chosen and for a definite earth-space path, the free space loss, L_{FS} is also fixed.

However, the atmospheric loss term, L_A, sometimes get dominant and we do not have much control over this term. For fixed satellite-ground station geometry, care is taken such that the effective L_A remains as low as possible. This is done by using techniques like diversity, etc. When the location of the satellite and that of the ground station is fixed, the elevation angle is preset and then the statistics of the attenuation can be obtained using the long-term statistical models discussed in previous chapters. The models give us either the probability of occurrence of different values of the attenuation, etc. or exceedance values of those impairments for a given percentage of time over a complete year. Now, suppose we need a_v% of availability over a year, for the service that the satellite system is designated to provide. This means the maximum acceptable outage is $p = (1 - a_v)$% of the year. Now, using the model, we can find out from this probability distribution of the attenuation value which is exceeded only for p% of time over the year, such that fade more than this value occurs only for p% of time. Let this value be L_P. So, we have to arrange for compensation of this amount of atmospheric loss, L_P, in addition to compensating other losses and satisfying the other requirements, to achieve the required availability. Randomly varying channel gains leading to fading are taken into account by adding some margin depending on the anticipated severity of its effects.

Therefore, knowing all these, the system design parameters EIRP and G_R/T values can be established. In equation Eqs (9.7b), putting to the left side, the terms which are estimated a priori as a requirement for establishing the link, we get,

$$\frac{E_b}{N_0} + L_{FS} + L_A + k + R = \text{EIRP} + \frac{G_R}{T} \text{ in dB} \tag{9.8}$$

Consequently, the terms that can be configured to establish and maintain the link, i.e. EIRP and G_R/T, remain to the right of the equation. Therefore, the EIRP and the G_R/T are to be so chosen that they are equated to the terms on the left hand side of the equation, where we have all the requirement terms (plus a constant). It is the resource availability at receive and transmit end and also the designer's wisdom that apportions the required value in these items. Setting of EIRP or G/T to a value over the minimum requirement leads to the advantage of having overall link margin. This helps in compensating any unaccounted or unforeseen losses.

For certain applications, due to large demand in the availability, or due to large values of atmospheric loss soaring the value of L_A, the estimated requirement on the left of Eq. (9.8) gets exceedingly high. This large demand is not possible to be readily implemented with the available system. Further, as the enhanced values of atmospheric loss L_A occur infrequently, only for certain small period of the year, using fixed resources to compensate these events leads to huge wastage of resources. In such cases, various adaptive mitigation and postprocessing techniques are utilized to meet the requirements. Each of these techniques adds to an effective gain of the system. Thus, one or more such techniques may be used to get the necessary overall gain to maintain the performance under the constrained conditions. Therefore, the occasional enhanced demand of the L_A may be negotiated by the techniques mentioned below. Nevertheless, in most of the cases, it may lead to a compromise in the available throughput.

a. Change the modulation, such that the E_b/N_0 requirement for the same BER reduces
b. Introduce the FEC so that, due to the coding gain availed, the E_b/N_0 requirement for the same BER reduces
c. Reduce the bit rate R so that even for reduced C/N_0, the E_b/N_0 remains the same
d. Use mitigating techniques such as diversity, etc. so that effective probability of occurrence of L_A reduces.

In the design of satellite link, two main sections, namely Uplink and Downlink, are considered separately. In each of these, the above affecting factors are to be determined separately. C/N_0 ratio calculation is simplified by the use of link budgets which is done for individual transponder and for each section of the link. When a bent pipe transponder is used, the uplink and down link C/N_0 ratios must be combined to give an overall C/N_0.

For the downlink design, the satellite downlink frequency, is kept typically lower. As mentioned previously, the downlink EIRP is determined from available power and considering maximum signal power flux density permissible. Given the transmission power or EIRP, the received power and the C/N_0 may be estimated for any size of the receiver antenna with definite G_R/T. The necessary G_R/T may be obtained using Eq. (9.8).

The uplink design is rather easier than the downlink in most cases. The requirement of presenting an accurately specified carrier power flux density at the satellite can be satisfied in this case by using much higher power transmitters at earth stations and larger antennas, compared to those which can be used on a satellite in most cases. In uplink analysis, the uplink EIRP is determined from the calculation of the power level at the input to the transponder so that the necessary uplink C/N_0 ratio can be maintained with a margin. Care is taken that the received power does not saturate the amplifier at the satellite which may lead to effects due to nonlinearity. However, the situation worsens as the uplink antenna diameter is reduced, like in the case of small and mobile terminals. In such cases, typically for satellite mobile applications, when the ground receiver antennas are small, the satellite antenna is kept comparatively larger with higher G/T. Moreover, additional gain is obtained by using FEC, but at the cost of reduced throughput or by using lower bit rate for a given bandwidth.

In a complete link through a communication transponder, both the uplink and the downlink values come into the consideration. The overall C/N_0 of combined up and down link is given by

$$\left(\frac{C}{N_0}\right)_{Total} = \left[\left(\frac{C}{N_0}\right)_{Up}^{-1} + \left(\frac{C}{N_0}\right)_{Down}^{-1}\right]^{-1} \tag{9.9}$$

This equation is explained as following. The uplink signal is received at the satellites receiving antenna and amplified and frequency converted to that for the downlink and transmitted again through the downlink antenna. The noise present in the uplink signal still remains there while some additional noise is introduced in the downlink. In addition, the signal fade in the downlink also affects the final signal received on the ground. In the process, the total C/N_0 becomes a function of both the C/N_0 of the uplink and downlink as well. From Eq. (9.9) we can see that, the resultant C/N_0 is lower than the lowest of the C/N_0 values among the uplink and downlink. Therefore, if one of these values falls below the threshold for the particular receiving system, the total link is sure to fall below the minimum level and cannot be compensated just by raising the C/N_0 of the other side. Therefore, to keep the total C/N_0 above the threshold, both the uplink and the downlink components are individually to be maintained above it. When both the uplink and downlink C/N_0 are equal, the resultant total C/N_0 is half the value of each, while if one of them is kept sufficiently higher value compared to the other, the total link is determined by the C/N_0 of the other side. In such a condition, with all other parameters remaining constant, the overall C/N_0 will vary linearly with the attenuation of the determining path.

Focus 9.1 Link Calculation for Satellite

Our objective in this focus is to understand the link parameters in a satellite navigation system and also in a Communication system. First let us consider a navigation system with satellites in MEO at the zenith height of 20,000 Km. The signals are at 1.5 GHz. We shall try to estimate the minimum EIRP required to provide a performance defined by the BER of 10^{-7}. For this, we prepare the following table with typical values

Navigation system

Link components	Units	Values
Frequency	GHz	1.57542
Wavelength	m	0.1904
Lowest elevation angle	degrees	10
Path loss	dB	191.00
Atmospheric attenuation	dB	1.00
Sky noise temperature	K	250.00
LNA noise temperature	K	80.00
Equivalent system noise temp	K	330.00
Equivalent system noise temp	dB_K	25.20
Boltzman constant	dB	−228.60
Antenna gain, G (highest deviation from boresight)	dB	−3
Data rate @ 50 bps	dB	17

As a result, the link equation for such a navigation system becomes

$$\text{EIRP} - 191 - 1 - 3 - 25.2 + 228.6 - 17 = E_b/N_0$$

Now, considering a standard BPSK modulation for the signal and a required BER of 10^{-7}, the necessary E_b/N_0 to achieve the required performance is 11 dBHz for BPSK signal. So, rewriting the above equation, we get

$$\text{EIRP} = 11 + 191 + 1 + 3 + 25.2 + (-228.6) + 17$$
$$\text{Or, EIRP} = 19.6 \approx 20.0$$

Therefore, a minimum EIRP of 20 dB will be sufficient to obtain the required BER of 10^{-7}. This value of EIRP results in the received C/N_0 of $11 + 17 = 28$ dBHz and is just sufficient for acquisition of the signal. However, an additional 2–4 dB of excess power margin is added to EIRP to mitigate varying fading conditions, scintillation and to meet other requirements and ensure unperturbed operation. Therefore, the total EIRP of the signal becomes ~24 dBW.

Communication system

We shall here, however, for the sake of simplicity estimate the necessary link parameters towards the Hub side. As an example, let the gateway hub has an EIRP of 90 dB_W, while its antenna size is 9 m. in diameter. For a Ka band signal towards the hub side, the antenna gain thus obtained is 63 dB (this implies 27 dB_w transmission power) and G/T of about 39 dB/K. Let the satellite has a G/T of 18 dB/K and an EIRP of 56 dBw. The downlink at 20 GHz and the uplink at 30 GHz can experience a fade of around 15 and 28 dB, respectively, for 0.01% of time of year with rain rate exceedance value of 30 mm/h.

The forward uplink, taking a pragmatic value of the free space path loss of 210 dB, and considering that 28 dB of rain fade has to be compensated to get 99.99% availability, the link will have a worst case of C/N_0 of

$$(C/N_0)_{UP} = 90 - 210 - 28 + 18 + 228.6 = 98.6 \, \text{dBHz}$$

Considering an E_b/N_0 requirement of 12 dBHz for BER of 10^{-5} with BPSK modulation, if the data rate it can handle be R, then, with 99.99% of confidence, the system can transmit with bit rate of R, such that

$$10 \log R = 98.6 - 12 = 86.6$$

So, the value of R becomes = 457 MHz.

Similarly, the reverse downlink towards the hub will have a C/N_0 of

$$(C/N_0)_{DN} = 56 - 210 - 15 + 39 + 228.6 = 98.6 \, \text{dBHz}$$

Accordingly, it can support the same data rate. If the satellite is connected with four such gateways with dual polarizations, the total bit rate that it has to support is

$$B_{Total} = 4 \times 2 \times 457 \, \text{MHz} = 3.5 \, \text{GHz}$$

9.3 PERFORMANCE

9.3.1 LONG-TERM PERFORMANCE

Long term performance of a satellite link depends upon many factors. Some of them are long term stability of the transmitter clock, long term gain variations, deterioration of the noise characteristics due to ageing, etc. Further, the propagation impairments also vary in long term. For example, the rain attenuation, which is a functions of the meteorological factors, vary with time, not only over the days of a year, but also from one year to another. Therefore, the estimates of the attenuation and other impairments calculated at one time have finite probability of failing in other years. So, there is a possibility that the predicted variations do not comply with the pragmatic variation of the meteorological factors. Hence, there should be a way for assessing the risk parameters associated with the variation of rain attenuation statistics.

The knowledge of the variability of propagation phenomena in addition to other parameters, is required to access the performance variability. The performance variability may be used to analyse and quantify the system reliability, availability of the service. Here, we shall discuss a prediction procedure to estimate risk parameters associated with the variation of rain attenuation statistics in which the risk parameter is obtained from the statistics of the year-to-year variations.

9.3.1.1 Risk associated with propagation margin

Rain attenuation models are used to estimate the amount of degradation (or fading) of the signal when passing through rain. Consider that, we have designed a link by considering a calculated value of attenuation A_p dB, as per the required availability figure of a_v. A_p corresponds to the exceedance probability '$p\%$', where $p = 100 - a_v$. If we set this value as our link margin, we expect that the link fade will only exceed the margin for $p\%$ of time giving availability of a_v.

However, this number is purely statistical, and hence, due to the variability of the meteorological conditions, there is a finite chance that the true occurrence of attenuation A_p or above happens for more than $p\%$. If the true probability of occurrence is $p_R\%$, then the occurrence probability of A_p and above, in excess of p, is $(p_R - p)$ in percentage and this will be equal to the depreciation in the availability from a_v.

The associated risk R in terms of probability, representing the probability that the yearly exceedance probability attached to A_p will be p_R instead of p, with p_R $(0 < p_R < 1)$ satisfies:

$$R = Q[(p_R - p)/\sigma(p)] \tag{9.10}$$

Where p_R and p are in factors and $\sigma^2(p)$ is the variance of estimation that considers both the variance of modelling and the interannual climatic variances. In other words, Eq. (9.10) can be stated as, for a desired location, the interannual fluctuations of rain attenuation statistics around the long-term value 'p' is normally distributed with mean p and yearly variance σ^2 so that (ITU-678, 2015)

$$\sigma^2(p) = \sigma_c{}^2(p) + \sigma_E{}^2(p) \tag{9.11}$$

where, σ_E^2 is the variance of estimation and σ_c^2 is the interannual climatic variance. These values can be derived from long-term historical database. Here $Q(x) = 1/\sqrt{(2\pi)} \int_x^\infty \exp\left(t^2/2\right) dt$. It is important to note that $p_R = p$ leads, as expected, to $R = 0.5$ which means it is equiprobable that the estimated exceedance values will be met or not in reality.

9.3.2 SHORT-TERM PERFORMANCE

The matter of this section could be well-discussed in Chapter 7. But, since these factors are also related to implementation in addition to theory, we prefer to discuss it here. Propagation effects experienced by the signal in their paths quantitatively determine the link performance. Propagation loss, i.e. the power loss in the signal, due to one or more effects, manifests as the depreciations in terms of C/N_0. For bent pipe transponder, the received signal power is determined by both the uplink and the downlink components, as given in Eq. (9.8). In a star network with a centralized gateway hub and a numbers of terminals, all connected connected to the hub through the satellite, the complete link consists of the forward link and the reverse link. The forward link is the overall communications link from the gateway to the terminal and consists of the gateway uplink and the terminal downlink. The Reverse link, on the other hand, is the overall communications link from the terminal to the gateway. It consists of the terminal uplink and the gateway downlink.

The effects of weather in terminal locations are generally more significant than the effects of weather in gateway locations. So, the forward link performance is dominated by the terminal downlink, while the reverse link performance is dominated by the terminal uplink. This is because the gateway can be designed to have large EIRP as well as large antenna gain such that in the overall C/N_0 expression, the C/N_0 of the gateway side is large compared to that of the terminal side. Hence, the link performance is dominated by the behaviour of the terminal side of the link, which is limited by the antenna size and power constraints. This is evident from Eq. (9.9). Also, the gateway location can often be chosen to avoid regions with the worst weather conditions, but the terminals must be located where service is needed, so that the terminals do not have the option to avoid the fade characteristics of that region.

Many rain events will result in short-term fades, lasting for few seconds. These fades can be mitigated using various network protocols like by retransmitting the missing data or by compensating for the missing information otherwise. A very effective technique used in such implementation is by buffering data. This stores the most recent data received in the buffer and allows the system to serve the application from the buffer while waiting for missing data to be resent, in case of any such event. Yet, for longer duration of the fades, this may lead to a short-term loss of data. This situation can often be overcome by the correct choice of FMT. As described in Chapter 7, the Fade Mitigation Techniques aim at adapting the operating conditions like the modulation, coding and power level of the link. Therefore, the deep and long fading events have a lower impact on the overall system performance. Broadband Satellite Systems take advantage of a combination of the Fade Mitigation Techniques, most important of which are discussed below.

9.3.2.1 Use of ULPC

While fades of very short interval can be mitigated using network protocols of retransmitting and buffering of data, comparatively longer term fades, lasting for several seconds to several minutes, require more active mitigation to maintain an adequate quality of service. One of such adaptive mitigation techniques is the Uplink Power Control. In ULPC, the terminals and gateways increase the transmit power to compensate for uplink fades only, with mitigation capability ranging from few to several units of decibels. Adaptive power control is mainly applicable to the reverse uplink, and hence, is mostly limited by the power economics of the terminals and typically remains <10 dB. Since weather fades on the uplink are larger than on the downlink, adaptive uplink power control is useful for mitigating fades.

9.3.2.2 Use of MODCOD

While adaptive power control can compensate only a limited value of the fade, changing modulation and coding can play a significant role in the system by compensating the excess weather-induced fades. When the maximum ULPC compensation range is reached, the residual signal power reduction due to fading is counteracted by means of appropriate coding and modulation. In this technique, during the event of uncompensated fade, the FEC and the modulation is adaptively changed to the level that matches the link conditions.

The MODCOD, i.e. the combination of modulation type and forward error correction code, of each terminal is adaptively tuned to meet the current terminal requirements, determined by channel conditions. The goal of adaptation is to give each terminal the highest possible data rate that the link may support, while providing the operating margin enough to compensate short-term fluctuations as required.

To explain how the method works, we shall first start by explaining the concept of a 'modcod point' and then we'll describe how adaptive modulation can compensate for fades, even if the fade is as deep as it is found in Ka band satellite signals.

A MODCOD is a combination of Modulation (MOD) type and FEC coding (COD) rate used to send information. As an example, one modulation among the choice of BPSK, QPSK, 8 PSK and 16 APSK along with a particular choice of FEC rate among 1/2, 2/3, 3/4, 4/5 and 5/6 rate Turbo code form a definite MODCOD combination or a modcod point. The coded units are called 'symbols' to distinguish them from the pure information carrying bits. For better understanding, a subset of modcod points defined in the DVB-S2 satellite communications standard for forward link communications is shown in Table 9.2. The table shows the modulation type and FEC rate with corresponding normalized data rate and the symbol energy to noise density ratio, E_s/N_0 required for that modcod symbols to be received reliably. By using the listed modulations, the modulation type can be adjusted to change the effective data rate by a factor of 9 as the required E_s/N_0 changes by about 18 dB. As per this table, when there exist a clear sky condition, let the received E_s/N_0 is such that the system can communicate with 16 APSK modulation with 8/9 rate Turbo code. It has 3.5 times the information bit rate compared to a normal BPSK with no FEC, which is our reference for comparison here. This is because the system can now send 4 times the symbols over the same channel than BPSK due to the modulation, while 8/9 rate Turbo coding will make the information bits 8/9 of $4 = 3.55$ times of it. Again, due to lesser numbers of redundant bits, the coding gain is less and hence the required modcode needs about 13 dB higher value of E_s/N_0 than the reference. On the other hand, on choosing QPSK with 1/4 rate Turbo code, although the system can now send symbols at 2 times the BPSK rate, due to the large redundancy in the code, the actual bit rate is 1/4 part of it, i.e. 1/2 of BPSK rate. This large redundancy allows the system to work with very low E_s/N_0 of about -2 dB. Hence, under large fade condition, the system can resort down to this modcode combination; however, sacrificing throughput during the period this choice persists. Here, we have considered that the effective bandwidth of the system remains unchanged.

Table 9.2 MODCOD points for DVB-S2 standard

Modulation	BW efficiency (w.r.t. BPSK)	FEC rate	Effective bit rate (w.r.t. BPSK)	E_s/N_0[a]
QPSK	2	1/4	0.50	−2.35
QPSK	2	1/3	0.67	−1.24
QPSK	2	2/5	0.80	−0.30
QPSK	2	1/2	1.00	1.00
QPSK	2	3/5	1.20	2.23
QPSK	2	2/3	1.33	3.10
QPSK	2	3/4	1.50	4.03
QPSK	2	4/5	1.60	4.68
QPSK	2	5/6	1.67	5.18
QPSK	2	8/9	1.78	6.20
QPSK	2	9/10	1.80	6.42
8PSK	3	3/5	1.80	5.50
8PSK	3	2/3	2.00	6.62
8PSK	3	3/4	2.25	7.91
8PSK	3	5/6	2.50	9.35
8PSK	3	8/9	2.67	10.69
8PSK	3	9/10	2.70	10.98
16APSK	4	2/3	2.67	8.97
16APSK	4	3/4	3.00	10.21
16APSK	4	4/5	3.20	11.03
16APSK	4	5/6	3.33	11.61
16APSK	4	8/9	3.56	12.89
16APSK	4	9/10	3.60	13.13
32APSK	5	3/4	3.75	12.73
32APSK	5	4/5	4.00	13.64
32APSK	5	5/6	4.17	14.28
32APSK	5	8/9	4.44	15.69
32APSK	5	9/10	4.50	16.05

[a]For BER $\sim 10^{-10}$ resulting in packet error rate per $< 10^{-7}$ considering 188 byte in a packet.

Choice of modcode is thus based on requirement threshold, given a fading condition in a communication configuration. Thresholds are defined based on the Satellite System operational scenario and on the performance obligation of the services. The possible ways for choosing different modcod point and configuring the system can be divided in three distinct categories, as described below:

Constant coding and modulation

In constant coding and modulation (CCM), all terminals use the same modcod point, for all the time and for all services. This is an approach, typically used for one-way satellite broadcast systems, where all terminals receive the same signal. CCM is commonly used in Ku-band satellite systems.

The choice of the modcod is based upon the basis of the highest expected fades in the service area such that the system can reliably transmit to all the terminals through any weather event. So, the data rate is effectively matched to the needs of the worst effect terminal. Consequently, most of the time—when there is no fade or for the terminals with lesser fades—the system will be operating at a lower data rate than it could otherwise support. Hence, the overall throughput will be lower and the system runs at reduced efficiency.

In a CCM operation, suppose the observed statistics shows that, within the area of service, for 99.99% of time of the year, the maximum fade experienced is such that, for the worst hit terminal, the estimated E_s/N_0 (estimated C/N_0 with available options of FEC/corresponding symbol rate) can drop during the fade to a minimum of 4.5 dB and will eventually recover. The selected CCM modcode is hence QPSK rate 3/4, which delivers a normalized data rate of 1.5 bits/Hz. The terminal will continue functioning with the same performance metric, even during the deepest part of the fade, not considering the 0.01% of outage time, however. Although the fade occurs to the deepest level only for a very small period of time and only at few of the worst hit terminals, the related modcode is used throughout the service period and for all terminals. Thus the general performance of the system is reduced.

CCM is simple to implement and works well in systems that don't experience large variations in signal level. Since many Ku-band systems, which don't generally experience deep weather-induced fades, use CCM.

Variable coding and modulation

In variable coding and modulation (VCM), different services (and different terminals or terminal types) may use different modcod points, but the modcod selected for a given service (and, a given terminal or terminal type) will remain constant at all times. That is, the variation of the selected modcod is across the terminals, but not over the time. The modcod point selected for each service is determined by considering the data rate requirements against the quality required even under the worst case of weather fade for that particular service. VCM can mitigate weather effects by allocating more robust modcod providing larger margin to terminals subject to more extreme weather effects, while allocating a modest modcod accepting lower fade margin to terminals subject to less fading. The data rate also changes accordingly. In this way, VCM partially avoids a limitation of CCM. A disadvantage of VCM is that each terminal must accept a lower data rate at congenial conditions than the link could otherwise support in consideration for providing a suitable weather margin at worst times.

VCM is not much more complicated to implement than CCM, and it offers the network means to achieve higher throughput for services that can inherently tolerate more and longer outages. Nevertheless, it still forces the users of the networks to accept compromises by using the configuration for the worst condition of the year. In brief, it can take care of terminal-wise variation of the fade conditions, but not for temporal variation of the same for a given terminal.

Dynamic/adaptive coding and modulation

The above two implementations of modcode are fixed in time. In adaptive coding and modulation (ACM), the modcod for each terminal is adaptively tuned over time to meet the current requirements of the terminal. As channel conditions change, such as the fade varying during a rain and nonrainy period, the modcod adjusts accordingly to become just adequate to compensate. Terminals in the same beam are likely to use

---

Box 9.1 MATLAB Exercise

The objective of this matlab exercise is to understand the operation of an ACM system and its effects. Here, we shall run a matlab programme ACM.m that selects different ACM modcodes on the basis of different fade values predicted for forthcoming fade. The prediction part is not used here and a predetermined value of fade time series is used instead. Assuming that the prediction is for 10 s and the selected modcod is valid for at least for this period, until the next prediction arrives, the average bit rate is also calculated. Note that, for smaller fades up to a preset value of 'U', the fade is compensated by ULPC. Comparison of Average throughput over the event duration is also done for ACM. The corresponding figure is shown below. The upper panel shows the channel efficiency, or simply, the factor by which the actual bit rate is more than that in BPSK without coding. The lower panel shows the fade that the ACM has to compensate.

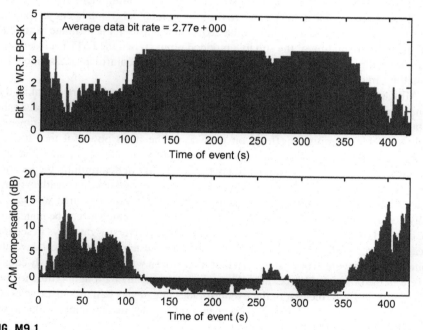

FIG. M9.1

Effective bit rate and magnitude of ACM compensation.

Notice in the lower panel that for a period of time, the ACM compensation value is negative. It represents the case when the fade is already compensated by the ULPC. The data bit rate during this period is therefore clamped to the highest rate of 3.5 times the BPSK. The overall rate, considering the whole fade period, is given with the plot in the top panel. Run the programme ACM.m to obtain similar plots with different distribution of modcod points. Change the values of the ULPC threshold 'U' to any other positive value and notice the change in the values derived.

different modcods during a weather event, because the fade conditions are very localized and varies greatly even within the footprint of one beam. The adaptation is targeted to give each terminal the highest possible data rate that the link will support at those individual terminals, while preserving some operating margin to accommodate short-term fluctuations. In this way, ACM overcomes the limitation of VCM.

In an ACM operation, before the fade starts, the terminal operates with a very modest modcod providing a high data rate. As the fade progresses, the modcode adapts to a more stringent choice to meet the current requirements of the channel. Such change in the modcod point reduces the instantaneous data rate of the terminal. However, the terminal gets back to its initial modcod through other intermediate ones as the fade reduces down and finally vanishes. Since, at the final application level, data may not be used at the transmission rate, the user may remain unaware of this variation.

ACM is more complicated to implement than VCM or CCM, but it is a powerful tool for mitigating the large weather-induced fades, typically experienced in Ka-band. A sufficiently large set of available modcode points can accommodate very deep fades. For example, the modcodes shown in Table 9.2 can accommodate more than 18 dB variation. When coupled with adaptive power control, ACM can compensate for even the large fades experienced in Ka-band (Petranovich, 2012).

An ACM has to add a margin due to the system loop delay and latency. That is, it has to take care of the fade variation that takes place between the estimate of the fade and the implementation of the suitable modcod point. When the FMT is closed loop, the loop delay should include the overall end to end control process, including signalling transmission, propagation and processing.

Mitigation of weather effects in the reverse downlink and forward uplink can be done using the resources at the hub. A deep fade in a gateway location could cause outages to all the terminals serviced by that gateway. However, there are powerful tools available to mitigate effects on gateways. The most powerful tool for it is 'appropriate planning'. The gateways may be located outside of the spot beams of the terminals they are servicing. Utilizing this flexibility in determining the location, the gateways can be placed in areas where occurrences of rain or other weather-related fades are less probable. Further, each gateway is resourcefully enriched, and specifically, the available power and the size of the antenna at the gateway can be matched to the demands of its location. It is also possible to deploy diversity gateways in separate locations to reduce the effect of the fade on the overall system performances. If weather conditions are severe in one gateway, it is unlikely that it will be simultaneously equally severe in the other, and hence, the alternative gateway can be brought online to compensate for it. Nevertheless, in the event of fades, a gateway can use adaptive power control and adaptive modulation to compensate for fades in a manner similar to that described in the previous section.

To conclude our discussion in this chapter, and to summarize, there is no doubt that propagation effects present a significant challenge to the satellite applications. However, a technologically advanced system design and implementation is mostly able to overcome these challenges. The understanding of the propagation, the associated impairments and consequently the effective mitigation of them through

FIG. 9.2

Scenarios when a satellite service system is properly designed and when it is not.

proper designing of the system can make the difference, not merely in the performance of the satellite services, but in turn in the quality of life for the generations to come. The following Fig. 9.2, showing the contrasting scenarios, however, is only illustrative.

CONCEPTUAL QUESTIONS

1. What should be the factors of consideration while choosing a link frequency and polarization for deep space communication? Does ACM has any role to play here?
2. Can the gateway estimate both the uplink and downlink fade of a forward link without any auxiliary information from the terminals? Is it possible for the reverse link?
3. Establish a relation between the E_s/N_0 and effective E_b/N_0 for any particular modcode scheme.
4. If ACM provides highest throughput, then why some systems still use CCM?

REFERENCES

ITU, 2015. Recommendation ITU-R P.678-3: Characterization of the Variability of Propagation Phenomena and Estimation of the Risk Associated With Propagation Margin. International Telecommunication Union, Geneva.
ITU, 2013. Recommendation ITU-R P.531-12: Ionospheric Propagation Data and Prediction Methods Required for the Design of Satellite Services and Systems. International Telecommunication Union, Geneva.

ITU, 2017. http://www.itu.int/en/ITU-R/terrestrial/broadcast/Pages/Bands.aspx, Accessed 4 January 2017.

Petranovich, J., 2012. Mitigating the Effect of Weather on Ka-band High-Capacity Satellites. ViaSat Inc, Carlsbad, CA.

FURTHER READING

Atayero, A.A., Luka, M.K., Alatishe, A.A., 2011. Satellite link design: a tutorial. Int. J. Electr. Comput. Sci 11 (4), 1–6.

European Telecommunications Standards Institute, 2009. European Standard (Telecommunications series) Digital Video Broadcasting (DVB); Second Generation Framing Structure, Channel Coding and Modulation Systems for Broadcasting, Interactive Services, News Gathering and Other Broadband Satellite Applications (DVB-S2), ETSI EN 302 307 V1.2.1 (2009-08).

Index

Note: Page numbers followed by *f* indicate figures, *t* indicate tables, and *b* indicate boxes.

Printed in the United States
By Bookmasters